T0231015

Heat Transfer Enhancement with Nanofluids

EDITORS
**Vincenzo Bianco
Oronzio Manca
Sergio Nardini
Kambiz Vafai**

CRC Press
Taylor & Francis Group
Boca Raton London New York

CRC Press is an imprint of the
Taylor & Francis Group, an **informa** business

CRC Press
Taylor & Francis Group
6000 Broken Sound Parkway NW, Suite 300
Boca Raton, FL 33487-2742

First issued in paperback 2017

ISBN-13: 978-1-4822-5400-6 (hbk)
ISBN-13: 978-1-138-74948-1 (pbk)

Library of Congress Cataloging-in-Publication Data

Heat transfer enhancement with nanofluids / editors, Vincenzo Bianco, Oronzio Manca, Sergio Nardini, and Kambiz Vafai.
    pages cm
    Includes bibliographical references and index.
    ISBN 978-1-4822-5400-6 (alk. paper)
    1. Heat exchangers--Fluid dynamics. 2. Nanofluids. 3. Microfluidics. I. Bianco, Vincenzo, editor. II. Manca, Oronzio, editor. III. Nardini, Sergio, editor. IV. Vafai, K. (Kambiz), editor.

TJ263.H427 2015
621.402'2--dc23
                                                  2014043704

Visit the Taylor & Francis Web site at
http://www.taylorandfrancis.com

and the CRC Press Web site at
http://www.crcpress.com

Contents

Preface

AIM AND SCOPE

Convective heat transfer is very important in many industrial heating and cooling applications. This mode of heat transfer can be passively enhanced by changing flow geometry, boundary conditions, or fluid thermophysical properties. An innovative way of improving the thermal conductivity of the fluid is to suspend small solid particles within it.

This idea is not new. Maxwell, at the end of nineteenth century, demonstrated the possibility of increasing the thermal conductivity of a liquid mixture by adding solid particles to it. At that time, because of the limited manufacturing capabilities, only particles with dimensions on the order of micrometers could be produced. But suspensions with these types of particles caused abrasion of the tube wall and a substantial increase in the wall shear stress.

Today, it is possible to readily manufacture nanometer-sized particles and disperse them within a base fluid such as water. These types of suspensions have been referred to as nanofluids. It appears that Choi from Argonne Laboratory coined this word in 1995.

Many researchers from all over the world have focused their attention on nanofluids, as they offer the possibility of increasing the heat transfer in various applications including electronic cooling.

The aim of this book is to bring together a number of recent contributions regarding nanofluids from several researchers from all over the world. The latest research results, innovations, and methodologies are reported in the book in order to support discussion and to circulate ideas and knowledge about heat transfer enhancement in nanofluids without a preconceived opinion, either positive or negative, on this topic.

This book covers basic topics such as analysis and measurements of thermophysical properties, convection, heat exchanger performance as well as issues such as convective instabilities, nanofluids in porous media, and entropy generation in nanofluids. Moreover, a chapter is dedicated to the analysis of the anomalous heat exchange associated with nanofluid convection and another is devoted to the innovative topic of gas-based nanofluids, which can be explored further in the future.

More specifically, Chapter 1 discusses the possible mechanisms of thermal conduction enhancement and introduces main results obtained with experimental, numerical, and theoretical models available in the literature. In Chapter 2, a theoretical analysis to determine the anomalous enhancement of heat transfer in nanofluid flow is reported and the results obtained are compared with those in the existing literature. Chapter 3 reports different approaches to model thermal conductivity enhancement of nanofluids, whereas Chapter 4 focuses on the experimental methodologies to determine thermophysical properties of nanofluids. Chapters 5 and 6 analyze forced convection heat transfer in nanofluids in both laminar and turbulent convection, with reference to possible practical applications. The main features and methodologies to analyze nanofluids flow are introduced together with an in-depth discussion of experimental results available in the literature. Chapters 7 and 8 focus on the application of nanofluids in heat exchangers and microchannels, respectively. Different applications are introduced and discussed. Chapters 9 and 10 introduce the effect of buoyancy by analyzing mixed and natural convection of nanofluids flow. The modeling of these kinds of flows is introduced and comparisons are performed with data available in the literature. Chapters 11 and 12 discuss the utilization of nanofluids in porous media and how to model this kind of complex flow. Problems regarding instability are also treated. In Chapter 13, the boiling of nanofluids is introduced. Pool and flow boiling are treated by analyzing the effect of the nanoparticles on these complex phenomena. Future research directions are also indicated to further develop this area of knowledge.

This book is intended to be a reference for researchers and engineers working in the field of heat transfer within academic institutions or industries. The editors acknowledge all the authors for their help and availability during the preparation of the book. They specially thank the staff of Taylor & Francis Group for their helpful support in preparing this book.

Vincenzo Bianco
Università degli Studi di Genova

Oronzio Manca
Seconda Università degli Studi di Napoli

Sergio Nardini
Seconda Università degli Studi di Napoli

Kambiz Vafai
University of California, Riverside

Contributors

Eiyad Abu-Nada
Department of Mechanical
  Engineering
Khalifa University of Science
  Technology and Research
Abu Dhabi, United Arab Emirates

Khaled Al-Salem
Department of Mechanical
  Engineering
College of Engineering
King Saud University
Riyadh, Kingdom of Saudi Arabia

Antonio Barletta
Department of Industrial
  Engineering
Alma Mater Studiorum Università
  di Bologna
Bologna, Italy

Sergio Bobbo
Istituto per le Tecnologie della
  Costruzione—Consiglio
  Nazionale delle Ricerche
Padova, Italy

Matthias H. Buschmann
Institut für Strömungsmechanik
Technische Universität Dresden
Dresden, Germany

Michele Celli
Department of Industrial
  Engineering
Alma Mater Studiorum Università
  di Bologna
Bologna, Italy

Lixin Cheng
Department of Engineering
Aarhus University
Aarhus, Denmark

Massimo Corcione
DIAEE Sezione Fisica Tecnica
Università di Roma "La Sapienza"
Rome, Italy

Laura Fedele
Istituto per le Tecnologie della
  Costruzione—Consiglio
  Nazionale delle Ricerche
Padova, Italy

Ehsan B. Haghighi
Department of Energy
  Technology
KTH Royal Institute of
  Technology
Stockholm, Sweden

Seok Pil Jang
School of Aerospace
and Mechanical Engineering
Korea Aerospace University
Goyang, Republic of Korea

Ali Kianifar
Department of Mechanical
Engineering
Ferdowsi University of
Mashhad
Mashhad, Iran

Clement Kleinstreuer
Department of Mechanical
and Aerospace Engineering
North Carolina State
University
Raleigh, North Carolina

Andrey V. Kuznetsov
Department of Mechanical
and Aerospace Engineering
North Carolina State
University
Raleigh, North Carolina

Seung-Hyun Lee
School of Aerospace and
Mechanical Engineering
Korea Aerospace University
Goyang, Republic of Korea

Wenhao Li
Department of Mechanical
Engineering
Shizuoka University
Hamamatsu, Japan

Giulio Lorenzini
Department of Industrial
Engineering
University of Parma
Parma, Italy

Omid Mahian
Department of Mechanical
Engineering
Ferdowsi University
of Mashhad
Mashhad, Iran

Akira Nakayama
Department of Mechanical
Engineering
Shizuoka University
Hamamatsu, Japan

and

School of Civil Engineering
and Architecture
Wuhan Polytechnic
University
Hubei, China

Donald A. Nield
Department of Engineering
Science
University of Auckland
Auckland, New Zealand

Hakan F. Oztop
Department of Mechanical
Engineering
Fırat University
Elazig, Turkey

Andrzej W. Pacek
School of Chemical Engineering
University of Birmingham
Birmingham, United Kingdom

Björn E. Palm
Department of Energy Technology
KTH Royal Institute of Technology
Stockholm, Sweden

Samuel Paolucci
Department of Aerospace and
    Mechanical Engineering
University of Notre Dame
Notre Dame, Indiana

Gianluca Puliti
Department of Aerospace and
    Mechanical Engineering
University of Notre Dame
Notre Dame, Indiana

Alessandro Quintino
DIAEE Sezione Fisica Tecnica
Università di Roma "La Sapienza"
Rome, Italy

Gilles Roy
Department of Mechanical
    Engineering
Université de Moncton
Moncton, New Brunswick,
    Canada

Ahmet Z. Sahin
Department of Mechanical
    Engineering
King Fahd University of
    Petroleum & Minerals
Dhahran, Kingdom of Saudi Arabia

Eugenia Rossi di Schio
Department of Industrial
    Engineering
Alma Mater Studiorum
    Università di Bologna
Bologna, Italy

Bengt Sundén
Department of Energy
    Sciences
Lund University
Lund, Sweden

Adi T. Utomo
School of Chemical Engineering
University of Birmingham
Birmingham, United Kingdom

Wesley C. Williams
Craft and Hawkins Department
    of Petroleum Engineering
Louisiana State University
Baton Rouge, Louisiana

Somchai Wongwises
Fluid Mechanics, Thermal
    Engineering and Multiphase
    Flow Research Lab
Department of Mechanical
    Engineering
King Mongkut's University
    of Technology Thonburi
Bangkok, Thailand

Zan Wu
Department of Energy Sciences
Lund University
Lund, Sweden

Zelin Xu
Department of Mechanical
and Aerospace Engineering
North Carolina State
University
Raleigh, North Carolina

Chen Yang
Department of Mechanical
Engineering
Shizuoka University
Hamamatsu, Japan

CHAPTER 1

Properties of Nanofluid

Samuel Paolucci and Gianluca Puliti

CONTENTS

1.1 INTRODUCTION

1.1.1 Why Nanofluids?

Nanofluids are colloidal suspensions of nanosized solid particles in a liquid. Recently conducted experiments have indicated that nanofluids tend to have substantially higher thermal conductivity than the base fluids (Keblinski, Eastman, and Cahill 2005). Among the many advantages of nanofluids over conventional solid–liquid suspensions, the following are worth mentioning (Choi 1995; Saidur, Leong, and Mohammad 2011): higher specific surface area, higher stability of the colloidal suspension, lower pumping power required to achieve the equivalent heat transfer, reduced particle clogging compared to conventional colloids, and higher level of control of the thermodynamics and transport properties by varying the particle material, concentration, size, and shape.

Though attempts have been made to explain the physical reasons for such enhancement in nanofluids, there are still many conspicuous inconsistencies (Keblinski, Prasher, and Eapen 2008). There are at least four reasons why a definitive theory on nanofluids still does not exist (Das et al. 2007; Wang and Fan 2010):

- The thermal behavior is too different from solid–solid composites or standard solid–liquid suspensions.

- The thermal transport in nanofluids, besides being surprisingly efficient compared to standard solid–liquid suspensions, depends on nontraditional variables, such as particle size, shape, and surface treatment.

- The understanding of the physics behind nanofluids requires a multidisciplinary approach.

- Probably, the most daunting difficulty is related to multiscale issues. In fact, nanofluids involve at least four scales (Wang and Fan 2010): the molecular scale, the microscale, the mesoscale, and the macroscale. The main difficulty is in the methods chosen to correlate and optimize the interplay among these scales.

1.1.2 Scientific and Engineering Significance

There is a large number of engineering applications that can benefit from a better understanding of the thermal conductivity enhancement of nanofluids.

One example is ionic liquids, which are salts that are liquid at room temperature. However, ionic liquids do not have a very high thermal conductivity compared to (say) water, and if this could be improved by the addition of nanoparticles, the liquid would be better suited for heat transfer applications such as in absorption refrigeration (Sen and Paolucci 2006) or cooling circuits.

Liquid cooling with high thermal conductivity fluids would also address many other heat dissipation problems. For instance, microelectromechanical systems generate large quantities of heat during operation and require high-performance coolants to mitigate the large heat flux. Such a system requires precise temperature control, and a higher conductive fluid would allow for more efficient heat transfer control (Escher et al. 2011; Nguyen et al. 2007; Saidur, Leong, and Mohammad 2011).

There are also many everyday applications in which nanofluids could be suitable for, such as in the automotive industry. The high thermal conductivity enhancement observed in ethylene glycol-based nanofluids (Das et al. 2007) suggests that this common antifreeze could have better performance simply with a nanoparticle suspension.

1.2 EXISTING STUDIES OF NANOFLUIDS

Three possible approaches have been pursued for the study of nanofluids: experimental, empirical, and numerical. Although the number of experimental works has been constantly increasing since 1993, very few works have been published on empirical or numerical studies of nanofluids (Das et al. 2007). The current lack of understanding of the basic mechanism of energy transport at the nanoscale makes the published empirical works on nanofluids extremely case dependent. In other words, researchers have been fitting data to experiments, rather than obtaining fundamental understanding. Numerical works aimed at basic understanding are sporadic in the literature, and the only papers published recently (Ghosh et al. 2011; Li, Zhang et al. 2008; Sankar, Mathew, and Sobhan 2008; Sarkar and Selvam 2007; Shukla and Dhir 2005) oversimplify the physics. In particular, two of these works use molecular dynamics (MD) with a simple Lennard–Jones (LJ) potential to study a solid copper particle in liquid argon. No experiment with such materials is available; thus, validating the results is impossible. Furthermore, such a simple potential is incapable of capturing surface effects (liquid layering, particle clustering, etc.) properly for polar fluids, such as water, which has been extensively used in experimental studies.

Figure 1.1 shows the rapid growth of the number of papers on nanofluids since 1993 worldwide.

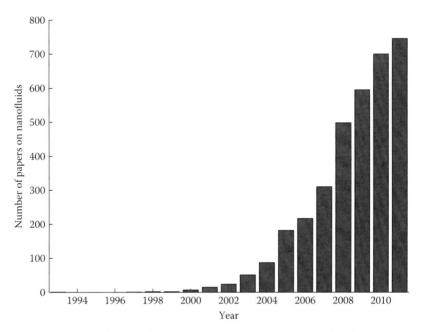

FIGURE 1.1   Growth of publications on *nanofluids*. (Reprinted with permission from Puliti, G. et al., *Appl. Mech. Rev.*, 64, 3, 030803, 2012a.)

1.2.1 Possible Mechanisms of Thermal Conduction Enhancement

Based on experimental results collected in the literature since 1993, many explanations have been given by different research groups for the thermal conduction enhancement observed in their own experiments (see Table 1.1). Mechanisms proposed to explain thermal conductivity enhancement include Brownian motion of nanoparticles, liquid layering of the base fluid surrounding nanoparticles, and nanoparticle aggregation. However, which, if any, of these possible mechanisms is mainly responsible for the thermal conductivity enhancement is still under debate (Chandrasekar and Suresh 2009; Keblinski, Prasher, and Eapen 2008; Prevenslik 2009). Each of these mechanisms is discussed in more detail in Sections 1.2.1.1 through 1.2.1.3.

1.2.1.1 Brownian Motion

Nanoparticles move through the molecules of the base fluid and sometimes collide with each other by means of Brownian motion (Brown 1828; Einstein 1906b). Particularly, when two particles collide, the solid–solid heat transfer mode could increase the overall thermal

TABLE 1.1   Possible Transport Mechanisms and Relevant References

Transport Mechanism	References
Brownian motion	Wang, Xu, and Choi (1999)
	Keblinski et al. (2002)
	Yu et al. (2003)
	Patel et al. (2003)
	Das, Putra, Thiesen et al. (2003)
	Koo and Kleinstreuer (2004)
	Bhattacharya et al. (2004)
	Jang and Choi (2004)
	Kumar et al. (2004)
	Patel et al. (2005)
	Prasher et al. (2005)
	Ren, Xie, and Cai (2005)
	Prasher, Bhattacharya, and Phelan (2006)
	Evans, Fish, and Keblinski (2006)
	Beck, Sun, Teja et al. (2007)
	Shukla and Dhir (2008)
	Nie, Marlow, and Hassan (2008)
	Godson et al. (2010b)
	Kondaraju, Jin, and Lee (2010)
Liquid layering	Keblinski et al. (2002)
	Yu and Choi (2003, 2004)
	Xue et al. (2004)
	Eastman et al. (2004)
	Xie, Fujii, and Zhang (2005)
	Shukla and Dhir (2005)
	Evans et al. (2008)
	Chandrasekar et al. (2009)
Nanoparticle aggregation	Xuan, Li, and Hu (2003)
	Wang, Zhou, and Peng (2003)
	Prasher, Bhattacharya, and Phelan (2006)
	Zhu et al. (2006)
	Prasher, Bhattacharya, and Phelan (2006)
	Feng et al. (2007)
	Evans et al. (2008)
	Karthikeyan, Philip, and Raj (2008)
	Xu, Yu, and Yun (2006)
	Li, Zhang et al. (2008)
	Philip, Shima, and Raj (2008)

conductivity of the nanofluid. The effect of Brownian motion is a diffusive process (Einstein 1906a) with a diffusion constant D given by the Stokes–Einstein formula:

$$D = \frac{k_B T}{3\Pi\eta d} \tag{1.1}$$

where:
k_B is the Boltzmann constant
T is the temperature
η is the viscosity of the fluid
d is the particle diameter

As we note, the higher the temperature, the higher the diffusivity, and thus the higher the thermal conductivity. In some cases (Kumar et al. 2004; Nie, Marlow and Hassan 2008), the thermal conductivity measured experimentally followed the T/η behavior suggested by the Stokes–Einstein formula for diffusion.

Keblinski et al. (2002) estimated the effect of Brownian motion on the thermal conductivity of a nanofluid by comparing the timescale of the diffusion of a nanoparticle with that of heat diffusion in the fluid. The characteristic timescale to cover a distance equal to the particle diameter d due to Brownian diffusion is

$$\tau_D = \frac{d^2}{6D} \tag{1.2}$$

where D is the diffusion constant defined in Equation 1.1. Similarly, the timescale for heat to move in the fluid by the same distance is

$$\tau_H = \frac{d^2}{6\chi} \tag{1.3}$$

where χ is the thermometric conductivity, $k/\rho c_p$ is defined as the ratio of thermal conductivity to the heat capacity per unit volume of the fluid, or $\chi = k/\rho c_p$. For a 10-nm particle in water at 300 K, Equations 1.2 and 1.3 yield $\tau_D \simeq 2 \times 10^{-7}$ s and $\tau_H \simeq 4 \times 10^{-10}$ s. Hence, the ratio τ_D/τ_H is approximately 500 and drops to 25 when the particle size is reduced to the atomic size, showing that the thermal diffusion is much faster than Brownian diffusion, even when the particle is extremely small. In other words, although there is some evidence in the literature that higher temperatures would raise the thermal conductivity of the base fluid (Patel et al. 2003; Prasher, Bhattacharya, and Phelan 2006; Wang, Xu, and Choi 1999), according to Keblinski et al. (2002), the motion of nanoparticles due to Brownian motion is too slow to transport a significant amount of heat through a nanofluid.

Thus far, Brownian motion has only been related to diffusive effects in nanofluids, and convection has not been mentioned. A comparison between Brownian motion-based diffusion and Brownian motion-based convection was performed by Prasher, Bhattacharya, and Phelan (2006). They claimed that the reason why Keblinski et al. (2002) ignore the role that Brownian motion plays in thermal conductivity enhancement of nanofluids is based only on the fact that energy transport due to Brownian diffusion is 2 orders of magnitude smaller than that due to conduction in the fluid. However, Keblinski et al. (2002) did not account for energy transport due to convection caused by Brownian motion. Prasher, Bhattacharya, and Phelan (2006) compared the timescale of Brownian diffusion to that of Brownian convection and showed that the effect of convection is felt almost instantaneously compared to the Brownian diffusion of the nanoparticle. Prasher, Bhattacharya, and Phelan (2006) also developed a semiempirical model for predicting the thermal conductivity of nanofluids (Puliti, Paolucci, and Sen 2012a).

A Brownian-based convection model did not appear in the literature until 2003 when Yu, Hull, and Choi (2003) developed a one-dimensional drift velocity model for thermal conductivity of a nanofluid. In their model, they superimposed a thermophoretical particle drift on top of their Brownian motion, and claimed that only a few fluid particles would be dragged along in the process. This model failed to predict the effects of particle size. One year later, Koo and Kleinstreuer (2004) extended the model developed by Yu, Hull, and Choi (2003) to include the effects of the fluid particles dragged along by the nanoparticles. In the same year, Jang and Choi (2004) proposed the concept of nanoconvection, purely induced by Brownian motion of nanoparticles without thermophoretically drifting velocity. This model was able to predict temperature- and size-dependent thermal conductivity (Puliti, Paolucci, and Sen 2012a). Prasher, Bhattacharya, and Phelan (2005) extended the concept of nanoconvection by including the effects of multiparticle convection and developed a semiempirical model to show that Brownian motion is a leading contributor to thermal conductivity enhancement in nanofluids. Patel et al. (2005)

developed a microconvection model to evaluate the thermal conductivity of nanofluids by taking into consideration Brownian motion-induced nanoconvection and the specific surface area of nanoparticles. Ren, Xie, and Cai (2005) obtained a model that couples kinetic theory-based microconvection, liquid layering, and fluid and particle conduction.

Shukla and Dhir (2008) provided a detailed understanding of the effects of Brownian motion on the thermal conductivity of nanofluids from a different perspective. They divided the net heat flux due to Brownian motion into two components: a kinetic and an interaction part. They showed that for a typical nanofluid (with 10-nm-diameter nanoparticles with a 0.01 volume fraction in water at 300 K), the effect of the kinetic contribution to Brownian motion is negligible. On the contrary, the contribution of the interaction potential seemed to play a major role in the thermal conductivity enhancement of nanofluids. However, they did not provide a fundamental method for predicting the surface potential and the Debye length, and their model included only two-body interactions (Chandrasekar and Suresh 2009).

A more recent experimental work conducted by Beck, Sun, and Teja (2007), however, confirmed that Brownian motion models are not needed to describe the dependence of thermal conductivity enhancement on temperature, because they simply follow the temperature behavior of the base fluid. Even more recently, Nie, Marlow, and Hassan (2008) have reported that they finally found the order of magnitude of the thermal conductivity enhancement of nanofluids due to Brownian motion. It also happened to be proportional to T/η; however, its magnitude was on the order of 10^{-15} W/mK, negligible if compared with the thermal conductivity of water.

Whether Brownian motion plays a key role in the thermal conductivity enhancement of nanofluids is still under strong debate (Bastea 2005; Das et al. 2007; Sergis and Hardalupas 2011): A large group of researchers agrees that Brownian motion is important (Bhattacharya et al. 2004; Jang and Choi 2004, 2007; Koo and Kleinstreuer 2004, 2005; Kumar et al. 2004; Prasher, Bhattacharya, and Phelan 2005; Shukla and Dhir 2008), whereas others claim that it does not play a major role in the thermal conductivity enhancement (Bastea 2005; Beck, Sun and Teja 2007; Evans, Fish and Keblinski 2006; Keblinski and Cahill, 2005; Keblinski et al. 2002; Nie, Marlow, and Hassan 2008; Wang, Xu, and Choi 1999).

1.2.1.2 Liquid Layering

A liquid in contact with a solid interface is more ordered than the bulk liquid (Xie, Fujii, and Zhang 2005). The interaction between the atoms of the liquid and the solid generates an oscillatory behavior in the liquid density profile in the direction normal to the interface. The strength of the solid–liquid bonding determines the magnitude of the layering, which can even extend up to several molecular distances for sufficiently strong interactions. As the bonding gets stronger, a crystal-like structure develops in the liquid surrounding the particle. These structural changes in the liquid structure have been shown to have significant effects on various properties, such as liquid–solid phase transition, flow, and tribological properties. Other mechanical properties have been shown to be affected, such as viscosity and forces between bodies with thick liquid layering (Xue 2003; Xue et al. 2004). Such a change in the interfacial liquid ordering must also have an effect on thermal transport properties. In fact, lattice vibrations (phonons) are responsible for the highly efficient heat transfer mechanism in solids. Phonons have the exceptional ability to travel ballistically over long distances (phonon mean free path) before being scattered, either from crystal structure defects or from other lattice excitations. On the contrary, the absence of any sort of order in the liquid structure shrinks the effective collision mean free path to the order of one atomic distance. This is the reason why liquids do not have comparable heat transfer properties to solids. However, the increased order generated by liquid layering surrounding nanoparticles may have the ability to increase the liquid mean free path through which phonons can travel, causing an increase in thermal conductivity.

The possible enhancement of thermal conductivity by liquid layering is negatively affected by the interfacial resistance at the solid–liquid interface. This interfacial resistance R_K, also known as the Kapitza resistance (Kapitza 1941), can arise from differences in phonon spectra in the two phases and from scattering at the interface between the phases (Eastman et al. 2004). In the presence of a heat flux, this phonon scattering generates a temperature difference ΔT at the interface. The heat flux J is related to this temperature difference by

$$J = G\Delta T \tag{1.4}$$

where $G \equiv 1/R_K$ is the interfacial conductance. Thus, the thermal conductivity κ is related to the interfacial conductance by

$$h = \frac{\kappa}{G} \tag{1.5}$$

where h is the Kapitza radius, or the equivalent thickness of the crystal, defined as the distance over which the temperature drop is the same as that at the interface (Eastman et al. 2004). The value of h is usually small if the contact between phases is good and decreases with temperature. It can usually be neglected at room temperature for large, grain-sized materials (Swartz and Pohl 1989). For a solid–liquid interface, it is strongly affected by the properties of the absorbed layer of liquid (Nakayama 1985) and is usually still small at a macroscopic scale. However, for smaller lateral dimensions, such as for nanoparticles, the value of h can play a bigger role in the overall nanofluid heat transfer, because it becomes comparable to the size of the microstructure (particle size and interparticle distance) (Eastman et al. 2004). Jang and Choi (2007) calculated the Kapitza resistance per unit area and reported that it was on the order of 10^{-7} cm^{2}K/W; the Kapitza radius was not given. Hence, the Kapitza resistance is not dominant and the liquid layer cannot act as a barrier to heat transfer.

Shukla and Dhir (2005) performed an MD study of liquid layering in nanofluids using simplified LJ force fields and showed that when nanoparticles were at a distance of 2 nm from each other, the observed thermal conductivity of the layer was more than twice as the thermal conductivity of the bulk fluid. However, when the distance between nanoparticles was 27 nm, no thermal conductivity enhancement was observed in the liquid layer. This is due to the fact that the liquid layer extends only up to two or three atomic layers, and with a low interparticle distance of 2 nm, the contribution of the layer in enhancing energy transfer will be more compared to a larger interparticle spacing of 27 nm.

More recently, Evans, Fish, and Keblinski (2006) have reported that there was no thermal conduction enhancement in nanofluids when the radius of the nanoparticles becomes equal to the Kapitza radius. Thus, the Kapitza resistance might become an issue only for very small nanoparticles. Chandrasekar et al. (2009) estimated that the liquid layer thickness becomes important for the thermal transport in nanofluids only at higher concentrations.

Puliti, Paolucci, and Sen (2011) studied the effects of confinement in the case of a gold nanolayer immersed in water and reported a more ordered water structure in the neighborhood of the metal surface. This ordering is manifested by an increase of water density near the layer surfaces and extends to approximately 0.7 nm away from the metal–liquid interface, in agreement with similar experimental evidence (Li, Zhang et al. 2008). At least two water layers were identified near the metal nanolayer. This observed liquid layering was further confirmed by the study of the radial distribution function (RDF) of the confined water. In fact, in this case, the RDF shows higher peaks than pure water, and hence more ordered liquid structure up to about 0.7 nm away from the metal surfaces. Furthermore, the second and third coordination shells appear shifted to the right, indicating a nonstandard water structure. A count of the number of water molecules at different distances from the metal interface confirms the double layering with peaks at 0.25 and 0.75 nm.

Based on today's knowledge, conclusions on the importance of liquid layering for the thermal conductivity enhancement of nanofluids cannot be drawn. One of the factors that makes this task so difficult is the absence of a method to experimentally estimate more accurately the nanolayer thickness and conductivity, even though some attempts have been made (Lee 2007; Tillman and Hill 2007).

1.2.1.3 Nanoparticle Aggregation

One of the most controversial heat transfer mechanisms in nanofluids is perhaps nanoparticle clustering. Nanoparticles have been experimentally observed to agglomerate into clusters when suspended in the liquid (Hong, Hong, and Yang 2006; Yoo, Hong, Hong et al. 2007). In theory, nanoparticle clustering into percolating patterns creates paths of lower thermal resistance that would have a major effect on the overall thermal conductivity (McLachlan, Blaszkiewicz, and Newnham 1990; Shih et al. 1990) and viscosity (Gharagozloo and Goodson 2010). The effect on thermal conductivity enhancement would, however, be negated for low particle volume fractions, because there would be particle-free areas in the liquid.

Several MD studies have shown a substantial thermal conductivity enhancement with increasing nanoparticle clustering (Evans et al. 2008; Feng et al. 2007; Philip, Shima, and Raj 2008; Prasher, Phelan, and Bhattacharya 2006a, 2006b), despite the overwhelming opposite opinion of experimentalists (Hong, Hong, and Yang 2006; Karthikeyan, Philip, and Raj 2008; Li, Qu, and Feng 2008; Xu, Yu, and Yun 2006; Xuan, Li, and Hu 2003; Yoo, Hong, Hong et al. 2007). It is obvious that particle clustering must have some effect on the thermal conductivity, but its magnitude is still unknown (Özerinç, Kakaç, and Yazıcıoğlu 2009).

1.2.2 Experiments

1.2.2.1 Thermal Conductivity

Since the first experimental evidence (Masuda et al. 1993) with alumina nanoparticles in water, the curiosity and interest in the scientific community has grown. Many different materials have been tested for nanoparticles with different base fluids, and the results are astonishingly different for various combinations. The vast majority of thermal conductivity experiments in nanofluids are conducted using the transient hot-wire method (Beck et al. 2009; Choi et al. 2001; Lee et al. 1999; Liu, Lin, Huang et al. 2006; Liu, Lin, Tsai et al. 2006; Murshed, Leong, and Yang 2006; Özerinç, Kakaç, and Yazıcıoğlu 2009; Zhang, Gu, and Fujii 2006). Other common techniques in the literature include the temperature oscillation method (Das, Putra, Thiesen et al. 2003), the optical beam deflection technique (Das, Putra, Thiesen et al. 2003), the 3-ω method (Putnam and Cahill 2004), and the thermal wave technique (Wang 2009). Some of these methods might not necessarily be valid when applied to nanofluids, because none of them was initially designed and intended for nanosuspensions. It has been reported (Buongiorno et al. 2009) that some systematic differences in the measured thermal conductivity of nanofluids were observed for different measurement techniques. However, as long as the same measurements were taken at the same temperature, the thermal conductivity enhancement was consistent between the various experimental techniques. A summary of the most relevant nanofluids used in the literature is shown in Table 1.2. It is important to note that this table just shows the two most prominent base fluids used: water and ethylene glycol. Other base fluids, including toluene, oils, and ionic liquids, have also been used, but these are not shown in the table for the sake of brevity. During the first 10 years of nanofluid investigation, particular emphasis was given to the relationship between thermal conduction enhancement, and particle size and volume fraction.

Masuda et al. (1993) in Japan were the first to report thermal conductivity enhancement with nanofluids. They showed that as little as 4.3% particle volume fraction of silica, alumina, and other oxides in water increased the nanofluid thermal conductivity by 30%, although the friction factor almost quadrupled. Two years later, Choi (1995) at Argonne National Laboratories, Lemont, Illinois, coined the term nanofluids to denote this new class of engineering fluids. Wang, Xu, and Choi (1999), a few years later, reported thermal conductivity enhancements for the same oxides that Masuda

TABLE 1.2   Some of the Most Prominent Nanofluid Experiments

	Base Fluids	
	Water	**Ethylene Glycol**
	Nanoparticle Oxides	
Alumina	Masuda et al. (1993)	Lee et al. (1999)
	Pak and Cho (1998)	Wang, Xu, and Choi (1999)
	Wang, Xu, and Choi (1999)	Eastman et al. (2001)
	Lee et al. (1999)	Prasher, Bhattacharya, and
	Das, Putra, Thiesen et al. (2003)	Phelan (2006)
	Yoo et al. (2007)	Timofeeva et al. (2007)
	Timofeeva et al. (2007)	Murshed, Leong, and Yang (2008)
	Tavman et al. (2010)	
Copper oxide	Lee et al. (1999)	Lee et al. (1999)
	Das, Putra, Thiesen et al. (2003)	Wang, Xu, and Choi (1999)
	Karthikeyan, Philip, and Raj (2008)	Eastman et al. (2001)
	Mintsa et al. (2009)	Kwak and Kim (2005)
		Karthikeyan, Philip, and Raj (2008)
Silica	Masuda et al. (1993)	Kang, Kim, and Oh (2006)
	Kang et al. (2006)	
	Hwang et al. (2006)	
	Duangthongsuk and Wongwises (2009)	
	Tavman et al. (2010)	
Titanium dioxide	Masuda et al. (1993)	Hong, Yang, and Choi (2005)
	Pak and Cho (1998)	Kim, Kang, and Choi (2007)
	Kim, Kang, and Choi (2007)	
	Tavman et al. (2010)	
Zirconia	Chopkar et al. (2007)	Chopkar et al. (2007)
	Metals	
Copper	Xuan and Li (2000)	Eastman et al. (2001)
	Xuan, Li, and Hu (2003)	Hong, Hong, and Yang (2006)
	Jana, Salehi-Khojin, and Zhong (2007)	Garg et al. (2008)
	Li, Zhang et al. (2008)	
Gold	Patel et al. (2003)	–
	Kumar et al. (2004)	
	Jana, Salehi-Khojin, and Zhong (2007)	
Iron	Hong, Yang, and Choi (2005)	Hong, Hong, and Yang (2006)
		Yoo, Hong, and Yang (2007)
Silver	Patel et al. (2003)	–
	Kang, Kim, and Oh (2006)	

considered not only in water but also in ethylene glycol. Wang's group showed that the thermal conductivity enhancement increases as the size of the particles decreases and is almost proportional to the volume fraction of the particles. A maximum thermal conductivity enhancement of 12% was observed when alumina particles were used with a volume fraction of just 3%. At the same time, the viscosity increased by 20%–30%. An interesting conclusion of Wang, Xu, and Choi (1999) is that the viscosity ratio (that of the nanofluid to the base fluid) has a quadratic dependence on the volume fraction of the nanoparticles, although the thermal conductivity has a linear dependence. A similar study was performed by Pak and Cho (1998), for alumina with the same volume fraction, but they obtained viscosity results 3 times larger than those of Wang, Xu, and Choi (1999). This is an indication that something is different in the particle dispersion, and that perhaps the size of the particles, volume fraction, shape, temperature, and so on must be taken into account while stabilizing each nanofluid dispersion. For this reason, further experiments were performed to study the particle size dependency. Eastman et al. (2001) reported that 10-nm copper particles dispersed in ethylene glycol enhanced the thermal conductivity by 40% with less than 0.3% particle volume fraction. However, with 35-nm copper oxide particles the enhancement was just 20% for a particle loading factor of 4%. Therefore, the size of particles is obviously related to the thermal conductivity enhancement. A factor that has not been considered so far, and discussed very little in the literature as a whole (Eapen, Li, and Yip 2004), is quantum effects, which are not insignificant at the nanoscale. In fact, Patel et al. (2003) demonstrated that thermal conduction enhancement is different for particles of different materials, even at the same surface-to-volume ratio. In particular, they showed that, for the same volume fraction, thermal conductivity enhancement is higher for copper oxide than for alumina. Furthermore, Patel's group noticed a linear increase in the conductivity ratio with temperature for 4-nm gold particles in toluene. Similar work was performed by Das, Putra, Thiesen et al. (2003) 2 years later, and they noticed the same linear trend for alumina and copper oxide in water. However, as the particle volume fraction increases, the trend of conductivity enhancement varies more like the square root of temperature. Choi et al. (2001) reported a 150% enhancement in the thermal conductivity of poly(α-olefin) oil with a suspension of 1% volume fraction of multiwalled carbon nanotubes (MWCNTs). Similar results were reported by Yang (2006), where only 0.35% by volume of MWCNT increased the thermal conductivity of poly(α-olefin) oil by 200%. A 38% increase in the thermal conductivity of water was reported by Assael et al. (2004) with 0.6% by volume of MWCNTs. They also noted that the thermal conductivity enhancement increased with the length-to-diameter ratio of the nanotubes.

These first 10 years of nanofluid research suggested that the thermal conductivity enhancement in nanofluid must be a function of temperature, particle size, and concentration, but not much was said on the interfacial effects of the liquid–solid interaction. An interesting conclusion from Patel's group (Patel et al. 2003) was that nanoparticles with a coating, such as thiolate on gold particles, were less effective in terms of thermal conductivity enhancement than uncoated gold particles. To produce the same effects as the uncoated nanoparticles, a larger volume fraction of gold/thiolate was needed. This suggests that part of the thermal conduction enhancement must be related to the particle surface's properties. After this experiment, many more results were published and many of them have taken into account not only temperature, particle size, and concentration, as in the past, but also nanoparticle clustering and surface treatment, besides other liquid/solid interfacial effects. It was 2001 when Eastman et al. (2001) added 1% volume fraction of thioglycolic acid to aid in the dispersion of nanoparticles, and they observed that this yielded a greater thermal conductivity than the same concentration of copper nanoparticle in ethylene glycol without the dispersant. Jana, Salehi-Khojin, and Zhong (2007) repeated a similar experiment (same 0.3% volume fraction of 10-nm copper nanoparticles), except that the base fluid was water and the dispersant was laurate salt. The thermal conductivity enhancement observed by Jana, Salehi-Khojin, and Zhong (2007) was now 70%, indicating that the chemical property of the dispersant had a large impact on the thermal conductivity enhancement. Kang, Kim, and Oh (2006) reported a 75% enhancement in thermal conductivity for ethylene glycol with a volume fraction of 1.2% diamond nanoparticles between 30 and 50 nm in diameter. Hong, Hong, and Yang (2006) showed that the thermal conductivity of an iron–ethylene glycol nanofluid was strictly connected to the nanoparticle clustering. In fact, they observed a nonlinear relation between the thermal conductivity and the nanoparticle volume fraction, and they attributed this nonlinearity to the increase of the nanoparticle cluster sizes. Another interesting observation in their study regards a comparison of Fe and Cu nanofluids. In fact, the thermal conductivities of the lowest concentrated Fe nanofluids of their study were fit to a linear function and showed a higher thermal conductivity than the Cu nanofluid. This is quite surprising because the dispersion of a higher conductive material is less efficient in improving the thermal conductivity. All this shows that the intrinsic properties of a nano-fluid are to some extent uncorrelated to those of the bulk materials due to size confinement and surface effects. Furthermore, the thermal conductivity of the nanofluid is related to the Kapitza interfacial resistance at the interface between the particles and the fluid (Hong, Hong, and Yang 2006). Following Hoang et al.'s hypothesis, Yoo, Hong, Hong et al. (2007) investigated the effects of nanoparticle clustering on thermal conductivity. Yoo's group controlled the size of the nanoparticle clusters by changing the pH of the liquid and analyzed how the overall thermal conductivity enhancement was affected. They dispersed alumina in water with pH values of 7 (neutral), 9.65, and 10.94, and noticed that as the pH increased, the particle clusters were getting smaller and the thermal conductivity of the nano-fluid was increasing. Therefore, they claimed that nanoparticle aggregations, due to van der Waals interaction, result in a reduction of the overall thermal conductivity of the nanofluid. It was now evident that the thermal conductivity of the nanofluids should also be a function of nanoparticle clustering, besides temperature, particle size, and concentration. Similar work was done by Li, Zhang et al. (2008), but using copper, rather than alumina, and a surfactant [sodium dodecylbenzenesulfonate (SDBS)] together with pH control. The main function of the surfactant was to stabilize and make more homogeneous the nanoparticle suspension. They reported that water with a dispersion of 25-nm copper nanoparticles had a noticeably higher thermal conductivity than the base fluid without nanoparticles. In particular, a 0.1% weight fraction of copper in water increased the overall thermal conductivity by 10.7% with an optimal pH value (pH 8.5–9.5) and SDBS concentration. Therefore, they again showed that the pH strongly affects the thermal conductivity. As the pH increases, the surface charge of the nanoparticles increases due to the more frequent attacks to the surface hydroxyl groups and phenyl sulfonic group by potential-determining ions (H+, OH, and phenyl sulfonic group) (Li, Zhang et al. 2008), stabilizing the colloidal particles, and consequently increasing the thermal conductivity.

At this point, it is obvious that interfacial effects and nanoparticle clustering both play roles in determining the effective thermal conductivity of nanofluids, together with temperature, particle size, and concentration. Meanwhile, other works in the literature still continued the investigation of how particle size and concentration affect the thermal conductivity. In particular, Kumar et al. (2004) showed that just 0.00013% by volume of 4.5-nm gold nanoparticles in water yielded up to 20% thermal conductivity enhancement. They concluded that the enhancement is inversely proportional to the nanoparticle size and linearly proportional to the particle concentration, at least for small concentrations. His group also observed some temperature dependence of thermal conductivity due to Brownian motion. Chopkar et al. (2007) reported a thermal conductivity increase of about 100% with only 1.5 volume fraction of 25-nm zirconia particles in ethylene glycol. They also noticed a strong dependence of thermal conductivity on particle size and shape. Anomalous behavior of nanofluids was again observed by Garg et al. (2008) who investigated copper in ethylene glycol. This group reports a thermal conductivity enhancement that is twice the prediction of a Maxwell model (Puliti, Paolucci, and Sen 2012a), but much lower than the one reported by Eastman et al. (2001). At the same time, the increase in viscosity with particle concentration was 4 times the prediction of the Stokes–Einstein model (Puliti, Paolucci, and Sen 2012a). Mintsa et al. (2009) concluded that the thermal conductivity of nanofluids also depends on the nanoparticle size and temperature. Vajjha and Das (2009a) agreed with Mintsa et al. (2009) that the thermal conductivity enhancement depends not only on the nanoparticle concentration but also on the temperature; in particular, nanofluids seemed to be more effective if used in high-temperature applications.

In some cases (Buongiorno et al. 2009; Escher et al. 2011; Putnam et al. 2006; Shalkevich et al. 2010; Tavman et al. 2010; Turanov and Tolmachev 2009), experimental evidence showed little or no enhancement in the thermal conductivity of nanofluids. Buongiorno et al. (2009) conducted an international nanofluid benchmark exercise, where several experimental datasets were analyzed by 34 different organizations around the world. They were able to use classical effective medium theory for well-dispersed particles (Puliti, Paolucci, and Sen 2012a) to accurately reproduce experimental data, providing evidence that there is no anomalous enhancement in the thermal conductivity of their limited sets of nanofluids.

Sergis and Hardalupas (2011) conducted a detailed statistical analysis of several nanofluid properties presented in the literature. Even though the overall trends they reported are simply the result of statistical tools rather than purely scientific techniques, they are still worth mentioning. With regard to the effective thermal conductivity of nanofluids, they showed that the enhancement usually increases with increasing nanofluid temperature and nanoparticle concentration, but there is a slight trend showing the enhancement increasing also with increasing nanoparticle size.

Many of the results discussed in this section and a few others are reflected in Figures 1.2 and 1.3, where k is the thermal conductivity.

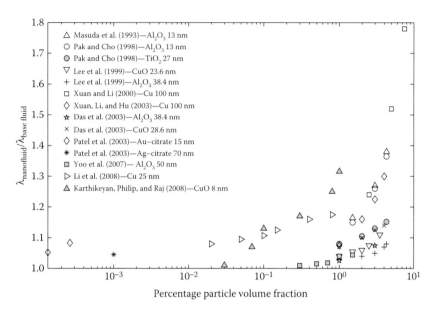

FIGURE 1.2 Experimental results of thermal conductivity enhancement of water-based nanofluids. (Reprinted with permission from Puliti, G. et al., *Appl. Mech. Rev.*, 64, 3, 030803, 2012a.)

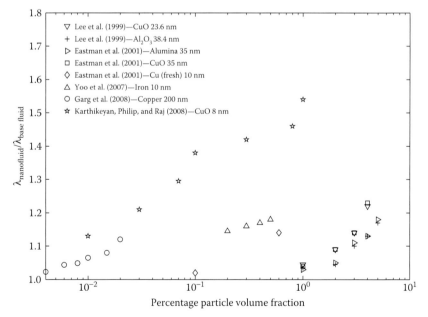

FIGURE 1.3 Experimental results of thermal conductivity enhancement of ethylene glycol-based nanofluids. (Reprinted with permission from Puliti, G. et al., *Appl. Mech. Rev.*, 64, 3, 030803, 2012a.)

1.2.2.2 Viscosity

Even though the literature on heat convection in nanofluids is limited compared to that in thermal conductivity, the results and approaches in the field are quite diverse and worth mentioning (Das et al. 2007). However, the understanding of the issues of convection is strictly related to the viscosity of the nanofluids, which we address here.

In the first place, we need to understand that whether nanofluids are Newtonian or shear thinning is important. This question was addressed by the first work ever done on heat convection in nanofluids in 1998 by Pak and Cho (1998). They reported that the nanofluids behaved as Newtonian when 13- and 27-nm nanoparticles of γAl_2O_3 and TiO_2 were suspended in water, but only for very low particle volume fractions. Shear-thinning behavior was, however, detected with an increase of particle volume fraction. In particular, the Al_2O_3–water nanofluid started showing shear-thinning behavior at 3% particle volume fraction onward, whereas TiO_2–water behaved as Newtonian up to 10%. Pak and Cho also observed a large increase in the viscosity of both nanofluids and showed that this behavior was not predicted by standard empirical models for suspension viscosities, such as the one developed by Batchelor in 1977 (Batchelor 1977; Das et al. 2007). Lee et al. (1999) reported similar substantial increase in viscosity for water and ethylene glycol-based nanofluids with Al_2O_3 nanoparticles. They also mentioned that the confirmation of this finding might offset the enhanced heat transfer observed in nanofluids.

In 2003, however, Das, Putra, and Roetzel (2003) conducted a similar study for a water–Al_2O_3 nanofluid, but reported no observation of shear-thinning behavior. In particular, they confirmed Newtonian behavior of the nanofluid with 1%–4% concentrations by volume. They also showed a linear trend of viscosity versus shear rate, even though viscosity values are still higher than those of pure water. Viscosity is also reported to increase with nanoparticle loading but remains Newtonian in nature.

As already discussed for thermal conductivity, nanofluids with carbon nanotubes (CNTs) behave quite differently than their nanoparticle counterparts, and this applies also to viscosity (Das et al. 2007). Ding et al. (2006) showed an interesting linear shear-thinning behavior for water–CNT nanofluids at lower shear rates, whereas the base fluid had a nonlinear trend.

Hence, there are some discrepancies in the literature on the nature of the rheology of nanofluids: some believe that nanofluids are Newtonian (Das, Putra, and Roetzel 2003; Prasher, Song et al. 2006; Wang, Xu, and Choi 1999; Xuan and Li 2000), whereas the others found non-Newtonian shear-thinning behavior (Ding et al. 2006; He et al. 2007; Kwak and Kim 2005; Studart et al. 2006; Tseng and Lin 2003). In the hope to explain these discrepancies, Chen, Ding, and Tan (2007) conducted both experimental and theoretical analyses of ethylene glycol-based nanofluids with TiO_2 nanoparticles at concentrations from 0.5% to 8.0% by weight at 20°C–60°C. They observed Newtonian behavior under their conditions, with shear viscosity being a strong function of temperature and nanoparticle loading. However, the relative shear viscosity (obtained by normalizing with respect to the shear viscosity of the base liquid) depends nonlinearly on the nanoparticle loading and is surprisingly independent of temperature. The high shear viscosity of nanofluids was shown to be in line with the prediction of the Krieger–Dougherty equation (Krieger 1959) if the nanoparticle volume fraction is replaced by the concentration of nanoparticle clusters. The shear-thinning behavior sometimes reported in the literature is mainly attributed to three factors: (1) the effective particle loading, (2) the range of shear rate, and (3) the viscosity of the base fluid. Such non-Newtonian behavior can be characterized by a characteristic shear rate that decreases with increasing nanoparticle loading, increasing viscosity of the base fluid, or increasing nanoparticle cluster size. Chen, Ding, and Tan (2007) claim that this might explain the reported controversy in the literature. At ambient temperatures, the relative high shear viscosity is independent of temperature due to a reduced Brownian diffusion in comparison with convection in high shear flows. Nevertheless, the characteristic shear rate can still have a strong temperature dependence, thus affecting the shear-thinning behavior of the nanofluid. The rheological nature of nanofluids can be categorized into four groups: (1) dilute nanofluids ($0 \leq \phi \leq 0.001$, with ϕ being the nanoparticle volume fraction) with well-dispersed nanoparticles—the nanofluid shows no shear-thinning behavior and the viscosity increase fits the Einstein model (Einstein 1906a, 1911); (2) semidilute ($0.001 \leq \phi \leq 0.05$) with nanoparticle clusters—the nanofluid viscosity is more in line with the modified Krieger–Dougherty equation and still shows Newtonian behavior; (3) semiconcentrated ($0.05 \leq \phi \leq 0.1$) with nanoparticle clusters—the viscosity still fits the modified Krieger–Dougherty equation but is obviously non-Newtonian; and (4) concentrated nanofluids ($\phi \leq 0.1$) with interpenetration of particle aggregation, but it is out of the standard nanoparticle loading range of nanofluids.

Tavman et al. (2010) investigated TiO_2, SiO_2, and Al_2O_3 nanoparticles in water and also reported a dramatic increase in nanofluid viscosity with an increase in the nanoparticle concentration; they also showed that classical empirical theories, such as the Einstein model (Einstein 1906a, 1911), were unable to predict the correct viscosity increase in nanofluids. Kole and Dey (2010) presented experimental results for viscosity measurements of a car engine coolant (50% water and 50% propylene glycol) with a dispersion of alumina nanoparticles. The nominal diameter of the nanoparticles was below 50 nm, whereas the volume fraction was between 0.001 and 0.015. They concluded that the addition of such a small amount of nanoparticles transforms the Newtonian behavior of the pure engine coolant to a non-Newtonian one, unlike what Chen, Ding, and Tan (2007) predicted, and it behaves as a Bingham plastic with small yield stress. In addition, yield stress versus shear strain rate data showed a power-law dependence on the particle volume fraction. The viscosity of the nanofluid, μ_{nf}, appears to depend on temperature T according to an empirical correlation of the type $\log(\mu_{nf}) = A\exp(-BT)$, where A and B are curve-fitting parameters (Namburu et al. 2007).

Kole and Dey also showed good agreement with the model derived by Masoumi, Sohrabi, and Behzadmehr (2009) with regard to the viscosity dependence on the nanoparticle concentration. Considering the fact that the model Masoumi, Sohrabi, and Behzadmehr (2009) derived is mainly based on the effects of Brownian diffusion on the nanofluid viscosity, this agreement might signify that Brownian motion plays a key role in the viscosity increase in nanofluids. A year later, Kole and Dey (2011) reported similar results, but for 40-nm-diameter spherical CuO nanoparticles in gear oil. Viscosity showed a strong dependence on both nanoparticle concentration and temperature between 10°C and 80°C. They reported an increase in viscosity of approximately 300% of the base fluid for a CuO volume fraction of 0.025. However, viscosity decreased significantly with increasing temperature. The Newtonian nature of gear oils appeared to change to non-Newtonian with an increase in nanoparticle loading, and shear thinning was observed for a CuO volume fraction above 0.005. CuO nanoparticles also appeared to aggregate into clusters with an average size of about 7 times the nanoparticle diameter.

The modified Krieger and Dougherty model (Wang, Zhou, and Peng 2003), derived by taking into account nanoparticle clustering, appeared to predict well the viscosity increase in the nanofluid. The temperature variation of the viscosity follows the modified Andrade equation (Andrade 1934; Chen, Ding and He et al. 2007).

The statistical analysis of viscosity trends in the literature conducted by Sergis and Hardalupas (2011) shows that the viscosity of nanofluids increases with decreasing temperature and increasing nanoparticle loading. In addition, there is a slight hint of an effective viscosity increase with decreasing nanoparticle size.

The significance of investigating the viscosity of nanofluids has been emphasized recently (Kole and Dey 2011), but much more still needs to be done (Saidur, Kazi et al. 2011). Viscosity is as critical as thermal conductivity in assessing adequate pumping power as well as the heat transfer coefficient in engineering systems with fluid flows (Kole and Dey 2011). In fact, Abu-Nada et al. (2010) discussed the important correlation between viscosity and the Nusselt number. Higher nanoparticle volume fractions cause the fluid to become more viscous, thus attenuating the velocity and resulting in a reduced convection. The reduction of velocity and convection will cause an increase of the thermal boundary layer thickness, thus reducing the temperature gradients and lowering the Nusselt number. Abu-Nada et al. (2010) mentioned that most of the numerical studies in heat convection of nanofluids underestimate the effective viscosity of nanofluids, despite the fact that it plays a major role in their overall heat transfer behavior. For instance, Escher et al. (2011) demonstrated that the relative thermal conductivity enhancement must be larger than the relative viscosity increase in order to gain any benefit in using nanofluids for electronic cooling.

1.2.2.3 Heat Convection

Although an enhanced thermal conductivity in nanofluids is an encouraging feature for possible application in heat transfer devices, it is not necessarily a sufficient condition. In fact, nanofluids should also be examined for performance under convective modes. Some of the most interesting experimental results in this area are discussed in this section.

Pak and Cho (1998) were the first to address convection in nanofluids and reported a large increase in the heat transfer coefficient in the turbulent flow regime when their nanofluids (13- and 27-nm nanoparticles of Al_2O_3 and TiO_2 in water, respectively) were used inside a 10.66-mm-diameter tube. In particular, the increase was 45% with 1.34% volume fraction of Al_2O_3 in water and 75% with a nanoparticle loading of 2.78%. However, they observed a 3%–12% decrease in the heat transfer coefficient for the case of a constant average velocity. Because of the peculiar behavior of thermal conductivity of nanofluids, Pak and Cho and many other researchers (Das et al. 2007) prefer to deal with a dimensional heat transfer coefficient rather than the Nusselt number, which includes the thermal conductivity of the nanofluids.

Xuan and Li (2003), however, reported results that contradicted the ones of Pak and Cho. In fact, they showed an increase of as much as 40% in the heat transfer coefficient of the nanofluid, at the same constant average velocity. The main difference between their experiments was in the choice of nanoparticles, which in the case of Xuan and Li were 100-nm copper particles. Xuan and Li justified Pak and Cho's results by assuming that, because of the large viscosity they reported, some of the turbulence might have been suppressed, consequently lowering the heat transfer coefficient. Based on these last results, it is now clear that the heat transfer coefficient must depend not only on the particle volume fraction but also on the particle size and material. Furthermore, the standard empirical equation for predicting the heat transfer coefficient such as the Dittus–Boelter equation (Das et al. 2007; Dittus and Boelter 1930) cannot be applied to nanofluids. In fact, convection in nanofluids is a function of untraditional variables, such as gravity, Brownian force, diffusion, and drag on the nanoparticles, and such empirical equation is 39% off for water with 2% copper nanoparticles (Das et al. 2007; Xuan and Li 2003).

In 2004, further progress in convection studies was accomplished by Wen and Ding (2004). They were the first to study entry-length effects (before the flow is fully developed) and reported a higher heat transfer coefficient in those regions, mainly because the boundary layer is thinner. In particular, they measured the heat transfer coefficient along a tube when the water–Al_2O_3 nanofluid was in the laminar flow regime. The enhancement in the heat transfer coefficient was substantial, larger at the entry-length region, and rising with increasing nanoparticle concentration. This discovery leads to the speculation that the thermal convection enhancement can be kept as high as it is in the entry-length region, by creating artificial entrance regions with boundary layer interruption techniques. The entry length of nanofluids also appears to be longer than that for pure fluids. Nanoparticle migration, and anisotropic thermal conductivity and viscosity might be the causes of the reduced boundary layer thickness and the consequent enhancement of the heat transfer coefficient.

Yang et al. (2005) studied heat convection of four different nanofluids and reported that nanoparticle concentration, material, temperature, and base fluid affected the heat transfer coefficient. Their experiment is worth mentioning mainly because they were the first to realize that electrical heating, commonly used for heat transfer measurements, might affect the motion of the nanoparticles, because they are likely to become electrically charged, as more recently discussed by Kabelac and Kuhnke (2006). Consequently, Yang et al. (2005) used hot water for heating purposes.

Heris, Etemad, and Nasresfahany (2006) reached similar conclusions as Xuan and Li (2003). They tested two different water-based nanofluids inside a 6-mm-diameter copper tube, the first with CuO and the other with Al_2O_3 nanoparticles. They used steam for heating purposes rather than the controversial electrical heating. Large enhancement was reported for both nanofluids, despite the use of nonmetallic nanoparticles, but it was more substantial for the water–Al_2O_3 nanofluid. They also concluded that the heat transfer coefficient increased with a higher nanoparticle volume fraction and at higher Peclet numbers. A similar conclusion was obtained by the numerical work of Vajjha, Das, and Namburu (2010). They analyzed the heat transfer performance of Al_2O_3 and CuO nanofluids in the flat tubes of a radiator and reported that the convective heat transfer and the local and averaged friction factors showed a remarkable improvement over the base fluid, and the enhancement increased with increasing nanoparticle concentration. An increase of convective heat transfer with increasing nanoparticle concentration was also reported in the numerical works of Arefmanesh and Mahmoodi (2011) and Gherasim et al. (2011).

Even though all the experiments in nanofluids heat convection discussed thus far seem to qualitatively agree with each other for the most part, quantitatively speaking, the trends of the data are quite diverging between Pak and Cho (1998) and Yang et al. (2005) on the one hand, and Xuan and Li (2003), Wen and Ding (2004), and Heris, Etemad, and Nasresfahany (2006) on the other (Das et al. 2007). The major differences between all these experiments, and the potential reasons for those divergent trends, are in the particle source, preparation method, nanoparticle size distribution, dispersion technique, and pH value, among others. Escher et al. (2011), however, showed that standard empirical correlations could be successfully used to estimate the convective heat transfer of a water–silica nanofluid, and no anomalous thermal conductivity or heat convection enhancements were observed.

While a significant number of studies have been performed in forced convection (Daungthongsuk and Wongwises 2007; Kakaça and Pramuanjaroenkij 2009), as discussed up to this point, not much is available for natural convection of nanofluids (Corcione 2011). The first work was done in 2003 by Putra, Roetzel, and Das (2003) with two water-based nanofluids: one with 131.2-nm Al_2O_3 nanoparticles and the other with 87.3-nm CuO particles. They observed that the natural convective heat transfer in nanofluids is lower than pure water, as the nanoparticle volume fraction increases. The deterioration was greater for the CuO nanofluid than for Al_2O_3–water nanofluid.

A decreasing effect of natural convection of nanofluids was also observed by Wen and Ding (2005) for a TiO_2–water nanofluid. They attributed this deterioration to the gradients in the nanoparticle concentration, to the particle–surface and particle–particle interaction, and to the modification of the properties of the dispersion.

However, Abu-Nada, Masoud, and Hijazi (2008; Abu-Nada 2009) showed that an enhancement of the heat transfer coefficient in the natural convection of a Al_2O_3–water nanofluid is possible, depending on a balance between the Rayleigh number, the particle concentration, and the aspect ratio of the enclosure. In particular, for a unity aspect ratio enclosure, the Nusselt number was slightly enhanced with increasing nanoparticle concentration at low Rayleigh number (Eastman et al. 2004), but it was reduced if the nanoparticle volume fraction exceeded 5% at a higher Rayleigh number (Evans et al. 2008; Xie, Fujii, and Zhang 2005). However, for a CuO–water nanofluid, the Nusselt number was always deteriorated, because the viscosity of CuO is relatively large compared to Al_2O_3. This means that viscosity dominates, enhancing thermal conduction due primarily to the high thermal conductivity of nanoparticles, consequently lowering the Nusselt number.

Corcione (2011), assuming that nanofluids behave more like single-phase fluids, analyzed their heat transfer performance for single-phase natural convection in bottom-heated enclosures. The heat transfer enhancement that derives from the dispersion of nanosized particles into the base fluid was shown to increase up to an optimal nanoparticle loading. This optimal loading increases as the temperature increases, but it may increase or decrease with increasing nanoparticle size depending on whether the flow is laminar or turbulent. Also, when different combinations of solid nanoparticles and base fluids are considered, the heat transfer enhancement and the optimal nanoparticle concentration depend much more on the base fluid than on the nanoparticle material (Corcione 2011). The heat transfer enhancement followed the same trend in the case of natural convection in annular spaces between horizontal concentric cylinders (Cianfrini, Corcione, and Quintino 2011).

More recent studies in the literature investigated more exotic factors in relation to heat convection, such as the effect of the presence of a magnetic field (Ghasemi, Aminossadati, and Raisi 2011; Hamad 2011), and how the nanoparticles in nanofluid flows affect entropy generation (Moghaddami, Mohammadzade, and Esfehani 2011). Some specific analytical investigations have also been reported using nanofluids such as the solution of the Falkner–Skan equation (Yacob, Ishak, and Pop 2011), or of the Cheng–Minkowycz problem (Nield and Kuznetsov 2011).

Sergis and Hardalupas (2011) conducted their statistical analysis also for trends in heat convection studies for nanofluids and reported that the convective mode seems to be enhanced with increasing temperature, increasing nanoparticle volumetric concentration, and decreasing particle size.

Unfortunately, experiments on convection only covers oxide nanoparticles and high nanoparticle concentrations (>1% by volume). More heat convection experiments are needed to fill the gaps, especially with metallic particles (that have much higher thermal conductivity) and lower concentrations (<1% by volume) (Godson et al. 2010a; Saidur, Leong, and Mohammad 2011).

1.2.2.4 Specific Heat
Literature on experimental specific heats of nanofluids is very limited and has just lately been getting more attention.

Namburu et al. (2009) reported that several ethylene glycol-based nanofluids exhibit lower specific heat than their respective base fluids. Similarly, Bergman (2009) reported experimental evidences that a water–alumina nanofluid appeared to have enhanced thermal conductivity, but lower specific heat, relative to the base fluid. Vajjha and Das (2009b) studied nanofluids with 2%–10% by volume of Al_2O_3, SiO_2, and ZnO nanoparticles in a 60:40 ethylene glycol–water mixture. They reported that the specific heat of all their nanofluids decreases substantially as the volumetric concentration of nanoparticles increases, and it increases moderately with temperature. They were not able to predict the specific heat of nanofluids using existing empirical models (Puliti, Paolucci, and Sen 2012a). Lower specific heat has also been reported in the numerical work of Puliti, Paolucci, and Sen (2011). In one case (Escher et al. 2011), no anomalous specific heat was observed, and classic theories were able to predict the nanofluid density and specific heat within a 10% deviation (Puliti, Paolucci, and Sen 2012a).

Conflicting results to the ones mentioned above, however, have also been reported. Nelson, Banerjee, and Ponnappan (2009) observed that the specific heat of polyalphaolefin is increased by 50% with a dispersion of exfoliated graphite nanoparticles at a concentration of 0.6% by weight. Similarly, Shin and Banerjee (2011a) reported a 26% enhancement of the specific heat of a molten salt (eutectic of 62% lithium carbonate and 38% potassium carbonate) with a suspension of silica nanoparticles at a concentration of 1% by weight. However, other contradictory results were reported by Zhou and Ni (2008); they observed that the specific heat of an alumina–water nanofluid decreased by 40%–50% at a concentration of 21.7% by volume of nanoparticles. Shin and Banerjee (2011a, 2011b) commented on the results by Zhou and Ni, pointing out that alumina nanoparticles tend to cluster, and the authors did not report if the nanoparticles were well dispersed. In fact, agglomerated alumina nanoparticles tend to precipitate out of the water solution, consequently degrading the thermal properties of the nanofluid. More recently, Shin and Banerjee (2011a) have reported a 14.5% enhancement in the specific heat of a chloride salt eutectic, using a dispersion of 1% by weight of silica nanoparticles (20–30 nm nominal diameter). In this case, the dispersion behavior of the nanoparticles was confirmed by scanning electron microscopy.

Shin and Banerjee (2011a) proposed three independent thermal transport mechanisms to explain the unusual enhancement of the specific heat they observed:

- *Mode 1*: The specific heat is enhanced due to higher specific surface energy of the surface atoms of the nanoparticles, compared to the bulk material. The surface energy is higher because of the low vibrational frequency and higher amplitudes of the vibrations at the surface of the nanoparticles.

- *Mode 2*: The enhancement of the specific heat can also be due to additional thermal storage mechanisms generated by interfacial interactions between nanoparticles and the liquid molecules, which act as virtual spring–mass systems. This interfacial effect is present due to the extremely high specific surface area of the nanoparticles.

- *Mode 3*: A third mechanism potentially involved is liquid layering, already discussed earlier with regard to the thermal conductivity enhancement (see Section 1.2.1.2). Solid-like liquid layers adhering to the nanoparticles are more likely to have an enhanced specific heat due to a shorter intermolecular mean free path compared to the bulk fluid.

Because the specific heat is a key thermal property for many engineering applications, there is a great need for additional experimental results on this fundamental property for nanofluids (Saidur, Kazi et al. 2011; Shin and Banerjee 2011a).

1.2.3 Empirical Models of Properties

The nature of the thermal conductivity enhancement in nanofluids is still under debate. Although some authors derived empirical models to predict the thermal conduction enhancement in nanofluids (Jang and Choi 2004; Kumar et al. 2004; Prasher, Bhattacharya et al. 2006; Yu and Choi 2003, 2004), others claim strong evidence for the classical nature of the nanofluid thermal conductivity (Buongiorno et al. 2009; Eapen, Rusconi et al. 2010; Tavman et al. 2010). Newer models have been developed in the past few years, but most of them are simply variations of classical models developed for solid–solid composites or microparticle suspensions (Puliti, Paolucci, and Sen 2012a).

Empirical theories of rheology began with Einstein's study of infinitely dilute suspensions of hard spheres in the early 1900s (Einstein 1906a, 1911). Since then, a few other models have been developed for higher particle volume fractions. Some of these classical models have been applied to predict the viscosity of several nanofluids with spherical and rodlike nanoparticles (Chen and Ding 2009). However, each proposed model applies only to a specific nanofluid, and it does not reduce to the Einstein equation at very low particle volume concentrations, thereby lacking a rigorous physical foundation (Wang and Mujumdar 2008).

Empirical models for the heat transfer coefficient of suspensions are quite limited in the literature, but lately have been gaining attention from researchers (Wang and Mujumdar 2008). Newer empirical models are usually modified versions of some classical theory, such as Maxwell's for the thermal conductivity of nanofluids (Maxwell 1881), or Einstein's for their viscosity (Einstein 1906a, 1911). Similarly, empirical models for the heat transfer coefficient are modified versions of the traditional Dittus–Boetler equation (Dittus and Boelter 1930) or of the Gnielinski equation (Gnielinski 1976).

Models for the specific heat of nanofluids, mostly at constant pressure, are also limited in the literature, but lately have been gaining much more attention. Please refer to Puliti, Paolucci, and Sen (2012a) for more details on the various empirical methods available in the literature.

1.2.4 Numerical Studies

The main shortcoming of the previously discussed models is that although they can predict correctly some nanofluid properties, others must be obtained by fitting experimental data. Furthermore, the many experimental works discussed so far do not provide a fundamental understanding of heat transport in nanofluids. Considering all the various practical challenges in producing nanofluids, the substantial disagreements among experiments in the literature, and the large variety of experimental techniques used (and possibly abused!) in those experiments, it appears that only a detailed numerical model could perhaps fill the gaps and potentially answer some fundamental questions on nanofluids.

1.2.4.1 Existing Simplified Numerical Studies of Thermal Conductivity

Very few numerical works exist in the literature (Ghosh et al. 2011; Kang, Zhang, and Yang 2011; Li, Zhang et al. 2008; Rudyak, Belkin, and Tomilina 2010; Sankar, Mathew, and Sobhan 2008; Sarkar and Selvam 2007; Shukla and Dhir 2005; Wang et al. 2011), and they usually do not model realistic nanofluids. The most relevant and recent numerical works are the ones by Sarkar and Selvam (2007) and Li, Zhang et al. (2008). They both model the interaction of a copper nanoparticle in liquid argon using MD. Because of the simplicity of the materials used (copper and argon are both monatomic materials), a simple LJ 12-6 potential is sufficient to model both the fluid and the particle. The particle–fluid interaction is also modeled using an LJ 12-6 potential, with coefficient defined by Berthlot mixing rules (Allen and Tildesley 1997). Sarkar and Selvam (2007) use a Green–Kubo formalism to obtain the thermal conductivity of their nanofluid, and they conclude that the thermal conductivity enhancement observed is mostly due to the increased movement of the liquid atoms in the presence of nanoparticles. Because they just model one copper particle in their simulation box of liquid argon, particle clustering is neglected. Li, Zhang et al. (2008) however, calculate the thickness of the liquid layer formed on top of their copper particle. Through analysis of the density distribution of the liquid argon near the nanoparticle, they find that the thickness of the layering is approximately 0.5 nm at the given conditions of their simulations.

Another MD work is the one by Sankar, Mathew, and Sobhan (2008). This work makes use of more realistic potentials and investigates the thermal conductivity of a more realistic nanofluid (platinum nanoparticles in water). However, the water potential was simplified to a simple LJ due to restrictions on the application of the algorithm used to compute thermal conductivities with nonpairwise potentials. Furthermore, not having any experimental counterpart, they compared their results to a copper–water nanofluid.

One of the most recent numerical works on nanofluid is the one by Ghosh et al. (2011). They were able to predict the thermal conductivity enhancement of copper–water nanofluids by using a hybrid MD–stochastic model. Their results are close to present experimental values (within a 25% error), but their model uses a simplified LJ potential for copper, does not use a rigorous copper–water potential, and couples MD with stochastic theories involving continuum-like assumptions for the underlying physics. These factors make the model less fundamental in nature, and it becomes challenging to attribute any observation to a specific physical principle. Ghosh et al. (2011) attributed the 25% error to the large particle size used in the experiments (6–100 nm) compared to their model (4 nm) (Ghosh et al. 2011). However, the lack of rigorous state-of-the-art potentials and the various assumptions related to the stochastic component of their model might also be associated with such discrepancy.

The MD work by Wang et al. (2011) focused on the study of the anomalous viscosity increase in nanocolloidal dispersions. Once again, the LJ potential is used for every interaction. Even though this simplified force field did not allow for a quantitative validation of their work, they did come to the interesting conclusion that the shear viscosity of their nanofluids is strongly dependent on the particle size, a dependency not included in classical effective medium theories. Upon decomposing the stress tensor, the solid–solid interaction potential was found to be the key player in the anomalous viscosity increase they observed.

Even though these models raise interesting physical arguments regarding the possible mechanisms of heat transport in nanofluids, none of them simulates a nanofluid that has been experimentally reported. Furthermore, unlike argon, most realistic nanofluids use polar molecular liquids (water, ethylene glycol). In addition, along with realistic materials, to capture what actually happens at the interface between the particle and fluid, a more realistic surface interaction potential should be used.

Puliti, Paolucci, and Sen (2011) focused on the study of the thermodynamics and structure of a plane nanolayer mixture. Specifically, they treat the case of water confined between gold nanolayers and examine their interfacial interactions. This combination is mainly chosen for the surprisingly high thermal conductivity enhancement that it shows with an extremely small particle volume fraction (see Figure 1.2). The novelty of this work is in a new, fundamental, realistic, and comprehensive approach to the problem of understanding solid–fluid mixtures, confinement, and, consequently, nanofluids. MD simulations make use of accurate potentials to effectively model realistic materials: the quantum Sutton–Chen for Au–Au interactions (Daw and Baskes 1984; Kimura et al. 1998; Meineke et al. 2005; Qi et al. 1999), the extended simple point charge for water–water interactions (Alejandre, Tildesley, and Chapela 1995; Berendsen, Grigera, and Straatsma 1987; Kusalik and Svishchev 1994), and a modified Spohr potential for Au–water interactions (Dou et al. 2001; Spohr 1989). These potentials ensure that most of the physics is captured properly. Thermodynamic properties (energies, RDFs, densities, heat capacities, compressibility, and bulk modulus) were successfully computed for validation purposes for both pure water and pure gold (Puliti, Paolucci, and Sen 2011). The model was extended to a mixture of a gold nanolayer immersed in water and to a realistic gold–water nanofluid (Puliti et al. 2012b). It is interesting to note that although the thermodynamic properties of the mixture can also be predicted using ideal mixture theory, such predictions are generally poor. All properties computed by Puliti, Paolucci, and Sen (2011) are between 10% and 400% off from ideal mixture predictions. In other words, thermodynamic properties of nanolayer mixtures and the nanofluids do not seem to behave as ideal mixtures. Nevertheless, several works in the nanofluid literature make extensive use of such relations (Bergman 2009; Buongiorno 2006; Corcione 2011; Lee and Mudawar 2007; Pak and Cho 1998; Shin and Banerjee 2011a; Wang and Mujumdar 2008). The anisotropy induced by the gold–water interface and its effects appear responsible for such behavior.

1.3 CONCLUSION

Anomalous enhancement (not following classical effective medium theories) of thermal conductivity over the base fluids was observed in most cases. Bare metal nanoparticles and CNTs appeared to give the most enhancement. However, the vast discrepancies among the experiments undermine the ability of nanofluid researchers to come up with a theory for the prediction and control of such a thermal conductivity enhancement. Too many experimental techniques have been used for the measurements, and most of them were not originally designed to work for nanoparticle suspensions. In at least one occasion, the hot-wire method showed no anomalous enhancement, whereas major enhancements were reported from a multitude of other researchers using the same technique for similar nanofluids. Most of the discrepancy comes from poor characterization of the nanofluid in the test cell: nanoparticle clustering; settling; hard-to-control size distributions; and presence of surfactants, ions, and other products of the synthesis of nanoparticles are just some of the potential threats to a reproducible experiment.

Even though a bit more sporadic, experiments on nanofluid viscosity also show some anomalous trends. The viscosity tends to increase when nanoparticles are suspended in a fluid. However, nanofluids appear to behave as Newtonian for lower nanoparticle loadings and shear thinning for higher volume fractions.

Heat convection in nanofluids has not been studied as much as the aforementioned transport properties, even though it is of fundamental importance for most engineering applications. Many of the available works are in disagreement with each other, even though most agree on the fact that forced heat convection in nanofluid is enhanced over the base fluid. This enhancement, once again, does not seem to follow classical theories and increases with nanoparticle loading. On the contrary, natural heat convection in some cases appears to deteriorate in nanofluids, and the deterioration worsens with increasing nanoparticle volume fractions. More heat convection experiments are needed, especially with metallic nanoparticles and with lower concentrations.

Similar controversy surrounds the nanofluid specific heat. Even though it usually decreases in nanofluids compared to the base fluids, this is not always the case. Given the key role that specific heat has in many engineering applications, more research should be done in this area.

Several empirical models were developed, based on classical effective medium theories, to predict the thermal conductivity, viscosity, heat convection, and specific heats of nanofluids. Some of them are discussed in more detail in the work of Puliti, Paolucci, and Sen (2012a); however, all of them are able to only predict enhancements of the specific nanofluid for which they were experimentally fitted to and not for other nanofluids.

Considering the high cost of manufacturing nanofluids, the discrepancies among experimental evidence, and the lack of a unified and reliable empirical theory, numerical work (such as MD simulations) offers a valid alternative. Nevertheless, only a handful of numerical studies are available in the literature, and most of them oversimplify the physics or the choice of nanofluid materials to lower the complexity of the algorithms and the computational costs. Furthermore, it is important to understand that the usefulness of nanofluids for heat transfer applications depends not only on the thermal conductivity but also on other transport properties, such as viscosity and mass diffusion, and on thermodynamic properties, such as specific heats. The fundamental understanding of all properties will allow us to produce relevant physical and mathematical models of nanofluids and, more importantly, enable the control of their thermodynamic and transport properties by proper selection of particles and solvents.

Much more needs to be done before a theory of nanofluids can be developed and used with confidence. Experimental techniques, especially the most classical methods, should be thoroughly reviewed in light of what we know on nanofluids today and on the assumptions made to design those techniques. If a specific method needs to be appropriately modified to work with nanofluids, each experimental researcher should take advantage of the same modifications, once they are made available to the scientific community. Among other experimental details, ultrasonic mixers and their usage for each measurement should be described in greater detail. More details are also needed for the statistical properties of the nanoparticle size distribution, and all the chemical species present in a suspension should be made known. More experiments, especially in heat convection and thermodynamics, are encouraged. Among thermal conductivity experiments, it would be beneficial to have more measurements made with ionic liquids as base fluids. Considering the fact that ionic liquids can be manufactured in a laboratory and their properties controlled, it would be interesting and insightful to observe how the properties of nanofluids are affected by various characteristics of the base fluid. This might provide a different perspective toward a better understanding of nanofluids. Realistic verified and validated numerical works, perhaps, can fill the gaps in the experimental literature and provide guidance to future experiments.

Many are the issues and challenges of nanofluids that need to be addressed and overcome before this new field of study can be fully established. As observed in the previous sections, the most stringent issues to face are: (1) the disagreement between most of experimental data, (2) the poor characterization of the nanosuspensions, and (3) the lack of understanding of the complex physical phenomena responsible for the anomalous behavior of nanofluids (Saidur, Leong, and Mohammad 2011). Additional challenges are more practical in nature, and they initially involve financial and manufacturing difficulties in producing nanofluids and finally issues in some engineering applications of nanofluids because of their usually higher viscosity and, occasionally, lower specific heat, compared to the base fluids.

Nanofluids are very complex fluids, but with an extremely vast range of applications in every field of science and engineering. Investing on their understanding will certainly be repaid many times.

REFERENCES

Abu-Nada, E. "Effects of variable viscosity and thermal conductivity of Al₂O₃-water nanofluid on heat transfer enhancement in natural convection." *Int. J. Heat Fluid Fl.* 30, no. 4, 2009: 679–690.

Abu-Nada, E., Z. Masoud, and A. Hijazi. "Natural convection heat transfer enhancement in horizontal concentric annuli using nanofluids." *Int. Commun. Heat Mass* 35, no. 5, 2008: 657–665.

Abu-Nada, E., Z. Masoud, H. F. Oztop, and A. Campo. "Effect of nanofluid variable properties on natural convection in enclosures." *Int. J. Therm. Sci.* 49, no. 3, 2010: 479–491.

Alejandre, J., D. J. Tildesley, and G. A. Chapela. "Molecular dynamics simulation of the orthobaric densities and surface tension of water." *J. Chem. Phys.* 102, no. 11, 1995: 4574–4583.

Allen, M. P., and D. J. Tildesley. *Computer Simulation of Liquids.* Oxford: Oxford University Press, 1997.

Andrade, E. N. D. "A theory of the viscosity of liquids—Part I." *Philos. Mag.* 17, no. 112, 1934: 497–511.

Arefmanesh, A., and M. Mahmoodi. "Effects of uncertainties of viscosity models for Al₂O₃-water nanofluid on mixed convection numerical simulations." *Int. J. Therm. Sci.* 50, no. 9, 2011: 1706–1719.

Assael, M. J., C.-F. Chen, I. Metaxa, and W. A. Wakeham. "Thermal conductivity of suspensions of carbon nanotubes in water." *Int. J. Thermophys.* 25, no. 4, 2004: 971–985.

Bastea, S. "Comment on 'model for heat conduction in nanofluids.'" *Phys. Rev. Lett.* 95, no. 1, 2005: 19401.

Batchelor, G. K. "The effect of Brownian motion on the bulk stress in a suspension of spherical particles." *J. Fluid Mech.* 83, no. 01, 1977: 97–117.

Beck, M. P., T. Sun, and A. S. Teja. "The thermal conductivity of alumina nanoparticles dispersed in ethylene glycol." *Fluid Phase Equilibr.* 260, no. 2, 2007: 275–278.

Beck, M. P., Y. H. Yuan, P. Warrier, and A. S. Teja. "The effect of particle size on the thermal conductivity of alumina nanofluids." *J. Nanopart. Res.* 11, no. 5, 2009: 1129–1136.

Berendsen, H. J. C., J. R. Grigera, and T. P. Straatsma. "The missing term in effective pair potentials." *J. Phys. Chem.* 91, no. 24, 1987: 6269–6271.

Bergman, T. "Effect of reduced specific heats of nanofluids on single phase, laminar internal forced convection." *Int. J. Heat Mass Transf.* 52, no. 5–6, 2009: 1240–1244.

Bhattacharya, P., S. K. Saha, A. Yadav, P. E. Phelan, and R. S. Prasher. "Brownian dynamics simulation to determine the effective thermal conductivity of nanofluids." *J. Appl. Phys.* 95, no. 11, 2004: 6492–6494.

Brown, R. "A brief account of microscopical observations made in the comments of June, July and August, 1827, on the particles contained in the pollen of plants; and on the general existence of active molecules in organic and inorganic bodies." *Philos. Mag.* 4, 1828: 161–173.

Buongiorno, J. "Convective transport in nanofluids." *J. Heat Transf.* 128, no. 3, 2006: 240–250.

Buongiorno, J. et al. "A benchmark study on the thermal conductivity of nanofluids." *J. Appl. Phys.* 106, no. 9, 2009: 94312.

Chandrasekar, M., and S. Suresh. "A review on the mechanisms of heat transport in nanofluids." *Heat Transf. Eng.* 30, no. 14, 2009: 1136–1150.

Chandrasekar, M., S. Suresh, R. Srinivasan, and A. C. Bose. "New analytical models to investigate thermal conductivity of nanofluids." *J. Nanosci. Nanotechnol.* 9, no. 1, 2009: 533–538.

Chen, H., and Y. Ding. "Heat aransfer and rheological behaviour of nanofluids— A review." In *Advances in Transport Phenomena.* Edited by Liqiu Wang. Vol. 1. Berlin, Germany: Springer, 2009. 140–157.

Chen, H., Y. Ding, Y. He, and C. Tan. "Rheological behaviour of ethylene glycol based titania nanofluids." *Chem. Phys. Lett.* 444, no. 4–6, 2007: 333–337.

Chen, H., Y. Ding, and C. Tan. "Rheological behaviour of nanofluids." *New J. Phys.* 9, no. 10, 2007: 367.

Choi, S. U. S. "Enhancing thermal conductivity of fluids with nanoparticles." Edited by D. A. Siginer and H. P. Wang. *ASME International Mechanical Engineering Congress and Exposition Proceedings.* San Francisco, CA: ASME, 1995. 99–105.

Choi, S. U. S., Z. G. Zhang, W. Yu, F. E. Lockwood, and E. A. Grulke. "Anomalous thermal conductivity enhancement in nanotube suspensions." *Appl. Phys. Lett.* 79, no. 14, 2001: 2252–2254.

Chopkar, M., S. Kumar, D. R. Bhandari, P. K. Das, and I. Manna. "Development and characterization of Al_2Cu and Ag_2Al nanoparticle dispersed water and ethylene glycol based nanofluid." *Mater. Sci. Eng. B* 139, no. 2–3, 2007: 141–148.

Cianfrini, M., M. Corcione, and A. Quintino. "Natural convection heat transfer of nanofluids in annular spaces between horizontal concentric cylinders." *Appl. Therm. Eng.* 31, no. 17–18, 2011: 4055–4063.

Corcione, M. "Rayleigh-Bénard convection heat transfer in nanoparticle suspensions." *Int. J. Heat Fluid Fl.* 32, no. 1, 2011: 65–77.

Das, S. K., S. U. S. Choi, W. Yu, and T. Pradeep. *Nanofluids: Science and Technology.* Hoboken, NJ: John Wiley & Sons, 2007.

Das, S. K., N. Putra, and W. Roetzel. "Pool boiling characteristics of nano-fluids." *Int. J. Heat Mass Transf.* 46, no. 5, 2003: 851–862.

Das, S. K., N. Putra, P. Thiesen, and W. Roetzel. "Temperature dependence of thermal conductivity enhancement for nanofluids." *J. Heat Transf.* 125, no. 4, 2003: 567–574.

Daungthongsuk, W., and S. Wongwises. "A critical review of convective heat transfer of nanofluids." *Renew. Sust. Energ. Rev.* 11, no. 5, 2007: 797–817.

Daw, M. S., and M. I. Baskes. "Embedded-atom method: Derivation and application to impurities, surfaces, and other defects in metals." *Phys. Rev. B* 29, no. 12, 1984: 6443–6453.

Ding, Y., H. Alias, D. Wen, and R. Williams. "Heat transfer of aqueous suspensions of carbon nanotubes (CNT nanofluids)." *Int. J. Heat Mass Transf.* 49, no. 1–2, 2006: 240–250.

Dittus, W., and L. M. K. Boelter. "Heat transfer in automobile radiators of the tubular type." *Univ. Calif. Publ. Eng.* 2, no. 13, 1930: 443–461.

Dou, Y., L. V. Zhigilei, N. Winograd, and B. J. Garrison. "Explosive boiling of water films adjacent to heated surfaces: A microscopic description." *J. Phys. Chem. A* 105, no. 12, 2001: 2748–2755.

Duangthongsuk, W., and S. Wongwises. "Measurement of temperature-dependent thermal conductivity and viscosity of TiO_2-water nanofluids." *Exp. Therm. Fluid Sci.* 33, no. 4, 2009: 706–714.

Eapen, J., J. Li, and S. Yip. *Modeling Transport Mechanism in Nanofluids. Nano-to-Micro Transport Processes*. Technical report, MIT 2.57 Project Report, Cambridge, MA: Department of Nuclear Science and Engineering; Massachusetts Institute of Technology, 2004.

Eapen, J., R. Rusconi, R. Piazza, and S. Yip. "The classical nature of thermal conduction in nanofluids." *J. Heat Transf.* 132, no. 10, 2010: 102402.

Eastman, J. A., S. U. S. Choi, S. Li, W. Yu, and L. J. Thompson. "Anomalously increased effective thermal conductivities of ethylene glycol-based nanofluids containing copper nanoparticles." *Appl. Phys. Lett.* 78, no. 6, 2001: 718–720.

Eastman, J. A., S. R. Phillpot, S. U. S. Choi, and P. Keblinski. "Thermal transport in nanofluids." *Annu. Rev. Mater. Res.* 34, no. 1, 2004: 219–246.

Einstein, A. "A new determination of the molecular dimensions." *Ann. Phys.* 19, no. 2, 1906a: 289–306.

Einstein, A. "The theory of the Brownian motion." *Ann. Phys.* 19, no. 2, 1906b: 371–381.

Einstein, A. "Correction of my work: A new determination of the molecular dimensions." *Ann. Phys.* 34, no. 3, 1911: 591–592.

Escher, W. et al. "On the cooling of electronics with nanofluids." *J. Heat Transf.* 133, no. 5, 2011: 051401.

Evans, W., J. Fish, and P. Keblinski. "Role of Brownian motion hydrodynamics on nanofluid thermal conductivity." *Appl. Phys. Lett.* 88, no. 9, 2006: 93116.

Evans, W., R. Prasher, J. Fish, P. Meakin, P. Phelan, and P. Keblinski. "Effect of aggregation and interfacial thermal resistance on thermal conductivity of nanocomposites and colloidal nanofluids." *Int. J. Heat Mass Transf.* 51, no. 5–6, 2008: 1431–1438.

Feng, Y., B. Yu, P. Xu, and M. Zou. "The effective thermal conductivity of nanofluids based on the nanolayer and the aggregation of nanoparticles." *J. Phys. D Appl. Phys.* 40, no. 10, 2007: 3164–3171.

Garg, J. et al. "Enhanced thermal conductivity and viscosity of copper nanoparticles in ethylene glycol nanofluid." *J. Appl. Phys.* 103, no. 7, 2008: 74301.

Gharagozloo, P. E., and K. E. Goodson. "Aggregate fractal dimensions and thermal conduction in nanofluids." *J. Appl. Phys.* 108, no. 7, 2010: 074309.

Ghasemi, B., S. M. Aminossadati, and A. Raisi. "Magnetic field effect on natural convection in a nanofluid-filled square enclosure." *Int. J. Therm. Sci.* 50, no. 9, 2011: 1748–1756.

Gherasim, I., G. Roy, C. T. Nguyen, and D. Vo-Ngoc. "Heat transfer enhancement and pumping power in confined radial flows using nanoparticle suspensions (nanofluids)." *Int. J. Therm. Sci.* 50, no. 3, 2011: 369–377.

Ghosh, M. M., S. Roy, S. K. Pabi, and S. Ghosh. "A molecular dynamics-stochastic model for thermal conductivity of nanofluids and its experimental validation." *J. Nanosci. Nanotechnol.* 11, no. 3, 2011: 2196–2207.

Gnielinski, V. "New equations for heat and mass transfer in turbulent pipe and channel flow." *Int. Chem. Eng.* 16, no. 2, 1976: 359–368.

Godson, L., B. Raja, D. M. Lal, and S. Wongwises. "Enhancement of heat transfer using nanofluids—An overview." *Renew. Sust. Energ. Rev.* 14, no. 2, 2010: 629–641.

Godson, L., B. Raja, D. M. Lal, and S. Wongwises. "Experimental investigation on the thermal conductivity and viscosity of silver-deionized water nanofluid." *Exp. Heat Transf.* 23, no. 4, 2010b: 317–332.

Hamad, M. A. A. "Analytical solution of natural convection flow of a nanofluid over a linearly stretching sheet in the presence of magnetic field." *Int. Commun. Heat Mass* 38, no. 4, 2011: 487–492.

He, Y., Y. Jin, H. Chen, Y. Ding, D. Cang, and H. Lu. "Heat transfer and flow behaviour of aqueous suspensions of TiO_2 nanoparticles (nanofluids) flowing upward through a vertical pipe." *Int. J. Heat Mass Transf.* 50, no. 11–12, 2007: 2272–2281.

Heris, S. Z., S. Etemad, and M. Nasresfahany. "Experimental investigation of oxide nanofluids laminar flow convective heat transfer." *Int. Commun. Heat Mass* 33, no. 4, 2006: 529–535.

Hong, K. S., T.-K. Hong, and H.-S. Yang. "Thermal conductivity of Fe nanofluids depending on the cluster size of nanoparticles." *Appl. Phys. Lett.* 88, no. 3, 2006: 31901.

Hong, T.-K., H.-S. Yang, and C. J. Choi. "Study of the enhanced thermal conductivity of Fe nanofluids." *J. Appl. Phys.* 97, no. 6, 2005: 064311.

Hwang, Y. J. et al. "Investigation on characteristics of thermal conductivity enhancement of nanofluids." *Curr. Appl. Phys.* 6, no. 6, 2006: 1068–1071.

Jana, S., A. Salehi-Khojin, and W.-H. Zhong. "Enhancement of fluid thermal conductivity by the addition of single and hybrid nano-additives." *Thermochim. Acta* 462, no. 1–2, 2007: 45–55.

Jang, S. P., and S. U. S. Choi. "Role of Brownian motion in the enhanced thermal conductivity of nanofluids." *Appl. Phys. Lett.* 84, no. 21, 2004: 4316–4318.

Jang, S. P., and S. U. S. Choi. "Effects of various parameters on nanofluid thermal conductivity." *J. Heat Transf.* 129, no. 5, 2007: 617.

Kabelac, S., and J. F. Kuhnke. "Heat transfer mechanisms in nanofluids." *International Heat Transfer Conference—Keynote Papers.* Sydney, Australia: Begell House Inc., 2006.

Kakaça, S., and A. Pramuanjaroenkij. "Review of convective heat transfer enhancement with nanofluids." *Int. J. Heat Mass Transf.* 52, no. 13–14, 2009: 3187–3196.

Kang, H. U., S. H. Kim, and J. M. Oh. "Estimation of thermal conductivity of nanofluid using experimental effective particle volume." *Exp. Heat Transf.* 19, no. 3, 2006: 181–191.

Kang, H., Y. Zhang, and M. Yang. "Molecular dynamics simulation of thermal conductivity of Cu-Ar nanofluid using EAM potential for Cu-Cu interactions." *Appl. Phys. A* 103, no. 4, 2011: 1001–1008.

Kapitza, P. L. "The study of heat transfer in helium II." *J. Phys.* 4, no. 1–6, 1941: 181–210.

Karthikeyan, N. R., J. Philip, and B. Raj. "Effect of clustering on the thermal conductivity of nanofluids." *Mater. Chem. Phys.* 109, no. 1, 2008: 50–55.

Keblinski, P., and D. G. Cahill. "Comment on 'Model for heat conduction in nanofluids.'" *Phys. Rev. Lett.* 95, no. 20, 2005: 209401.

Keblinski, P., J. A. Eastman, and D. G. Cahill. "Nanofluids for thermal transport." *Mater. Today* 8, 2005: 36–44.

Keblinski, P., S. R. Phillpot, S. U. S. Choi, and J. A. Eastman. "Mechanisms of heat flow in suspensions of nano-sized particles (nanofluids)." *Int. J. Heat Mass Transf.* 45, no. 4, 2002: 855–863.

Keblinski, P., R. Prasher, and J. Eapen. "Thermal conductance of nanofluids: Is the controversy over?" *J. Nanopart. Res.* 10, no. 7, 2008: 1089–1097.

Kim, J., Y. T. Kang, and C. K. Choi. "Soret and Dufour effects on convective instabilities in binary nanofluids for absorption application." *Int. J. Refrig.* 30, no. 2, 2007: 323–328.

Kimura, Y., Y. Qi, T. Çağin, and W. A. Goddard. "The Quantum Sutton-Chen many-body potential for properties of fcc metals." Unpublished, 1998.

Kole, M., and T. K. Dey. "Viscosity of alumina nanoparticles dispersed in car engine coolant." *Exp. Therm. Fluid Sci.* 34, no. 6, 2010: 677–683.

Kole, M., and T. K. Dey. "Effect of aggregation on the viscosity of copper oxide-gear oil nanofluids." *Int. J. Therm. Sci.* 50, no. 9, 2011: 1741–1747.

Kondaraju, S., E. K. Jin, and J. S. Lee. "Direct numerical simulation of thermal conductivity of nanofluids: The effect of temperature two-way coupling and coagulation of particles." *Int. J. Heat Mass Transf.* 53, no. 5–6, 2010: 862–869.

Koo, J., and C. Kleinstreuer. "A new thermal conductivity model for nanofluids." *J. Nano. Res.* 6, no. 6, 2004: 577–588.

Koo, J., and C. Kleinstreuer. "Laminar nanofluid flow in microheat-sinks." *Int. J. Heat Mass Transf.* 48, no. 13, 2005: 2652–2661.

Krieger, I. M. "A mechanism for non-Newtonian flow in suspensions of rigid spheres." *J. Rheol.* 3, no. 1, 1959: 137–152.

Kumar, D., H. Patel, V. Kumar, T. Sundararajan, T. Pradeep, and S. K. Das. "Model for heat conduction in nanofluids." *Phys. Rev. Lett.* 93, no. 14, 2004: 4316.

Kusalik, P. G., and I. M. Svishchev. "The spatial structure in liquid water." *Science* 265, no. 5176, 1994: 1219–1221.

Kwak, K., and C. Kim. "Viscosity and thermal conductivity of copper oxide nanofluid dispersed in ethylene glycol." *Korea-Australia Rheol. J.* 17, no. 2, 2005: 35–40.

Lee, D. "Thermophysical properties of interfacial layer in nanofluids." *Langmuir* 23, no. 11, 2007: 6011–6018.

Lee, S., S. U. S. Choi, S. Li, and J. A. Eastman. "Measuring thermal conductivity of fluids containing oxide nanoparticles." *J. Heat Transf.* 121, no. 2, 1999: 280–289.

Lee, J., and I. Mudawar. "Assessment of the effectiveness of nanofluids for single-phase and two-phase heat transfer in micro-channels." *Int. J. Heat Mass Transf.* 50, no. 3–4, 2007: 452–463.

Li, Y.-H., W. Qu, and J.-C. Feng. "Temperature dependence of thermal conductivity of nanofluids." *Chinese Phys. Lett.* 25, no. 9, 2008: 3319–3322.

Li, L., W. Zhang, H. B. Ma, and M. Yang. "An investigation of molecular layering at the liquid-solid interface in nanofluids by molecular dynamics simulation." *Phys. Lett. A* 372, no. 25, 2008: 4541–4544.

Liu, M.-S., M. C.-C. Lin, I.-T. Huang, and C.-C. Wang. "Enhancement of thermal conductivity with CuO for nanofluids." *Chem. Eng. Technol.* 29, no. 1, 2006: 72–77.

Liu, M.-S., M. C.-C. Lin, C. Y. Tsai, and C.-C. Wang. "Enhancement of thermal conductivity with Cu for nanofluids using chemical reduction method." *Int. J. Heat Mass Transf.* 49, no. 17–18, 2006: 3028–3033.

Masoumi, N., N. Sohrabi, and A. Behzadmehr. "A new model for calculating the effective viscosity of nanofluids." *J. Phys. D: Appl. Phys.* 42, no. 5, 2009: 055501.

Masuda, H., A. Ebata, K. Teramae, and N. Hishinuma. "Alteration of thermal conductivity and viscosity of liquid by dispersing ultra-fine particles (dispersions of γ-Al_2O_3, SiO_2, and TiO_2 ultra-fine particles)." *Jpn. J. Thermophys. Prop.* 7, no. 4, 1993: 227–233.

Maxwell, J. C. *A Treatise on Electricity and Magnetism.* Vol. 1. 2nd ed. Oxford: Clarendon Press, 1881.

McLachlan, D. S., M. Blaszkiewicz, and R. E. Newnham. "Electrical resistivity of composites." *J. Am. Ceram. Soc.* 73, no. 8, 1990: 2187–2203.

Meineke, M. A., C. F. Vardeman, T. Lin, C. J. Fennell, and J. D. Gezelter. "OOPSE: An object-oriented parallel simulation engine for molecular dynamics." *J. Comput. Chem.* 26, no. 3, 2005: 252–271.

Mintsa, H., G. Roy, C. Nguyen, and D. Doucet. "New temperature dependent thermal conductivity data for water-based nanofluids." *Int. J. Therm. Sci.* 48, no. 2, 2009: 363–371.

Moghaddami, M., A. Mohammadzade, and S. A. V. Esfehani. "Second law analysis of nanofluid flow." *Energ. Convers. Manage* 52, no. 2, 2011: 1397–1405.

Murshed, S. M. S., K. C. Leong, and C. Yang. "Determination of the effective thermal diffusivity of nanofluids by the double hot-wire technique." *J. Phys. D Appl. Phys.* 39, no. 24, 2006: 5316–5322.

Murshed, S. M. S., K. C. Leong, and C. Yang. "Investigations of thermal conductivity and viscosity of nanofluids." *Int. J. Therm. Sci.* 47, no. 5, 2008: 560–568.

Nakayama, T. "New channels of energy transfer across a solid-liquid He interface." *J. Phys. C Solid State Phys.* 18, no. 22, 1985: L667–L671.

Namburu, P. K., D. K. Das, K. M. Tanguturi, and R. S. Vajjha. "Numerical study of turbulent flow and heat transfer characteristics of nanofluids considering variable properties." *Int. J. Therm. Sci.* 48, no. 2, 2009: 290–302.

Namburu, P. K., D. P. Kulkarni, D. Misra, and D. K. Das. "Viscosity of copper oxide nanoparticles dispersed in ethylene glycol and water mixture." *Exp. Therm. Fluid Sci.* 32, no. 2, 2007: 397–402.

Nelson, I. C., D. Banerjee, and R. Ponnappan. "Flow loop experiments using poly-alphaolefin nanofluids." *J. Thermophys. Heat Tr.* 23, no. 4, 2009: 752–761.

Nguyen, C. T., G. Roy, C. Gauthier, and N. Galanis. "Heat transfer enhancement using Al_2O_3-water nanofluid for an electronic liquid cooling system." *Appl. Therm. Eng.* 27, no. 8–9, 2007: 1501–1506.

Nie, C., W. H. Marlow, and Y. A. Hassan. "Discussion of proposed mechanisms of thermal conductivity enhancement in nanofluids." *Int. J. Heat Mass Transf.* 51, no. 5–6, 2008: 1342–1348.

Nield, D. A., and A. V. Kuznetsov. "The Cheng-Minkowycz problem for the double-diffusive natural convective boundary layer flow in a porous medium saturated by a nanofluid." *Int. J. Heat Mass Transf.* 54, no. 1–3, 2011: 374–378.

Özerinç, S., S. Kakaç, and A. G. Yazıcıoğlu. "Enhanced thermal conductivity of nanofluids: A state-of-the-art review." *Microfluid. Nanofluid.* 8, no. 2, 2009: 145–170.

Pak, B. C., and Y. I. Cho. "Hydrodynamic and heat transfer study of dispersed fluids with submicron metallic oxide particles." *Exp. Heat Transf.* 11, no. 2, 1998: 151–170.

Patel, H., S. K. Das, T. Sundararajan, N. A. Sreekumaran, B. George, and T. Pradeep. "Thermal conductivities of naked and monolayer protected metal nanoparticle based nanofluids: Manifestation of anomalous enhancement and chemical effects." *Appl. Phys. Lett.* 83, no. 14, 2003: 2931–2933.

Patel, H. E., T. Sundararajan, T. Pradeep, A. Dasgupta, N. Dasgupta, and S. K. Das. "A micro-convection model for thermal conductivity of nanofluids." *Pramana J. Phys.* 65, no. 5, 2005: 863–869.

Philip, J., P. D. Shima, and B. Raj. "Evidence for enhanced thermal conduction through percolating structures in nanofluids." *Nanotechnolgy* 19, no. 30, 2008: 305706.

Prasher, R., P. Bhattacharya, and P. E. Phelan. "Thermal conductivity of nanoscale colloidal solutions (nanofluids)." *Phys. Rev. Lett.* 94, no. 2, 2005: 25901.

Prasher, R., P. Bhattacharya, and P. E. Phelan. "Brownian-motion-based convective-conductive model for the effective thermal conductivity of nanofluids." *J. Heat Transf.* 128, no. 6, 2006: 588–595.

Prasher, R., P. E. Phelan, and P. Bhattacharya. "Effect of aggregation kinetics on the thermal conductivity of nanoscale colloidal solutions (nanofluid)." *Nano Lett.* 6, no. 7, 2006a: 1529–1534.

Prasher, R., P. E. Phelan, and P. Bhattacharya. "Effect of aggregation on thermal conduction in colloidal nanofluids." *Appl. Phys. Lett.* 89, no. 14, 2006b: 143119.

Prasher, R., D. Song, J. Wang, and P. Phelan. "Measurements of nanofluid viscosity and its implications for thermal applications." *Appl. Phys. Lett.* 89, no. 13, 2006: 133108.

Prevenslik, T. "Nanofluids by quantum mechanics." *ASME 2009 Second International Conference on Micro/Nanoscale Heat and Mass Transfer.* Vol. 1. Shanghai, People's Republic of China: ASME, 2009. 387.

Puliti, G., S. Paolucci, and M. Sen. "Thermodynamic properties of gold-water nanolayer mixtures using molecular dynamics." *J. Nanopart. Res.* 13, no. 9, 2011: 4277–4293.

Puliti, G., S. Paolucci, and M. Sen. "Nanofluids and their properties." *Appl. Mech. Rev.* 64, no. 3, 2012a: 030803.

Puliti, G., Paolucci, S., and M. Sen. "Thermodynamic properties of gold-water nanofluids using molecular dynamics." *J. Nanopart. Res.* 2012b: 1296.

Putnam, S. A., and D. G. Cahill. "Micron-scale apparatus for measurements of thermodiffusion in liquids." *Rev. Sci. Instrum.* 75, no. 7, 2004: 2368–2372.

Putnam, S. A., D. G. Cahill, P. V. Braun, Z. Ge, and R. G. Shimmin. "Thermal conductivity of nanoparticle suspensions." *J. Appl. Phys.* 99, no. 8, 2006: 084308.

Putra, N., W. Roetzel, and S. K. Das. "Natural convection of nano-fluids." *Heat Mass Transf.* 39, no. 8–9, 2003: 775–784.

Qi, Y., T. Çağın, Y. Kimura, and W. A. Goddard. "Molecular-dynamics simulations of glass formation and crystallization in binary liquid metals: Cu-Ag and Cu-Ni." *Phys. Rev. B* 59, no. 5, 1999: 3527–3533.

Ren, Y., H. Xie, and A. Cai. "Effective thermal conductivity of nanofluids containing spherical nanoparticles." *J. Phys. D Appl. Phys.* 38, no. 21, 2005: 3958–3961.

Rudyak, V. Y., A. A. Belkin, and E. A. Tomilina. "On the thermal conductivity of nanofluids." *Tech. Phys. Lett.* 36, no. 7, 2010: 660–662.

Saidur, R., S. N. Kazi, M. S. Hossain, M. M. Rahman, and H. A. Mohammed. "A review on the performance of nanoparticles suspended with refrigerants and lubricating oils in refrigeration systems." *Renew. Sust. Energ. Rev.* 15, no. 1, 2011: 310–323.

Saidur, R., K. Y. Leong, and H. A. Mohammad. "A review on applications and challenges of nanofluids." *Renew. Sust. Energ. Rev.* 15, no. 3, 2011: 1646–1668.

Sankar, N., N. Mathew, and C. B. Sobhan. "Molecular dynamics modeling of thermal conductivity enhancement in metal nanoparticle suspensions." *Int. Commun. Heat Mass* 35, no. 7, 2008: 867–872.

Sarkar, S., and R. P. Selvam. "Molecular dynamics simulation of effective thermal conductivity and study of enhanced thermal transport mechanism in nanofluids." *J. Appl. Phys.* 102, no. 7, 2007: 74302.

Sen, M., and S. Paolucci. "The use of ionic liquids in refrigeration." Paper No. IMECE 2006-14712. *ASME International Mechanical Engineering Congress and Exposition Proceedings.* Chicago, IL, 2006. 131–134.

Sergis, A., and Y. Hardalupas. "Anomalous heat transfer modes of nanofluids: A review based on statistical analysis." *Nanoscale Res. Lett.* 6, no. 1, 2011: 391.

Shalkevich, N., W. Escher, T. Bürgi, B. Michel, L. Si-Ahmed, and D. Poulikakos. "On the thermal conductivity of gold nanoparticle colloids." *Langmuir* 26, no. 2, 2010: 663–670.

Shih, W.-H., W. Y. Shih, S.-I. Kim, J. Liu, and I. A. Aksay. "Scaling behavior of the elastic properties of colloidal gels." *Phys. Rev. A* 42, no. 8, 1990: 4772–4779.

Shin, D., and D. Banerjee. "Enhancement of specific heat capacity of high-temperature silica-nanofluids synthesized in alkali chloride salt eutectics for solar thermal-energy storage applications." *Int. J. Heat Mass Transf.* 54, no. 5–6, 2011a: 1064–1070.

Shin, D., and D. Banerjee. "Enhanced specific heat of silica nanofluid." *J. Heat Transf.* 133, no. 2, 2011b: 024501.

Shukla, R. K., and V. K. Dhir. "Numerical study of the effective thermal conductivity of nanofluids." *Proceedings of ASME Summer Heat Transfer Conference*. San Francisco, CA: ASME, 2005. 1–5.

Shukla, R. K., and V. K. Dhir. "Effect of Brownian motion on thermal conductivity of nanofluids." *J. Heat Transf.* 130, no. 4, 2008: 042406.

Spohr, E. "Computer simulation of the water/platinum interface." *J. Phys. Chem.* 93, no. 16, 1989: 6171–6180.

Studart, A. R., E. Amstad, M. Antoni, and L. J. Gauckler. "Rheology of concentrated suspensions containing weakly attractive alumina nanoparticles." *J. Am. Ceram. Soc.* 89, no. 8, 2006: 2418–2425.

Swartz, E. T., and R. O. Pohl. "Thermal boundary resistance." *Rev. Mod. Phys.* 61, no. 3, 1989: 605–668.

Tavman, I., A. Turgut, M. Chirtoc, K. Hadjov, O. Fudym, and S. Tavman. "Experimental study on thermal conductivity and viscosity of water-based nanofluids." *Heat Transf. Res.* 41, no. 3, 2010: 339–351.

Tillman, P., and J. Hill. "Determination of nanolayer thickness for a nanofluid." *Int. Commun. Heat Mass* 34, no. 4, 2007: 399–407.

Timofeeva, E. V. et al. "Thermal conductivity and particle agglomeration in alumina nanofluids: Experiment and theory." *Phys. Rev. E* 76, no. 6, 2007: 061203.

Tseng, W., and K. Lin. "Rheology and colloidal structure of aqueous TiO_2 nanoparticle suspensions." *Mater. Sci. Eng. A* 355, no. 1–2, 2003: 186–192.

Turanov, A. N., and Y. V. Tolmachev. "Heat- and mass-transport in aqueous silica nanofluids." *Heat Mass Transf.* 45, no. 12, 2009: 1583–1588.

Vajjha, R. S., and D. K. Das. "Experimental determination of thermal conductivity of three nanofluids and development of new correlations." *Int. J. Heat Mass Transf.* 52, no. 21–22, 2009a: 4675–4682.

Vajjha, R. S., and D. K. Das. "Specific heat measurement of three nanofluids and development of new correlations." *J. Heat Transf.* 131, no. 7, 2009b: 071601.

Vajjha, R. S., D. K. Das, and P. K. Namburu. "Numerical study of fluid dynamic and heat transfer performance of Al_2O_3 and CuO nanofluids in the flat tubes of a radiator." *Int. J. Heat Fluid Fl.* 31, no. 4, 2010: 613–621.

Wang, Z. "Thermal wave in thermal properties measurements and flow diagnostics: With applications of nanofluids thermal conductivity and wall shear stress measurements." PhD dissertation, Oregon State University, Corvallis, OR, 2009.

Wang, L., and J. Fan. "Nanofluids research: Key issues." *Nanoscale Res. Lett.* 5, no. 8, 2010: 1241–1252.

Wang, X.-Q., and A. S. Mujumdar. "A review of nanofluids—Part I: Theoretical and numerical investigations." *Braz. J. Chem. Eng.* 25, no. 4, 2008: 613–630.

Wang, T., X. Wang, Z. Luo, M. Ni, and K. Cen. "Mechanisms of viscosity increase for nanocolloidal dispersions." *J. Nanosci. Nanotechnol.* 11, no. 4, 2011: 3141–3150.

Wang, X., X. Xu, and S. U. S. Choi. "Thermal conductivity of nanoparticle-fluid mixture." *J. Thermophys. Heat Tr.* 13, no. 4, 1999: 474–480.

Wang, B.-X., L.-P. Zhou, and X.-F. Peng. "A fractal model for predicting the effective thermal conductivity of liquid with suspension of nanoparticles." *Int. J. Heat Mass Transf.* 46, no. 14, 2003: 2665–2672.

Wen, D., and Y. Ding. "Experimental investigation into convective heat transfer of nanofluids at the entrance region under laminar flow conditions." *Int. J. Heat Mass Transf.* 47, no. 24, 2004: 5181–5188.

Wen, D., and Y. Ding. "Formulation of nanofluids for natural convective heat transfer applications." *Int. J. Heat Fluid Fl.* 26, no. 6, 2005: 855–864.

Xie, H. Q., M. Fujii, and X. Zhang. "Effect of interfacial nanolayer on the effective thermal conductivity of nanoparticle-fluid mixture." *Int. J. Heat Mass Transf.* 48, no. 14, 2005: 2926–2932.

Xu, J., B.-M. Yu, and M.-J. Yun. "Effect of clusters on thermal conductivity in nanofluids." *Chinese Phys. Lett.* 23, no. 10, 2006: 2819–2822.

Xuan, Y., and Q. Li. "Investigation on convective heat transfer and flow features of nanofluids." *J. Heat Transf.* 125, no. 1, 2003: 151.

Xuan, Y. M., and Q. Li. "Heat transfer enhancement of nanofluids." *Int. J. Heat Fluid Fl.* 21, no. 1, 2000: 58–64.

Xuan, Y. M., Q. Li, and W. F. Hu. "Aggregation structure and thermal conductivity of nanofluids." *AIChE J.* 49, no. 4, 2003: 1038–1043.

Xue, L., P. Keblinski, S. R. Phillpot, S. U. S. Choi, and J. A. Eastman. "Effect of liquid layering at the liquid-solid interface on thermal transport." *Int. J. Heat Mass Transf.* 47, no. 19–20, 2004: 4277–4284.

Xue, Q. "Model for effective thermal conductivity of nanofluids." *Phys. Lett. A* 307, no. 5–6, 2003: 313–317.

Yacob, N. A., A. Ishak, and I. Pop. "Falkner-Skan problem for a static or moving wedge in nanofluids." *Int. J. Therm. Sci.* 50, no. 2, 2011: 133–139.

Yang, Y. "Carbon nanofluids for lubricant application." PhD dissertation, University of Kentucky, Lexington, KY, 2006.

Yang, Y., Z. Zhang, E. Grulke, W. Anderson, and G. Wu. "Heat transfer properties of nanoparticle-in-fluid dispersions (nanofluids) in laminar flow." *Int. J. Heat Mass Transf.* 48, no. 6, 2005: 1107–1116.

Yoo, D.-H., K. S. Hong, T. E. Hong, J. A. Eastman, and H.-S. Yang. "Thermal conductivity of Al_2O_3/water nanofluids." *J. Korean Phys. Soc.* 51, no. 12, 2007: S84–S87.

Yoo, D., K. Hong, and H. Yang. "Study of thermal conductivity of nanofluids for the application of heat transfer fluids." *Thermochim. Acta* 455, no. 1–2, 2007: 66–69.

Yu, W., and S. U. S. Choi. "The role of interfacial layers in the enhanced thermal conductivity of nanofluids: A renovated Maxwell model." *J. Nanopart. Res.* 5, 2003: 167–171.

Yu, W., and S. U. S. Choi. "The role of interfacial layers in the enhanced thermal conductivity of nanofluids: A renovated Hamilton-Crosser model." *J. Nanopart. Res.* 6, 2004: 355–361.

Yu, W., J. H. Hull, and S. U. S. Choi. "Stable and highly conductive nanofluids—Experimental and theoretical studies." Paper TED-AJ03-384. *Proceedings of the 6th ASME-JSME Thermal Engineering Joint Conference.* New York, 2003.

Zhang, X., H. Gu, and M. Fujii. "Experimental study on the effective thermal conductivity and thermal diffusivity of nanofluids." *Int. J. Thermophys.* 27, no. 2, 2006: 569–580.

Zhou, S.-Q., and R. Ni. "Measurement of the specific heat capacity of water-based Al_2O_3 nanofluid." *Appl. Phys. Lett.* 92, no. 9, 2008: 093123.

Zhu, H., C. Zhang, S. Liu, Y. Tang, and Y. Yin. "Effects of nanoparticle clustering and alignment on thermal conductivities of Fe_3O_4 aqueous nanofluids." *Appl. Phys. Lett.* 89, no. 2, 2006: 23123.

CHAPTER 2

Exact Solutions and Their Implications in Anomalous Heat Transfer

Wenhao Li, Chen Yang, and Akira Nakayama

CONTENTS

2.1 INTRODUCTION

Masuda et al. (1993) carried out initiative measurements on thermal conductivities of nanofluids and found that their thermal conductivities are quite high even with small volume fraction of nanoparticles. This experimental finding was followed by a substantial number of investigations, most of which supported excellent thermal conductivity and heat transfer characteristics associated with nanofluids (Choi, 1995; Lee et al., 1999; Xuan and Roetzel, 2000; Eastman et al., 2001; Heris et al., 2007). In order to investigate the inconsistency observed in the thermal conductivity data, Buongiorno et al. (2009) conducted the International Nanofluid Property Benchmark Exercise and concluded that the thermal conductivity data from most organizations lie within a reasonably narrow band and that the Maxwell theory (Maxwell, 1881; Nan et al., 1997) is found to be in good accord with most of the data, suggesting that there is no such anomalous enhancement of thermal conductivity.

Many investigators such as Pak and Cho (1998) conducted forced convection experiments using nanofluids and reported abnormal convective heat transfer enhancement (namely, anomalous heat transfer enhancement), in which the measured heat transfer coefficient exceeds the level expected from the increase in the effective thermal properties of nanofluids. This claim has led to a controversial issue, namely, whether or not the anomalous convective heat transfer enhancement is possible in nanofluid convective heat transfer. Prabhat et al. (2012) carried out a detailed analysis examining the database reported in 12 nanofluid papers and found that in nanofluid laminar flows, in fact, there seems to be anomalous heat transfer enhancement in the entrance region. However, the turbulent flow data basically follow Dittus–Boelter's correlation, once the temperature dependence of viscosity was included in the prediction of the Reynolds number. However, all these findings are not conclusive, because most papers do not report information about the temperature dependence of the viscosity for their nanofluids. Excellent reviews on nanofluid heat transfer investigations for the past decade may be found elsewhere (Ding et al., 2007; Wang and Mujumdar, 2007).

Despite that there are a substantial number of theoretical attempts, a satisfactory theoretical explanation for possible heat transfer enhancement mechanism associated with nanofluids has not yet been fully provided. Some (Pak and Cho, 1998; Xuan and Roetzel, 2000; Wang and Mujumdar, 2007) claimed that convective heat transfer enhancement is mainly due to dispersion of the suspended nanoparticles and partly due to intensification of turbulence by nanoparticles (Xuan and Li, 2003). Buongiorno (2006) revealed that such dispersion is negligible and that turbulence is not affected by the presence of nanoparticles. He estimated relative magnitudes of the terms associated with all possible slip mechanisms, namely, inertia, Brownian diffusion, thermophoresis, diffusiophoresis, Magnus effect, fluid drainage, and gravity, and concluded that only the Brownian diffusion and thermophoresis are important in nanofluids. In this way, he was able to derive a two-component four-equation nonhomogeneous equilibrium model for mass, momentum, and heat transfer in nanofluids, which has been used by Tzou (2008) for the analysis of nanofluid Bernard convection, Hwang et al. (2009) for the analysis of laminar forced convection, and Nield and Kuznetsov (2009) for the study of thermal instability in a porous medium layer saturated by nanofluids. In recent years, a considerable number of computational fluid dynamics (CFD) papers on nanofluid convection have also been published in the open literature (Bianco et al., 2009; Akbari et al., 2011; Tahir and Mital, 2012).

In most of these previous investigations, the spatial variation of density is neglected and the governing equations are treated under incompressible flow assumptions. This incompressible flow treatment initially proposed by Buongiorno (2006) and followed by many investigators is inconsistent with the nanoparticle transport equation, which yields the spatial distribution of nanoparticle volume fraction. It is essential to take into account the effects of the nanoparticle volume fraction not only on the effective properties such as viscosity, thermal conductivity, and diffusion coefficient, but also on the nanofluid density, because all properties including the nanofluid density are naturally sensitive to the spatial distribution of the nanoparticle volume fraction. The anomaly observed in the heat transfer enhancement may be more or less 10% for some cases. This is the reason why the anomalous heat transfer enhancement has been controversial even after many years of experimental investigation. Only careful theoretical treatments that account for both Brownian diffusion and thermophoresis with the nanofluid density variation can reveal such a possibility related to the anomalous heat transfer enhancement associated with nanofluid convection.

In this chapter, such mathematical treatments based on the Buongiorno equations will be presented in detail, fully accounting for the effects of the nanoparticle distributions on all properties, including the effective density (Yang, Li, Sano et al., 2013; Yang, Li and Nakayama, 2013). Exact solutions are obtained for well-defined nanofluid convection cases, namely, hydrodynamically and thermally fully developed flows within straight conduits such as channels and tubes. Possible anomalous heat transfer enhancement associated with nanofluids is theoretically investigated, focusing on heat transfer characteristics of nanofluid-forced convection in a laminar flow regime, so as to clarify the controversial issue on anomalous convective heat transfer enhancement.

In this chapter, the temperature dependency of thermal properties of nanofluids will also be investigated thoroughly for the first time (Li and Nakayama, 2014). In most previous studies on nanofluid convection, the anomaly of heat transfer has been discussed without considering the temperature dependency of thermal properties. It has not yet been theoretically understood, whether or not the anomalous heat transfer is possible, when such dependency of thermal properties is fully taken into account. Recently, upon performing a regression analysis on a considerable number of experimental data available in the literature, Corcione (2011) has proposed what seems to be the most reliable set of empirical correlations, as function of both nanoparticle volume fraction and local temperature, for estimating the effective thermal conductivity and dynamic viscosity of nanofluids. Thus, another aim of this chapter is to investigate the effect of temperature-dependent thermophysical properties on the convective heat transfer characteristics in nanofluids, using the Buongiorno model along with Corcione's empirical correlations, so as to clarify the controversial issue of the anomalous convective heat transfer enhancement associated with nanofluids.

2.2 THERMOPHYSICAL PROPERTIES OF NANOFLUIDS

Thermophysical properties of nanofluids have been discussed over the past two decades in the literature. Because most classical models such as Brinkman (1952) are found unable to predict the thermal conductivity and dynamic viscosity of nanofluids, a substantial number of empirical correlations have been proposed. These can be classified into two groups: particle loading-dependent correlations and particle loading–temperature-dependent correlations. The aforementioned classical models and empirical sets of correlations giving both thermal conductivity and dynamic viscosity of alumina–water nanofluids are summarized in Table 2.1. Note that Pak and Cho's correlations for titania–water combinations (Pak and Cho, 1998) are also listed in Table 2.1 for the convenience of forthcoming calculations. Moreover, thermophysical properties of base fluid and nanoparticles are provided in Table 2.2.

A set of typical empirical correlations, namely, Pak and Cho's correlations for alumina–water combinations, is illustrated in Figure 2.1. The figure shows

TABLE 2.1   Correlations of Thermophysical Properties for Alumina–Water Nanofluids

Correlations	References	Expressions
Classical model	Brinkman, 1952; Maxwell, 1881	Maxell model $k = k_{bf}\{[k_p + 2k_{bf} - 2\phi(k_{bf} - k_p)]/[k_p + 2k_{bf} + \phi(k_{bf} - k_p)]\}$ Brinkman model $\mu = [\mu_{bf}/(1-\phi)^{2.5}]$
Pak and Cho's correlations	Pak and Cho, 1998	Alumina–water $k = k_{bf}(1 + 7.47\phi)$ $\mu = \mu_{bf}(1 + 39.11\phi + 533.9\phi^2)$ Titania–water $k = k_{bf}(1 + 2.92\phi - 11.99\phi^2)$ $\mu = \mu_{bf}(1 + 5.45\phi + 108.2\phi^2)$
Maiga's correlation	Maiga et al., 2005	$k = k_{bf}(1 + 2.72\phi + 4.97\phi^2)$ $\mu = \mu_{bf}(1 + 7.3\phi + 123\phi^2)$
Palm et al.'s correlation	Palm et al., 2006	$k = 0.003352T - 0.3708\ (\phi = 1\%)$ $k = 0.004961T - 0.8078\ (\phi = 4\%)$ $\mu = 3.4 \times 10^{-2} - 1.975 \times 10^{-4}T + 2.912 \times 10^{-7}T^2$ $(\phi = 1\%)$ $\mu = 4.051 \times 10^{-2} - 2.353 \times 10^{-4}T + 3.475 \times 10^{-7}T^2$ $(\phi = 4\%)$
Williams et al.'s correlation	Williams et al., 2008	$k(\phi, T) = k_{bf}(T)(1 + 4.5503\phi)$ $\mu(\phi, T) = \mu_{bf}(T)\exp[4.91\phi/(0.2092 - \phi)]$
Venerus et al.'s correlation	Venerus et al., 2010	$k = k_{bf}(1 + 4\phi)$ $\mu = \mu_{bf}(1 + 23.4\phi)$

TABLE 2.2   Thermophysical Properties of Base Fluids and Nanoparticles

Thermophysical Properties	Density (kb/m³)	Heat Capacity [J/(kg·K)]	Thermal Conductivity [W/(m·K)]	Dynamic Viscosity [kg/(m·s)]
Water	998.2	4182	0.597	9.93×10^{-4}
Alumina	3880	773	36	N/A
Titania	4175	692	8.4	N/A

the variations of the dimensionless nanofluid properties based on their base fluid counterparts. The ratio of nanofluid Prandtl number $Pr = \mu c/k$, where μ, c, and k are the viscosity, heat capacity, and thermal conductivity of the nanofluid, respectively, is also indicated. As can be seen from Figure 2.1, the density ratio ρ/ρ_{bf}, where ρ is the density and the subscript bf refers to the base fluid, is as sensitive as k/k_{bf} to the nanoparticle volume fraction ϕ. Therefore, the nanofluid density modification must be made, because the constancy of ρ adopted in the Buongiorno model may lead to substantial errors.

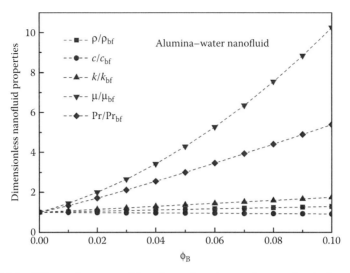

FIGURE 2.1   Effects of nanoparticle volume fraction on thermophysical properties of nanofluids.

2.3 CONSERVATION EQUATIONS FOR NANOFLUIDS

Buongiorno assumed incompressible flow, no chemical reactions, negligible external forces, dilute mixture, negligible viscous dissipation, negligible radiative heat transfer, and local thermal equilibrium between nanoparticles and base fluid. His two-component mixture model may be modified to allow the nanofluid density (ρ) variation in mass, momentum, and energy conservations as follows:

$$\frac{\partial \rho u_j}{\partial x_j} = 0 \tag{2.1}$$

$$\frac{\partial \rho u_i}{\partial t} + \frac{\partial \rho u_j u_i}{\partial x_j} = -\frac{\partial p}{\partial x_i} + \frac{\partial}{\partial x_j}\mu\left(\frac{\partial u_i}{\partial x_j} + \frac{\partial u_j}{\partial x_i}\right) \tag{2.2}$$

$$\frac{\partial \rho c T}{\partial t} + \frac{\partial \rho c u_j T}{\partial x_j} = \frac{\partial}{\partial x_j} k \frac{\partial T}{\partial x_j} + \rho_p c_p \left(D_B \frac{\partial \phi}{\partial x_j} + \frac{D_T}{T}\frac{\partial T}{\partial x_j}\right)\frac{\partial T}{\partial x_j} \tag{2.3}$$

$$\rho_p \left(\frac{\partial \phi}{\partial t} + \frac{\partial u_j \phi}{\partial x_j}\right) = \frac{\partial}{\partial x_j}\left(\rho_p D_B \frac{\partial \phi}{\partial x_j} + \frac{\rho_p D_T}{T}\frac{\partial T}{\partial x_j}\right) \tag{2.4}$$

where:
   T is the temperature of nanofluids
   t is the time

u is the velocity of nanofluids

x is the coordinate

D is the diffusion coefficient as defined in equation (2.6a) and (2.6b)

For convenience, it is first assumed that thermal conductivity and dynamic viscosity of nanofluids depend only on ϕ. Thus, the particle loading-dependent correlations of nanofluids' thermophysical properties (Pak and Cho, 1998) are employed here to obtain exact solutions. The properties ρ, *c*, μ, and *k* are expressed as the functions of ϕ as follows:

$$\rho = \phi \rho_p + (1-\phi)\rho_{bf} \tag{2.5a}$$

$$c = \frac{\phi \rho_p c_p + (1-\phi)\rho_{bf} c_{bf}}{\rho} \tag{2.5b}$$

$$\mu = \mu_{bf}(1 + a_\mu \phi + b_\mu \phi^2) \tag{2.5c}$$

$$k = k_{bf}(1 + a_k \phi + b_k \phi^2) \tag{2.5d}$$

The Brownian and thermophoretic diffusion coefficients are given by

$$D_B = \frac{k_{BO} T}{3\pi\mu_{bf} d_p} \tag{2.6a}$$

$$D_T = 0.26 \frac{k_{bf}}{2k_{bf} + k_p} \frac{\mu_{bf}}{\rho_{bf}} \phi \tag{2.6b}$$

where:

The subscript *p* refers to the nanoparticle

k_{BO} is the Boltzmann constant

d_p is the nanoparticle diameter, which can be anywhere on the order of 1–100 nm

Buongiorno proved that the heat transfer associated with the nanoparticle dispersion (i.e., nanoparticle diffusion flux), namely, the second term on the right-hand side of Equation 2.3, is always negligible compared with convection and conduction. As will be shown shortly, the term vanishes for the case of fully developed flows. Thus, the energy equation (2.3) becomes identical to that of a pure fluid, except that all properties are functions of ϕ. It should be noted that the nanoparticle continuity equation (2.4) must be solved simultaneously using Equations 2.1 through 2.3

for the other dependent variables, because the thermophysical properties strongly depend on the spatial distribution of ϕ.

The results naturally depend on a particular set of correlations employed for the thermophysical properties. However, Yang, Li, Sano et al. (2013) and Yang, Li and Nakayama (2013) studied the anomalous heat transfer using two asymptotic correlations of thermophysical properties to cover most of reliable correlations proposed so far and proved that all reliable correlations lead to similar results on the heat transfer anomaly. Thus, it is expected that the following conclusions for the cases of fully developed channel flow and tube flow drawn using Pak and Cho's correlations generally apply to all cases in reality.

2.3.1 Fully Developed Flow in a Channel

Referring to Figure 2.2a for hydrodynamically and thermally fully developed channel flow subject to constant heat flux, one may write the set of the conservation equations as follows:

$$\frac{d}{dy}\left(\mu\frac{du}{dy}\right) - \frac{dp}{dx} = 0 \tag{2.7}$$

$$\rho c u \frac{\partial T}{\partial x} = \frac{\partial}{\partial y}\left(k\frac{\partial T}{\partial y}\right) \tag{2.8}$$

$$\frac{\partial}{\partial y}\left(D_B\frac{\partial \phi}{\partial y} + \frac{D_T}{T}\frac{\partial T}{\partial y}\right) = 0 \tag{2.9}$$

where the origin of the vertical coordinate y is set on the wall. The boundary conditions are given by

$$y = 0 \text{ (on the wall):}$$

$$u = 0 \tag{2.10a}$$

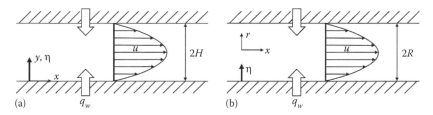

FIGURE 2.2 Physical models: (a) channel flow; (b) tube flow.

$$-k\frac{\partial T}{\partial y} = q_w \tag{2.10b}$$

$$D_B\frac{d\phi}{dy} + \frac{D_T}{T}\frac{d(T-T_w)}{dy} = 0 \tag{2.10c}$$

$y = H$ (along the center):

$$\frac{du}{dy} = 0 \tag{2.11a}$$

$$\frac{d(T-T_w)}{dy} = 0 \tag{2.11b}$$

$$\frac{d\phi}{dy} = 0 \tag{2.11c}$$

The nanoparticle continuity equation (2.9) indicates that the diffusion mass flux of the nanoparticles is constant across the channel. As the wall is impermeable, the boundary condition (2.10c) holds such that the Brownian diffusion flux and thermophoretic diffusion flux cancel out everywhere in the channel. This fact is already implemented to simplify the foregoing energy equation, which may be integrated over the lower half of the channel from $y = 0$ to H with the boundary conditions (2.10b) and (2.11b) to give

$$\langle \rho c u \rangle \frac{d T_B}{dx} = \frac{q_w}{H} \tag{2.12}$$

where:

$$\langle \phi \rangle \equiv \frac{1}{A}\int_A \phi dA = \frac{1}{H}\int_0^H \phi dy \tag{2.13}$$

denotes the average over the cross section such that

$$T_B \equiv \frac{\langle \rho c u T \rangle}{\langle \rho c u \rangle} \tag{2.14}$$

is the bulk mean temperature.

The foregoing considerations on both nanoparticle diffusion flux and axial temperature gradient are implemented to write the set of the governing equations in a dimensionless form as follows:

$$\frac{du^*}{d\eta} = \frac{1-\eta}{(\mu/\mu_w)} \tag{2.15}$$

$$\frac{d^2 T^*}{d\eta^2} = -\frac{1}{(k/k_w)}\left[\frac{\rho c u^*}{\langle \rho c u^* \rangle} + \frac{a_k + 2b_k\phi}{1+a_k\phi_w + b_k\phi_w{}^2}\left(\frac{d\phi}{d\eta}\right)\left(\frac{dT^*}{d\eta}\right)\right] \tag{2.16}$$

$$\frac{d\phi}{d\eta} = \frac{\phi}{N_{BT}\left(1-\gamma T^*\right)^2}\frac{dT^*}{d\eta} \tag{2.17}$$

with the boundary conditions:

$$\eta = 0 \,(\text{on the wall}):$$

$$u^* = 0 \tag{2.18a}$$

$$T^* = 0 \tag{2.18b}$$

$$\frac{dT^*}{d\eta} = 1 \tag{2.18c}$$

$$\phi = \phi_w \tag{2.18d}$$

where the dimensionless coordinate, velocity, and temperature are defined as

$$\eta = \frac{y}{H} \tag{2.19a}$$

$$u^* = \frac{u}{\{(H^2/\mu_w)[-(dp/dx)]\}} \tag{2.19b}$$

$$T^* = \frac{k_w(T_w - T)}{q_w H} \tag{2.19c}$$

and the dimensionless parameters as follows:

$$N_{BT} \equiv \frac{D_{Bw}T_w\phi_w k_w}{D_{Tw}q_w H} = \frac{D_{Bw}\phi_w}{D_{Tw}\gamma} = \frac{k_{B0}T_w^2\,\rho_{bf}k_w}{3\pi d_p\,\{0.26[k_{bf}/(2k_{bf}+k_p)]\}\,\mu_{bf}^2\,q_w H} \tag{2.20a}$$

$$\gamma \equiv \frac{q_w H}{k_w T_w} \tag{2.20b}$$

The ratio of Brownian and thermophoretic diffusivities $N_{BT}(\propto 1/d_p)$ can range over a wide range from 0.1 to 10 for typical cases of alumina and copper nanoparticles with $d_p \sim 10$ nm and $\phi_B \sim 0.01$, whereas the ratio of wall and fluid temperature difference to absolute temperature $\gamma \sim (T_w - T_B)/T_w$ is usually much smaller than unity, as estimated by Buongiorno (2006). In reality, the bulk mean particle volume fraction ϕ_B is prescribed instead of that at the wall ϕ_w. However, for the sake of computational convenience, ϕ_w is prescribed and ϕ_B is calculated later to find out ϕ_w as a function of ϕ_B. Note that the properties with the subscript w are based on the wall.

The symmetry boundary conditions (2.11a–c) are already implemented in the energy equation, such that the set of the ordinary equations can easily be solved as an initial value problem with all boundary values prescribed at the wall. Any standard integration scheme such as the Runge–Kutta–Gill method (Nakayama, 1995) can be used to determine the distributions of the dimensionless variables $u^*(\eta)$, $T^*(\eta)$, and $\phi(\eta)$ for given ϕ_w, N_{BT}, and γ. Once these distributions are known, the Nusselt number based on the bulk properties of nanofluid and hydraulic diameter can be evaluated as

$$\mathrm{Nu}_B \equiv \frac{h(4H)}{k_B} = \frac{q_w(4H)}{(T_w - T_B)k_B} = 4 \bigg/ \left[T_B^* \left(\frac{k_B}{k_w} \right) \right] \tag{2.21}$$

The bulk mean dimensionless temperature is defined as

$$T_B^* \equiv \frac{\langle \rho c u^* \, T^* \rangle}{\langle \rho c u^* \rangle} \tag{2.22}$$

The thermal properties with the subscript B are based on the bulk mean particle volume fraction:

$$\phi_B \equiv \frac{\langle u^* \phi \rangle}{\langle u^* \rangle} \tag{2.23}$$

such that $k_B = k(\phi_B)$ is the bulk mean thermal conductivity of the nanofluid.

The heat transfer enhancement may be evaluated in terms of the ratio of the nanofluid and base fluid heat transfer coefficients:

$$\frac{h}{h_{bf}} = \frac{\mathrm{Nu}_B}{8.24} \left(\frac{k_B}{k_{bf}} \right) = \frac{\mathrm{Nu}_B}{8.24} (1 + a_k \phi_B + b_k \phi_B^2) \tag{2.24}$$

Thus, one may infer possible nanofluid anomalous convective heat transfer enhancement, only if Nu_B is greater than the pure fluid Nusselt number, namely, $140/17 = 8.24$. Likewise, the dimensionless pressure gradient may be evaluated as

$$\frac{-(dp/dx)}{\mu_B\left[u_B/(4H)^2\right]} = \frac{\lambda}{2}\left(\frac{\rho_{bf}u_B 4H}{\mu_{bf}}\right) = \frac{16\rho_B}{\langle\rho u^*\rangle(\mu_B/\mu_w)} \tag{2.25}$$

2.3.2 Fully Developed Flow in a Tube

Hydrodynamically and thermally fully developed tube flow subject to constant heat flux may be treated by writing the governing equations in cylindrical coordinates (x, r) shown in Figure 2.2b as

$$\frac{1}{r}\frac{d}{dr}\left(r\mu\frac{du}{dr}\right) - \frac{dp}{dx} = 0 \tag{2.26}$$

$$\rho cu\frac{\partial T}{\partial x} = \frac{1}{r}\frac{\partial}{\partial r}\left(rk\frac{\partial T}{\partial r}\right) \tag{2.27}$$

$$\frac{1}{r}\frac{\partial}{\partial r}\left(D_B\frac{d\phi}{dr} + \frac{D_T}{T}\frac{\partial T}{\partial r}\right) = 0 \tag{2.28}$$

which can easily be transformed in a dimensionless form as

$$\frac{du^*}{d\eta} = \frac{1-\eta}{2(\mu/\mu_w)} \tag{2.29}$$

$$\frac{d^2 T^*}{d\eta^2} = -\frac{1}{(k/k_w)}\left\{2\frac{\rho cu^*}{\langle\rho cu^*\rangle} + \left[\frac{a_k + 2b_k\phi_w}{1 + a_k\phi_w + b_k\phi_w^2}\left(\frac{d\phi}{d\eta}\right) - \frac{(k/k_w)}{1-\eta}\right]\left(\frac{dT^*}{d\eta}\right)\right\} \tag{2.30}$$

$$\frac{d\phi}{d\eta} = \frac{\phi}{N_{BT}(1-\gamma T^*)^2}\frac{dT^*}{d\eta} \tag{2.31}$$

with the boundary conditions:

$$\eta = 0: \quad u^* = 0 \tag{2.32a}$$

$$T^* = 0 \tag{2.32b}$$

$$\frac{dT^*}{d\eta} = 1 \tag{2.32c}$$

$$\phi = \phi_w \tag{2.32d}$$

where the dimensionless coordinate, velocity, and temperature are defined as

$$\eta = \frac{(R-r)}{R} \tag{2.33a}$$

$$u^* = \frac{u}{\left[R^2/\mu_w \left(-dp/dx \right) \right]} \tag{2.33b}$$

$$T^* = \frac{k_w (T_w - T)}{q_w R} \tag{2.33c}$$

and the dimensionless parameters as follows:

$$N_{BT} \equiv \frac{k_{B0} T_w{}^2 \rho_{bf} k_w}{3\pi d_p \left\{ 0.26 \left[k_{bf} / \left(2k_{bf} + k_p \right) \right] \right\} \mu_{bf}{}^2 q_w R} \tag{2.34a}$$

$$\gamma \equiv \frac{q_w R}{k_w T_w} \tag{2.34b}$$

The average value $\langle \phi \rangle$ for the case of the tube is computed by

$$\langle \phi \rangle \equiv \frac{1}{A} \int_A \phi \, dA = \frac{1}{\pi R^2} \int_0^R 2\pi r \phi \, dr = 2 \int_0^1 (1 - \eta) \phi \, d\eta \tag{2.35}$$

Note that the singularity in Equation 2.30 at $\eta = 1$ may easily be removed by transforming it to

$$\frac{d^2 T^*}{d\eta^2} = -\frac{1}{2(k/k_w)} \left\{ 2 \frac{\rho c u^*}{\langle \rho c u^* \rangle} + \left[\frac{a_k + 2b_k \phi_w}{1 + a_k \phi_w + b_k \phi_w{}^2} \left(\frac{d\phi}{d\eta} \right) \right] \left(\frac{dT^*}{d\eta} \right) \right\} \tag{2.36}$$

The corresponding Nusselt number based on the bulk properties of nano-fluids and tube diameter can be evaluated as

$$Nu_B \equiv \frac{h(2R)}{k_B} = \frac{q_w (2R)}{(T_w - T_B) k_B} = 2 / \left[T_B^* \left(\frac{k_B}{k_w} \right) \right] \tag{2.37}$$

where T_B^* and ϕ_B are the bulk mean dimensionless temperature and particle volume fraction, respectively, already defined by Equations 2.22 and 2.23 along with Equation 2.35.

The heat transfer enhancement and pressure gradient may be evaluated as follows:

$$\frac{h}{h_{bf}} = \frac{Nu_B}{4.36} \left(\frac{k_B}{k_{bf}} \right) = \frac{Nu_B}{4.36} (1 + a_k \phi_B + b_k \phi_B^2) \tag{2.38}$$

$$\frac{-(dp/dx)}{\mu_B\left[u_B/(2R)^2\right]} = \frac{\lambda}{2}\left(\frac{\rho_{bf}u_B 2R}{\mu_{bf}}\right) = \frac{4\rho_B}{\langle\rho u^*\rangle(\mu_B/\mu_w)} \tag{2.39}$$

One may conclude that anomalous convective heat transfer enhancement is possible, only if Nu_B is greater than the pure fluid Nusselt number, namely, $48/11 = 4.36$.

2.4 DISTRIBUTIONS OF VELOCITY, TEMPERATURE, AND PARTICLE VOLUME FRACTION IN A CIRCULAR TUBE

The versatile software SOLODE based on the Runge–Kutta–Gill method (Nakayama, 1995) was used to integrate the set of ordinary differential equations. The effects of the bulk mean particle volume fraction ϕ_B on the velocity profile $u/u_B = u^*/(\langle\rho u^*\rangle/\rho_B)$ are shown in Figure 2.3a for the case of titania–water nanofluids in a tube with $N_{BT} = 0.2$ and $\gamma = 0$. Likewise, the temperature profiles are presented in Figure 2.3b in terms of $(T_w - T)/(T_w - T_B) = T^*/T_B^*$. Furthermore, the distributions of the particle volume fraction ϕ/ϕ_B are shown in Figure 2.3c, which indicates low volume fraction of particles near the wall and high volume fraction of particles in the core. Because of low and high volume fractions of nanoparticles, the viscosity near the wall surface as given by Equation 2.5c is much smaller

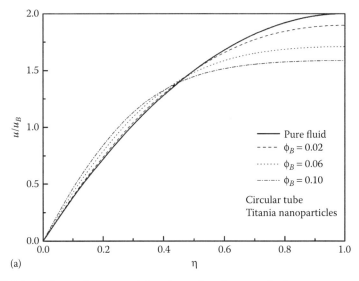

(a)

FIGURE 2.3 Effects of nanoparticle volume fraction on velocity, temperature, and volume fraction profiles in a tube with $N_{BT} = 0.2$ and $\gamma = 0$: (a) velocity. (*Continued*)

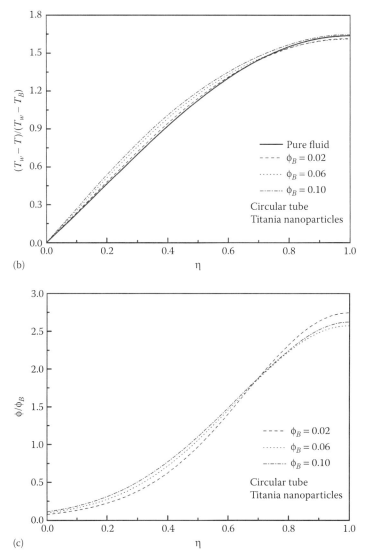

FIGURE 2.3 (Continued) Effects of nanoparticle volume fraction on velocity, temperature, and volume fraction profiles in a tube with $N_{BT} = 0.2$ and $\gamma = 0$: (b) temperature; (c) volume fraction.

than that in the core, which results in an increase in the velocity near the wall and a decrease in the velocity in the core, as can be confirmed in Figure 2.3a. This tendency is pronounced as adding more particles. Figure 2.3b consistently shows that the increase in the dimensionless velocity near the wall naturally makes the dimensionless temperature gradient steeper compared with the one with pure fluids ($\phi_B = 0$). The effects of γ on these profiles are found insignificant, as varying γ from 0 to 0.1. Thus, all calculations have been carried out with $\gamma = 0$.

2.5 HYDRAULIC CHARACTERISTICS IN CHANNELS AND TUBES

The dimensionless nanofluid pressure gradients in a channel and a tube are presented in Figure 2.4a and b, respectively. Both figures show that the pressure drops are less than what would be expected from the increase in the properties, because the dimensionless pressure gradients for the case of nanofluids are less than the pure fluid values. The decrease in the dimensionless pressure gradient is more pronounced for higher ϕ_B and smaller N_{BT} (i.e., larger particle diameter), especially for alumina–water nanofluids. Substantial decrease in the dimensionless nanofluid pressure gradients indicates that the pressure drops in nanofluids are much less than in pure fluids of the same bulk viscosity. However, the pressure drops of nanofluids for given u_B are never lower than those of the base fluids, because of the viscosity increase near the wall due to the particle addition. It is also interesting to note that the dimensionless pressure gradient of nanofluids asymptotically reaches to the level of the pure fluids, as increasing the ratio of Brownian and thermophoretic diffusivities N_{BT}.

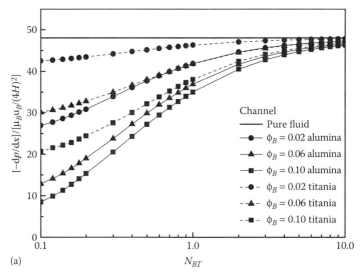

FIGURE 2.4 Effects of N_{BT} on the pressure gradients: (a) channel flow. (*Continued*)

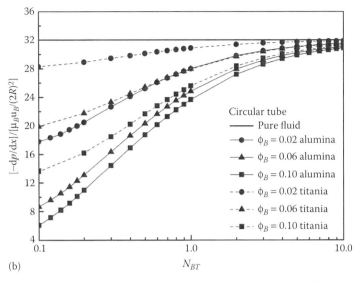

FIGURE 2.4 (Continued)   Effects of N_{BT} on the pressure gradients: (b) tube flow.

2.6 HEAT TRANSFER CHARACTERISTICS IN CHANNELS AND TUBES

The Nusselt number based on the hydraulic diameter and the nanofluid bulk thermal conductivity $k_B = k(\phi_B)$ for the channel flows is presented in Figure 2.5a for this case of alumina–water nanofluids and Figure 2.5b for this case of titania–water nanofluids, respectively. As shown in Figure 2.5a, Nu_B stays below the value of the pure fluid and asymptotically increases to that value 8.24, as increasing N_{BT}. Although the variations of Nu_B shown in Figure 2.5b are quite different from those shown in Figure 2.5a that Nu_B stays below the value of the pure fluid in a low range of N_{BT}, but overshoots its value as increasing N_{BT}, reaching to its maximum around $N_{BT} \cong 0.5$. This clearly suggests that the anomalous heat transfer enhancement takes place for titania–water nanofluids in a channel. This theoretical finding on the maximum can be quite useful for designing nanoparticle sizes, because one can capture the maximum Nusselt number around $N_{BT} \cong 0.5$.

The occurrence of the anomalous heat transfer enhancement also depends on the geometrical configurations of conduits. As shown in Figure 2.6a and b, Nu_B for the case of tubes exhibits the anomalous heat transfer enhancement even for the alumina nanoparticles, for which

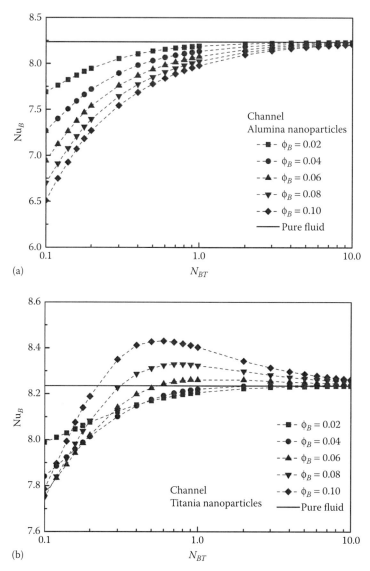

FIGURE 2.5   Heat transfer characteristics in alumina–water nanofluids: (a) channel flow; (b) tube flow.

the channel does not show any abnormality in heat transfer. This is due to the geometrical (radial) effects, namely, that events taking place near the peripheral walls, such as velocity and temperature changes, influence on bulk quantities more in a tube than in a channel. The maximum value of Nu_B is attained at $N_{BT} \cong 0.5$ as shown in Figure 2.6a.

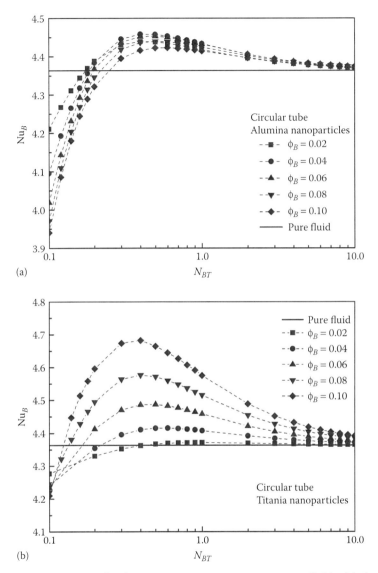

FIGURE 2.6   Heat transfer characteristics in titania–water nanofluids: (a) channel flow; (b) tube flow.

Figure 2.6b clearly indicates that the degree of the anomalous heat transfer enhancement in titania–water nanofluids is somewhat higher than in alumina–water nanofluids. However, as given by Equation 2.38 and shown in Figure 2.7, the heat transfer coefficient of the alumina–water nanofluids is higher than that of the titania–water nanofluids, simply due to the difference in k_B.

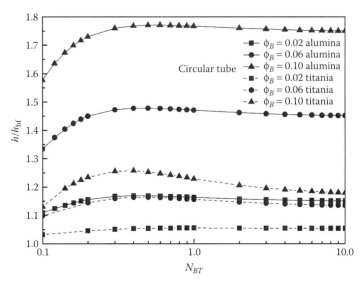

FIGURE 2.7   Comparison of the heat transfer coefficients of nanofluids in a tube.

2.7 TEMPERATURE DEPENDENCY OF NANOFLUID THERMOPHYSICAL PROPERTIES

As pointed out in Section 2.2, thermophysical properties of nanofluids depend not only on the loading of nanoparticles but also on the temperature of nanofluids (Das et al. 2003). Palm et al. (2006) provided temperature-dependent empirical correlations for specific concentrations of alumina–water nanofluid. Williams et al. (2008) presented temperature-dependent empirical correlations, assuming that temperature dependency only has influence on the thermal properties of the base fluid. However, all the correlations mentioned above have been obtained from a limited number of experimental data and their region of validity is somewhat limited. Thus, further efforts are needed to develop a general model to estimate thermophysical properties of nanofluids (Prabhat et al., 2012).

Upon exploiting a regression analysis, Corcione (2011) proposed a set of empirical correlations that match well with most experimental data available in the literature, substantiating a wide range of its validity. His correlation for the thermal conductivity of nanofluids runs as follows:

$$k = k_{\mathrm{bf}}(T)\left[1 + 4.4\,\mathrm{Re}^{0.4}\,\mathrm{Pr}^{0.66}\left(\frac{T}{T_{fr}}\right)^{10}\left(\frac{k_p}{k_{\mathrm{bf}}}\right)^{0.03}\phi^{0.66}\right] \qquad (2.40)$$

where the valid ranges of nanoparticle diameter, volume fraction, and temperature of nanofluids are 10–150 nm, 0.002–0.09, and 294–324 K, respectively. Pr is the Prandtl number of the base fluid, T is the absolute temperature of nanofluid, T_{fr} is the freezing point of the base fluid, and k_p and k_{bf} are the thermal conductivities of nanoparticle and base fluid, respectively. ϕ is the volume fraction of nanoparticles and $Re = (2\rho_{bf} k_{BO} T)/(\pi \mu_{bf}^2 d_p)$ is the nanoparticle Reynolds number, in which ρ_{bf} and μ_{bf} are the density and dynamic viscosity of the base fluids, respectively. k_{BO} is the Boltzmann's constant and d_p is the diameter of nanoparticle. Unlike the thermal conductivity, the dynamic viscosity ratio of nanofluid and base fluid is independent of the temperature within the range, although the viscosity of the base fluid varies with the temperature moderately.

$$\mu = \mu_{bf}(T)/\left[1 - 34.87(d_p/d_f)^{-0.3} \phi^{1.03}\right] \quad (2.41)$$

where the valid ranges of nanoparticle diameter, volume fraction, and temperature of nanofluids are 25–200 nm, 0.0001–0.071, and 293–323 K, respectively. $d_f = 0.1\left[(6M)/(N\pi\rho_{bf0})\right]^{1/3}$ is the equivalent diameter of a base fluid molecule, in which M is the molecule weight of the base fluid, N is the Avogadro number, and ρ_{bf0} is the density of the base fluid at $T_0 = 293$ K.

Figure 2.8 presents the effect of the temperature on the dimensionless nanofluid thermal conductivity k/k_{bf} with fixed bulk mean nanoparticle

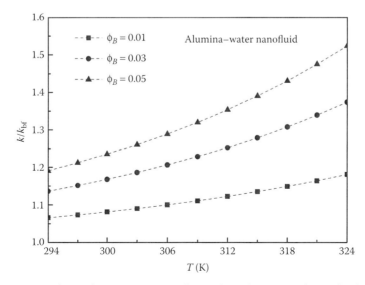

FIGURE 2.8 Effects of temperature on thermal conductivity of nanofluids.

volume fraction. As shown in Figure 2.8, the temperature significantly influences on the thermal conductivity ratio k/k_{bf} and naturally affects the convective heat transfer behavior.

2.8 CONSERVATION EQUATIONS FOR NANOFLUIDS WITH TEMPERATURE-DEPENDENT PROPERTIES

In this section, hydrodynamically and thermally fully developed forced convective tube flow subject to constant heat flux is considered to investigate the effect of temperature-dependent thermophysical properties on convective heat transfer. The derivation process of the exact solution accounting for the temperature dependency is identical to that of tube flow in Section 2.3 (i.e., Equations 2.26 through 2.39), except that the energy equation (2.30) should be replaced by

$$\frac{d^2 T^*}{d\eta^2} = -\frac{1}{(k/k_w)}\left\{2\frac{\rho c u^*}{\langle \rho c u^*\rangle} + \left[\left(\frac{\partial k}{\partial \phi}\frac{d\phi}{d\eta} + \frac{\partial k}{\partial T^*}\frac{dT^*}{d\eta}\right)/k_w - \frac{(k/k_w)}{1-\eta}\right]\left(\frac{dT^*}{d\eta}\right)\right\} \quad (2.42)$$

and Equation 2.36 replaced by

$$\frac{d^2 T^*}{d\eta^2} = -\frac{1}{2(k/k_w)}\left\{2\frac{\rho c u^*}{\langle \rho c u^*\rangle} + \left[\left(\frac{\partial k}{\partial \phi}\frac{d\phi}{d\eta} + \frac{\partial k}{\partial T^*}\frac{dT^*}{d\eta}\right)/k_w\right]\left(\frac{dT^*}{d\eta}\right)\right\} \quad (2.43)$$

because for nanofluids accounting for the temperature dependency, μ and k are considered as given functions of ϕ and T as indicated in Equations 2.40 and 2.41. The expressions of $\partial k/\partial \phi$ and $\partial k/\partial T^*$ in Equations 2.42 and 2.43 have to be given in terms of ϕ and T^*. Substituting Equations 2.33c and 2.34b into 2.40,

$$k = k_{bf}\left\{1 + 4.4\left[\frac{2\rho_{bf}k_{BO}T_w(1-\gamma T^*)}{\pi\mu_{bf}^2 d_p}\right]^{0.4}\mathrm{Pr}^{0.66}\left[\frac{T_w(1-\gamma T^*)}{T_{fr}}\right]^{10}\left(\frac{k_p}{k_{bf}}\right)^{0.03}\phi^{0.66}\right\} \quad (2.44)$$

The first-order partial derivatives of k with respect to ϕ and T^* can be obtained as

$$\frac{\partial k}{\partial \phi} = k_{bf}\left\{4.4\left[\frac{2\rho_{bf}k_{BO}T_w(1-\gamma T^*)}{\pi\mu_{bf}^2 d_p}\right]^{0.4}\mathrm{Pr}^{0.66}\left[\frac{T_w(1-\gamma T^*)}{T_{fr}}\right]^{10}\left(\frac{k_p}{k_{bf}}\right)^{0.03}0.66\phi^{-0.34}\right\} \quad (2.45)$$

and

$$\frac{\partial k}{\partial T^*} = k_{bf}\left[4.4\left(\frac{2\rho_{bf}k_{BO}T_w}{\pi\mu_{bf}^2 d_p}\right)^{0.4}\mathrm{Pr}^{0.66}\left(\frac{T_w}{T_{fr}}\right)^{10}10.4(-\gamma)(1-\gamma T^*)^{9.4}\left(\frac{k_p}{k_{bf}}\right)^{0.03}\phi^{0.66}\right] \quad (2.46)$$

Meanwhile, in order to assess the effect of temperature-dependent thermophysical properties on the convective heat transfer characteristics in nanofluids, another set of computations without accounting for the temperature dependency of thermophysical properties as a comparison is carried out, in which the following thermal conductivity correlation with T_B is used:

$$k = k_{bf}\left[1 + 4.4\left(\frac{2\rho_{bf}k_{BO}T_B}{\pi\mu_{bf}^2 d_p}\right)^{0.4}\mathrm{Pr}^{0.66}\left(\frac{T_B}{T_{fr}}\right)^{10}\left(\frac{k_p}{k_{bf}}\right)^{0.03}\phi^{0.66}\right] \quad (2.47)$$

where the bulk mean absolute temperature T_B is calculated from

$$T_B = T_w(1 - \gamma T_B^*) \quad (2.48)$$

where T_w and γ are prescribed and T_B^* is obtained from Equation 2.33 in the case with accounting for the temperature dependency of thermophysical properties.

Likewise, the derivation process without accounting for the temperature dependency is identical to that of tube flow in Section 2.3 (i.e., Equations 2.26 through 2.39), except that the expression of $\partial k/\partial \phi$ in terms of ϕ is given as

$$\frac{\partial k}{\partial \phi} = k_{bf}\left[4.4\left(\frac{2\rho_{bf}k_{BO}T_{B,d}}{\pi\mu_{bf}^2 d_p}\right)^{0.4}\mathrm{Pr}^{0.66}\left(\frac{T_{B,d}}{T_{fr}}\right)^{10}\left(\frac{k_p}{k_{bf}}\right)^{0.03}0.66\phi^{-0.34}\right] \quad (2.49)$$

One may conclude that anomalous convective heat transfer enhancement is possible, only if Nu_B is greater than the Nusselt number of nanofluid with uniform particle distribution (i.e., sufficiently large N_{BT}) $\mathrm{Nu}_{B,uniform} = \mathrm{Nu}_B(N_{BT} \to \infty)$.

2.9 EFFECTS OF TEMPERATURE DEPENDENCY OF THERMOPHYSICAL PROPERTIES ON ANOMALOUS HEAT TRANSFER

Figure 2.9 shows the Nusselt number Nu_B against N_{BT} for both cases with and without accounting for the temperature dependency under fixed $\gamma = 0.05$ and $\phi_B = 0.04$. Solid and dash straight lines in Figure 2.9 indicate the corresponding asymptotic cases with and without accounting for the temperature dependency for sufficiently large N_{BT} such that the Brownian diffusion dominates over the thermophoretic diffusion, resulting in a uniform distribution of the nanoparticles. The dashed line with the temperature dependency stays

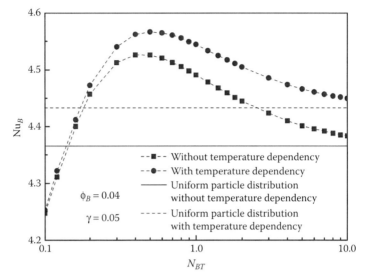

FIGURE 2.9   Comparison of Nu_B between with and without accounting for the temperature dependency under $\gamma = 0.05$ and $\phi_B = 0.04$.

higher than the solid line without the temperature dependency, because the thermal conductivity near the heated wall (i.e., wall heat transfer rate) is much higher than that in the core when the temperature dependency is accounted for. As can be seen from Figure 2.9, the Nusselt number Nu_B for both cases stays below the corresponding asymptotic values for a low range of N_{BT}, but overshoots them as increasing N_{BT}, reaching toward the values of the uniform particle distribution cases. Thus, anomalous heat transfer enhancement is possible for both cases with and without accounting for the temperature dependency of thermophysical properties. Furthermore, the anomaly takes place almost in the same range of N_{BT} and its maximum anomaly level is attained around $N_{BT} \cong 0.5$ for both cases. This clearly suggests that the effect of temperature dependency on thermophysical properties is only to alter the level of Nu_B (plotted against N_{BT}), following the asymptotic level corresponding with the uniform particle distribution case. The temperature dependency, however, has no significant influence on the occurrence of anomalous heat transfer enhancement. Hence, one can justify all discussions made on the anomaly by the previous researchers without accounting for the temperature dependency.

Finally, the dimensionless pressure gradient as defined by Equation 2.39 is plotted in Figure 2.10 against N_{BT} for both cases with and without accounting for the temperature dependency of thermophysical properties. The

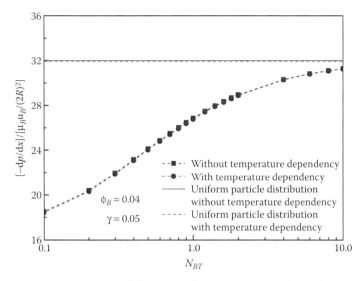

FIGURE 2.10 Comparison of dimensionless pressure gradients between with and without accounting for the temperature dependency under $\gamma = 0.05$ and $\phi_B = 0.04$.

dimensionless pressure gradient increases with N_{BT} approaching to the asymptotic value. However, both curves are almost identical, because the dynamic viscosity ratio of the nanofluid to the base fluid is independent of the temperature within the range.

2.10 CONCLUSIONS

The Buongiorno model equations for convective heat transfer in nanofluids have been fully solved accounting for the effects of nanoparticle volume fraction distributions on the continuity, momentum, and energy equations. Exact solutions for laminar fully developed nanofluid flows in channels and tubes have been obtained for the first time to clarify the controversial issue on nanofluid anomalous convective heat transfer enhancement. The anomalous heat transfer enhancement has been captured theoretically for the cases of titania–water nanofluids in a channel, alumina–water nanofluids in a tube, and also titania–water nanofluids in a tube. Comparatively low volume fraction of particles near the wall is responsible for the anomalous heat transfer, because it yields a relatively low viscosity field there, thus leading to high velocity and steep temperature gradient near the wall. The maximum Nusselt number based on the bulk mean nanofluid thermal conductivity is captured around at

$N_{BT} \cong 0.5$, which may be utilized for designing nanoparticles in view of heat transfer enhancement. Furthermore, it is found that the anomalous heat transfer enhancement is more likely to take place in a tube than in a channel due to the radial effects.

Two distinctive sets of the exact solutions were obtained for hydrodynamically and thermally fully developed nanofluid flows in a tube subject to constant heat flux with and without accounting for the temperature dependency of thermophysical properties. Alumina–water nanofluids are considered using available correlations, which take into account both nanoparticle volume fraction and temperature dependency of thermophysical properties of nanofluids. The study indicates that the temperature dependency on thermophysical properties only alters the level of the Nusselt number and that the anomaly takes place almost in the same range of N_{BT} and its maximum anomaly level is attained around $N_{BT} \cong 0.5$ for both two cases, namely, with and without accounting for the temperature dependency.

REFERENCES

Akbari, M., Galanis, N., Behzadmehr, A. 2011. Comparative analysis of single and two-phase models for CFD studies of nanofluid heat transfer. *Int. J. Therm. Sci.* 50: 1343–1354.

Bianco, V., Chiacchio, F., Manca, O., Nardini, S. 2009. Numerical investigation of nanofluids forced convection in circular tubes. *Appl. Therm. Eng.* 29: 3632–3642.

Brinkman, H. C. 1952. The viscosity of concentrated suspensions and solutions. *J. Chem. Phys.* 20: 571–581.

Buongiorno, J. 2006. Convective transport in nanofluids. *J. Heat Transf.* 128: 240–250.

Buongiorno, J. et al. 2009. A benchmark study on the thermal conductivity of nanofluids. *J. Appl. Phys.* 106: 1–14.

Choi, S. U. S. 1995. Enhancing thermal conductivity of fluids with nanoparticle. In *Developments and Applications of Non-Newtonian Flows*, eds. D. A. Soginer and H. P. Wang, 231: 99–105. ASME Fluids Engineering Division, San Francisco, CA.

Corcione, M. 2011. Empirical correlating equations for predicting the effective thermal conductivity and dynamic viscosity of nanofluids. *Energ. Convers. Manage.* 52: 789–793.

Das, S. K., Putra, N., Thiesen, P., Roetzel, W. 2003. Temperature dependence of thermal conductivity enhancement for nanofluids. *J. Heat Transf.* 125: 567–574.

Ding, Y. et al. 2007. Heat transfer intensification using nanofluids. *Powder Part.* 25: 23–38.

Eastman, J. A., Choi, S. U. S., Li, S., Yu, W., Thompson, L. J. 2001. Anomalously increased effective conductivities of ethylene glycol-based nanofluids containing nanoparticles. *Appl. Phys. Lett.* 78: 718–720.

Heris, S. Z., Esfahany, M. N., Etemad, S. G. 2007. Experimental investigation of convective heat transfer of Al_2O_3/water nanofluid in a circular tube. *Int. J. Heat Fluid Flow* 28: 203–210.

Hwang, K. S., Jang, S. P., Choi, S. U. S. 2009. Flow and convective heat transfer characteristics of water-based Al_2O_3 nanofluids in fully developed laminar flow regime. *Int. J. Heat Mass Transf.* 52: 193–199.

Lee, S., Choi, S. U. S., Li, S., Eastman, J. A. 1999. Measuring thermal conductivity of fluids containing oxide nanoparticles. *J. Heat Transf.* 121: 280–289.

Li, W., Nakayama, A. 2014. Effects of temperature-dependency of thermophysical properties on laminar forced convective heat transfer enhancement in nanofluids. *J. Thermophys Heat Transfer.*

Maiga, S. B., Palm, S. J., Nguyen, C. T., Roy, G., Galanis, N. 2005. Heat transfer enhancement by using nanofluids in forced convection flows. *Int. J. Heat Fluid Flow* 26: 530–546.

Masuda, H., Ebata, A., Teramae, K., Hishimura, N. 1993. Alteration of thermal conductivity and viscosity of liquid by dispersing ultra-fine particles. *Netsu Bussei (Japan)* 4: 227–233.

Maxwell, J. C. 1881. *A Treatise on Electricity and Magnetism.* Oxford: Oxford University Press.

Nakayama, A. 1995. *PC-Aided Numerical Heat Transfer and Convective Flow.* Boca Raton, FL: CRC Press.

Nan, C. W., Birringer, R., Clarke D. R., Gleiter, H. 1997. Effective thermal conductivity of particulate composites with interfacial thermal resistance. *Appl. Phys. Lett.* 81: 6692–6699.

Nield, D. A., Kuznetsov, A. V. 2009. Thermal instability in a porous medium layer saturated by a nanofluid. *Int. J. Heat Mass Transf.* 52: 5796–5801.

Pak, B. C., Cho, Y. 1998. Hydrodynamic and heat transfer study of dispersed fluids with submicron metallic oxide particles. *Exp. Heat Transf.* 11: 151–170.

Palm, S. J., Roy, G., Nguyen, C. T. 2006. Heat transfer enhancement with the use of nanofluids in radial flow cooling systems considering temperature dependent properties. *Appl. Therm. Eng.* 26: 2209–2218.

Prabhat, N., Buongiorno, J., Hu, L. W. 2012. Convective heat transfer enhancement in nanofluids: Real anomaly or analysis artifact? *J. Nanofluids* 1(1): 55–62.

Tahir, S., Mital, M. 2012. Numerical investigation of laminar nanofluid developing flow and heat transfer in a circular channel. *Appl. Therm. Eng.* 39: 8–14.

Tzou, D. Y. 2008. Thermal instability of nanofluids in natural convection. *Int. J. Heat Mass Transf.* 51: 2967–2979.

Venerus, D. C. et al. 2010. Viscosity measurements on colloidal dispersions (nanofluids) for heat transfer applications. *J. Appl. Rheol.* 20: 44582.

Wang, X. Q., Mujumdar, A. S. 2007. Heat transfer characteristics of nanofluids: A review. *Int. J. Therm. Sci.* 46: 1–19.

Williams, W., Buongiorno, J., Hu, L. W. 2008. Experimental investigation of turbulent convective heat transfer and pressure loss of alumina/water and zirconia/water nanoparticle colloids (nanofluids) in horizontal tubes. *J. Heat Transf.* 130, 042412.

Xuan, Y., Li, Q. 2003. Investigation on convective heat transfer and flow features of nanofluids. *J. Heat Transf.* 125: 151–155.

Xuan, Y., Roetzel, W. 2000. Conceptions for heat transfer correlations of nanofluids. *Int. J. Heat Mass Transf.* 43: 3701–3707.

Yang, C., Li, W., Nakayama, A. 2013. Convective heat transfer of nanofluids in a concentric annulus. *Int. J. Therm. Sci.* 71: 249–257.

Yang, C., Li, W., Sano, Y., Mochizuki, M., Nakayama, A. 2013. On the anomalous convective heat transfer enhancement in nanofluids: A theoretical answer to the nanofluids controversy. *J. Heat Transf.* 135, 054504.

CHAPTER 3

Mechanisms and Models of Thermal Conductivity in Nanofluids

Seung-Hyun Lee and Seok Pil Jang

CONTENTS

3.1 INTRODUCTION

Thermal conductivity of nanofluids has been widely investigated because of its great interest and potential to thermal scientists and engineers. As shown in Figure 3.1, a number of thermal conductivity data of nanofluids are beyond the prediction of classical Maxwell theory,[1] which is the conventional model. In particular, as shown in Figure 3.1b, the Maxwell theory quite underestimates the magnitude of the thermal conductivity enhancement of metal nanofluids. Moreover, as shown in Figure 3.2a and c, pioneering researchers experimentally discovered the novel behaviors of the heat conduction in nanofluids such as the temperature dependence,[2–6] the inverse size dependence,[4,6–9] and the pH dependence.[10,11] Thus, numerous mechanisms have been experimentally, theoretically, and numerically proposed to predict the magnitude and trend of the enhanced thermal conductivity of nanofluids. However, the thermal conduction mechanisms of nanofluids are focused on the subject of much uncertainty and debate, due to the inconsistencies in thermal conductivity data reported by each group. For example, as shown in Figure 3.2b and d, some researchers have recently reported that the effective thermal conductivity of nanofluids is not dependent on the temperature[12,13] and is directly proportional to the particle size.[14,15] The inconsistency of experimental data is because the various parameters affecting the thermal conductivity of nanofluids such as size, temperature, suspension quality, and pH, are not properly controlled or characterized. High uncertainty of measurement devices is also one of the causes. Although there are many contradictories and debates on the data of nanofluids, it is clearly shown that the conventional effective medium theories (EMTs) such as the Maxwell model are not sufficient to explain the novel thermal conduction behaviors of nanofluids.[16] Therefore, a number of new mechanisms and models are still proposed and discussed. In this chapter, the proposed mechanisms that can be divided into static mechanisms and dynamic mechanisms are introduced to explain the unique of conduction behavior in nanofluids. The static mechanisms, which include the classical EMT, the nanolayer, and the aggregation or clustering, are developed based on the postulation that the nanoparticles suspended in base fluids do not have mobility. The dynamic mechanisms, which consider the randomly moving nanoparticles in suspensions, include the nanoconvection model induced by the Brownian motion of nanoparticles. In addition, we separately discuss the combined models of static and dynamic mechanisms.

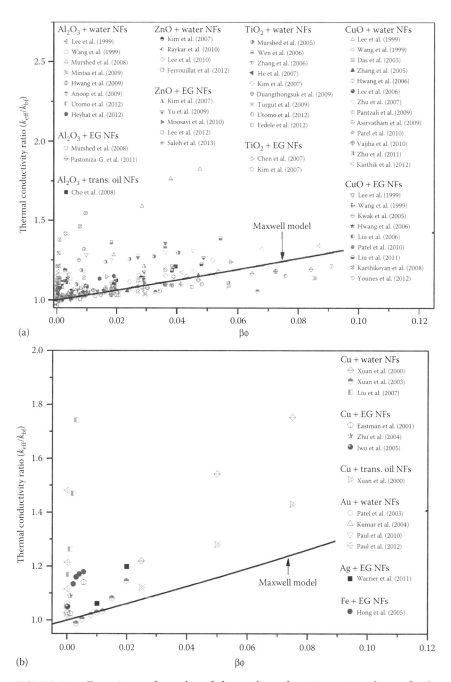

FIGURE 3.1 Experimental results of thermal conductivity ratio of nanofluids with respect to volume fractions, where $\beta = (k_p - k_{bf})/(k_p + 2k_{bf})$ and ϕ is the volume fraction: (a) metal oxide nanofluids; (b) metal nanofluids. EG, ethylene glycol; NFs, nanofluids; trans., transmission.

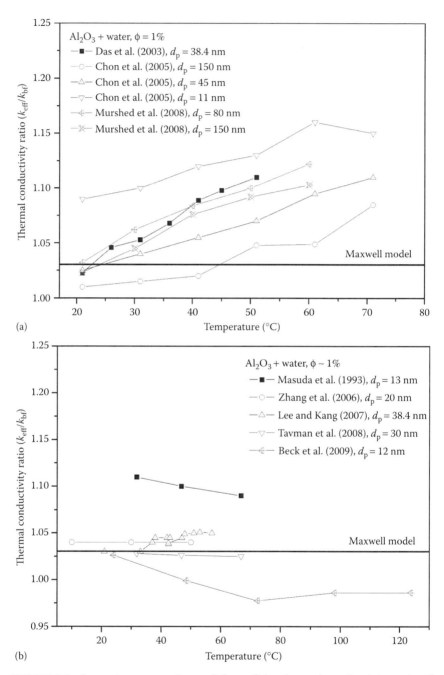

FIGURE 3.2 Inconsistent experimental data of the thermal conductivity ratio of nanofluids to basefluids according to the temperature and particle size: (a) temperature dependence results and (b) no temperature dependence results. (*Continued*)

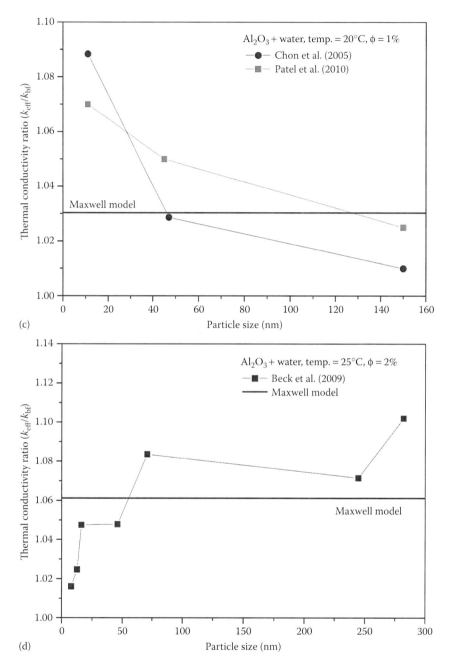

FIGURE 3.2 (Continued)   Inconsistent experimental data of the thermal conductivity ratio of nanofluids to basefluids according to the temperature and particle size: (c) inverse size dependence results; (d) size dependence results.

3.2 THERMAL CONDUCTIVITY MECHANISMS AND MODELS OF NANOFLUIDS

3.2.1 Static Mechanisms and Models

3.2.1.1 Classical EMT

Basically, the EMT or effective medium approximation is an analytical approach to predict the effective properties of a mixture with the relative fractions and the properties of its components. In the field of nanofluids, EMT-based models can be classified into static mechanisms because the models are based on the stationary dispersion of solid nanoparticles in a liquid medium as shown in Figure 3.3. As Maxwell[1] introduced the theory in 1873, modified models that can consider the various parameters such as the shape and size of inclusions, the interfacial thermal resistance, and the particle–particle interactions were presented. Therefore, the Maxwell model (or the Maxwell-Garnett model[17]), as a representative of EMT, has been widely used as a comparison model for the effective k of nanofluids as given by

$$k_{eff} = \frac{k_p + 2k_{bf} + 2\phi(k_p - k_{bf})}{k_p + 2k_{bf} - \phi(k_p - k_{bf})} k_{bf} \qquad (3.1)$$

where:

k and ϕ denote the thermal conductivity and the volume fraction of nanoparticles

The subscripts bf, eff, and p indicate the base fluid, the nanofluid, and the nanoparticle, respectively

This model can well estimate the effective properties for a dilute suspension of solid–liquid composite with the micro- and millimeter-sized particles.

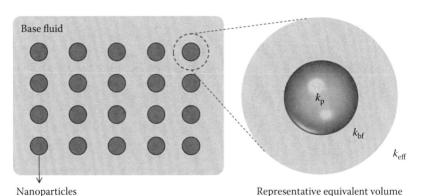

FIGURE 3.3 Schematic of classical effective medium theory.

In Equation 3.1, when the thermal conductivity ratio of particle to base fluid (k_p/k_{bf}) is high and the volume fraction is very low, the Maxwell model only depends on the volume concentration of nanoparticles. Bruggeman[18] developed a symmetrical EMT-based model to take into account the interactions between particles.

$$\phi\left(\frac{k_p - k_{eff}}{k_p + 2k_{eff}}\right) + (1-\phi)\left(\frac{k_{bf} - k_{eff}}{k_{bf} + 2k_{eff}}\right) = 0 \tag{3.2}$$

The Bruggeman model is more appropriate than the Maxwell model to predict the effective properties of a mixture at high concentrations. However, at low volume fractions, the difference between the Bruggeman model and the Maxwell model is not critical. Hamilton and Crosser[19] empirically presented an effective thermal conduction model based on the EMT by considering the arbitrary shape of inclusions. When the thermal conductivity ratio of discontinuous phase (k_p) to continuous phase (k_{bf}) is greater than 100, the effective thermal conductivity of a mixture is described by

$$k_{eff} = \left[\frac{k_p + (n-1)k_b - (n-1)(k_{bf} - k_p)\phi}{k_p + (n-1)k_{bf} + (k_{bf} - k_p)\phi}\right]k_{bf}, \quad n = \frac{3}{\psi} \tag{3.3}$$

where n and ψ are the shape factor and the particle sphericity, which is expressed by the ratio of the sphere surface area (with the same volume of the particle) to the given particle surface area. For sphere-shaped inclusions, the sphericity is 1 and the Hamilton–Crosser model approaches to the Maxwell model. Hashin and Shtrikman[20] developed the theoretical bounds of effective properties of a mixture based on the EMT. The Hashin and Shtrikman (HS) bounds is given by

$$k_{bf}\left[\frac{k_p + 2k_{bf} + 2\phi(k_p - k_{bf})}{k_p + 2k_{bf} - \phi(k_p - k_{bf})}\right] \leq k_{eff} \leq \left[\frac{3k_f + 2\phi(k_p - k_{bf})}{3k_p - \phi(k_p - k_{bf})}\right]k_p \tag{3.4}$$

For the $k_p > k_{bf}$, the HS lower bound is the same as the Maxwell model. With the HS bounds, Keblinski et al.[21] and Eapen et al.[22] demonstrated that almost all of thermal conductivity results of nanofluids fall within these bounds. Based on these results, they suggested that the thermal transport phenomena of nanofluids can be explained by the EMT theories without any new physics. However, the HS upper bound means that the base fluid is surrounded by the chain-clustered solid nanoparticles and this structure is not common in the well-dispersed dilute nanofluids.[16] Hasselman and Johnson[23] developed a model of effective thermal conductivity with the consideration of the interfacial thermal resistance and the various shapes such as sphere, cylinder, and flat plate. For the spherical shape, the effective thermal conductivity is expressed by

$$k_{\text{eff}} = \left\{ \frac{k_{\text{p}}(1+2\kappa)+2k_{\text{bf}}+2\phi\left[k_{\text{p}}(1-\kappa)-k_{\text{bf}}\right]}{k_{\text{p}}(1+2\kappa)+2k_{\text{bf}}-\phi\left[k_{\text{p}}(1-\kappa)-k_{\text{bf}}\right]} \right\} k_{\text{bf}} \tag{3.5}$$

where:

$\kappa = a_{\text{k}}/a_{\text{p}}$ is a dimensionless parameter

a_{k} is the Kapitza radius defined as $a_{\text{k}} = R_{\text{k}} k_{\text{bf}}$

a_{p} and R_{k} are the radius of nanoparticle and the Kapitza resistance

When the nanoparticles are perfectly in contact with the surrounding medium, the interfacial thermal resistance becomes 0 and Hasselman and Johnson model reduces to the Maxwell model. Nan et al.[24] derived a generalized form of EMT taking into account numerous parameters such as the arbitrarily shape and size of nanoparticles, the Kapitza resistance, and the orientation. Their model is as follows:

$$k_{\text{eff},11} = k_{\text{eff},22}$$

$$= \left\{ \frac{2+\phi\left[\beta_{11}(1-L_{11})\left(1+\langle\cos^2\theta\rangle\right)+\beta_{33}(1-L_{33})\left(1-\langle\cos^2\theta\rangle\right)\right]}{2-\phi\left[\beta_{11}L_{11}\left(1+\langle\cos^2\theta\rangle\right)+\beta_{33}L_{33}\left(1-\langle\cos^2\theta\rangle\right)\right]} \right\} k_{\text{bf}} \tag{3.6}$$

$$k_{\text{eff},33} = \left\{ \frac{1+\phi\left[\beta_{11}(1-L_{11})\left(1-\langle\cos^2\theta\rangle\right)+\beta_{33}(1-L_{33})\langle\cos^2\theta\rangle\right]}{1-\phi\left[\beta_{11}L_{11}\left(1-\langle\cos^2\theta\rangle\right)+\beta_{33}L_{33}\langle\cos^2\theta\rangle\right]} \right\} k_{\text{bf}}$$

where:

$$\langle\cos^2\theta\rangle = \frac{\int\rho(\theta)\cos^2\theta\cdot\sin\theta d\theta}{\int\rho(\theta)\sin\theta d\theta}$$

$$\beta_{ii} = \frac{k_{ii}^c - k_{\text{bf}}}{k_{\text{bf}} + L_{ii}\left(k_{ii}^c - k_{\text{bf}}\right)}$$

$$k_{ii}^{c} = k_{\text{IR}} \frac{k_{\text{IR}} + L_{ii}\left(k_{\text{p}} - k_{\text{IR}}\right)\left(1 - v\right) + v\left(k_{\text{p}} - k_{\text{IR}}\right)}{k_{\text{IR}} + L_{ii}\left(k_{\text{p}} - k_{\text{IR}}\right)\left(1 - v\right)},$$

$$v = \frac{a_1^2 a_3}{\left(a_1 + \delta_{\text{IR}}\right)^2 \left(a_3 + \delta_{\text{IR}}\right)}$$

where:

$$\begin{cases} L_{11} = L_{22} = \dfrac{p^2}{2(p^2 - 1)} - \dfrac{p}{2(p^2 - 1)^{3/2}} \cosh^{-1} p, \\[4mm] L_{33} = 1 - 2L_{11} \quad p = \dfrac{a_{33}}{a_{11}} \end{cases}$$

where:

a_i, L_{ii}, k_{ii}^{c}, and p are the radii of the ellipsoid, the geometrical factors dependent on the particle shape, equivalent thermal conductivities, and the aspect ratio of the ellipsoid

δ_{IR}, $\rho(\theta)$, and θ are the thickness of interface layer, the angle between the material axis and the local particle symmetry axis, and the distribution function of ellipsoidal particle orientation, respectively

For completely misoriented spherical inclusions, this model reduces to the Hasselman and Johnson model.[23] The above-mentioned EMT-based models are summarized in Table 3.1.

3.2.1.2 Nanoscale Layer between Fluid and Nanoparticle Interface

The nanolayer concept was introduced by Keblinski et al.[25] and Yu and Choi[26] based on the experimental findings.[27] As shown in Figure 3.4, the nanolayer is a well-ordered liquid structure adjacent to solid nanoparticle due to the interactions between liquid and solid molecules. Keblinski et al.[25] roughly presented the potential concept of enhanced thermal conductivity of nanofluids with the naonolayer. Yu and Choi[26] first proposed the effective thermal conductivity model with the nanolayer concept by renovating the Maxwell model. They assumed that the nanolayer works as

TABLE 3.1 Conventional EMT-Based Models

Authors	Formulation
Maxwell[1] and Maxwell-Garnett[17]	$k_{\text{eff}} = \left\{ \dfrac{\left[k_{\text{p}} + 2k_{\text{bf}} + 2(k_{\text{p}} - k_{\text{bf}})\phi \right]}{\left[k_{\text{p}} + 2k_{\text{bf}} - (k_{\text{p}} - k_{\text{bf}})\phi \right]} \right\} k_{\text{bf}}$
Bruggeman[18]	$\phi \left[\dfrac{(k_{\text{p}} - k_{\text{eff}})}{(k_{\text{p}} + 2k_{\text{eff}})} \right] + (1-\phi) \left[\dfrac{(k_{\text{bf}} - k_{\text{eff}})}{(k_{\text{bf}} + 2k_{\text{eff}})} \right] = 0$
Hamilton and Crosser[19]	$k_{\text{eff}} = \left\{ \dfrac{\left[k_{\text{p}} + (n-1)k_{\text{b}} - (n-1)(k_{\text{bf}} - k_{\text{p}})\phi \right]}{\left[k_{\text{p}} + (n-1)k_{\text{bf}} + (k_{\text{bf}} - k_{\text{p}})\phi \right]} \right\} k_{\text{bf}}$
Hashin and Shtrikman[20]	$k_{\text{bf}} \left\{ \dfrac{\left[k_{\text{p}} + 2k_{\text{bf}} + 2\phi(k_{\text{p}} - k_{\text{bf}}) \right]}{\left[k_{\text{p}} + 2k_{\text{bf}} - \phi(k_{\text{p}} - k_{\text{bf}}) \right]} \right\} \le k_{\text{eff}} \le \left\{ \dfrac{\left[3k_f + 2\phi(k_{\text{p}} - k_{\text{bf}}) \right]}{\left[3k_{\text{p}} - \phi(k_{\text{p}} - k_{\text{bf}}) \right]} \right\} k_{\text{p}}$
Hasselman and Johnson[23]	$k_{\text{eff}} = \left\{ \dfrac{\left[k_{\text{p}}(1+2\kappa) + 2k_{\text{bf}} + 2\phi(k_{\text{p}}(1-\kappa) - k_{\text{bf}}) \right]}{\left[k_{\text{p}}(1+2\kappa) + 2k_{\text{bf}} - \phi(k_{\text{p}}(1-\kappa) - k_{\text{bf}}) \right]} \right\} k_{\text{bf}}$
Nan et al.[24]	$k_{\text{eff},11} = k_{\text{eff},22}$ $= \left(\dfrac{\left\{ 2 + \phi\left[\beta_{11}(1-L_{11})\left(1+\langle \cos^2\theta \rangle\right) + \beta_{33}\left(1-L_{33}\right)\left(1-\langle \cos^2\theta \rangle\right) \right] \right\}}{\left\{ 2 - \phi\left[\beta_{11}L_{11}\left(1+\langle \cos^2\theta \rangle\right) + \beta_{33}L_{33}\left(1-\langle \cos^2\theta \rangle\right) \right] \right\}} \right) k_{\text{bf}}$ $k_{\text{eff},33} = \left(\dfrac{\left\{ 1 + \phi\left[\beta_{11}(1-L_{11})\left(1-\langle \cos^2\theta \rangle\right) + \beta_{33}\left(1-L_{33}\right)\langle \cos^2\theta \rangle \right] \right\}}{\left\{ 1 - \phi\left[\beta_{11}L_{11}\left(1-\langle \cos^2\theta \rangle\right) + \beta_{33}L_{33}\langle \cos^2\theta \rangle \right] \right\}} \right) k_{\text{bf}}$

a favorable thermal path between a fluid and a nanoparticle. The resulting expression is as follows:

$$k_{\text{eff}} = \frac{k_{\text{pe}} + 2k_{\text{bf}} + 2\left(k_{\text{pe}} - k_{\text{bf}} \right)(1-\beta)^3 \phi}{k_{\text{pe}} + 2k_{\text{bf}} - \left(k_{\text{pe}} - k_{\text{bf}} \right)(1+\beta)^3 \phi} k_{\text{bf}} \tag{3.7}$$

where:

$$k_{\text{pe}} = \frac{\left[2(1-\gamma) + (1+\beta)^3 (1+2\gamma) \right]\gamma}{-(1-\gamma) + (1+\beta)^3 (1+2\gamma)} k_{\text{p}} \quad \gamma = \frac{k_{\text{layer}}}{k_{\text{p}}} \text{ and } \beta = \frac{\delta}{a_{\text{p}}}$$

where k_{pe} and δ are the equivalent thermal conductivity and the nanolayer thickness, respectively. At $\gamma = 1$, the model is consistent with the Maxwell model. Yu and Choi[28] extended this concept to predict the enhancement of carbon nanotube (CNT) nanofluids' thermal conductivity by renovating

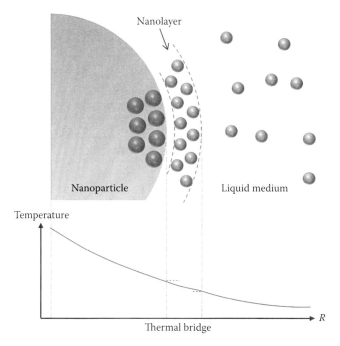

FIGURE 3.4   Schematic of nanolayer model at the solid–liquid interface.

the Hamilton and Crosser model.[19] Their model was well matched with the experimental data of polyalphaolefin (PAO)–CNT nanofluids, although they assumed the thickness and the thermal conductivity of nanolayer. Xue[29] derived a model based on the consideration of the nanolayer and the average polarization theory with the EMT. His model is described as follows:

$$
9\left(1-\frac{\phi}{\lambda}\right)\frac{k_{\mathrm{eff}}-k_{\mathrm{bf}}}{2k_{\mathrm{eff}}+k_{\mathrm{bf}}}
$$

$$
+\frac{\phi}{\kappa}\left[\frac{k_{\mathrm{eff}}-k_{c,x}}{k_{\mathrm{eff}}+B_{2,x}\left(k_{c,x}-k_{\mathrm{eff}}\right)}+4\frac{k_{\mathrm{eff}}-k_{c,y}}{2k_{\mathrm{eff}}+\left(1-B_{2,x}\right)\left(k_{c,y}-k_{\mathrm{eff}}\right)}\right]=0 \tag{3.8}
$$

where:

$k_{c,j}$ is the effective dielectric constant

$B_{2,x}$ is the depolarization factor along the x-symmetrical axis

$$
\lambda=\frac{abc}{[(a+\delta)(b+\delta)(c+\delta)]}
$$

where:

a, b, and c are half-radii of an elliptical shell

This model can explain a nonlinear behavior of thermal conductivity enhancement of oil-based CNT nanofluids. Also, Xue and Xu[30] derived a model with a nanolayer concept by modifying the Bruggeman model. The results are as follows:

$$\left(1 - \frac{\phi}{\chi}\right)\frac{k_{eff} - k_{bf}}{2k_{eff} + k_{bf}}$$

$$+ \frac{\phi}{\chi}\frac{\left(k_{eff} - k_{layer}\right)\left(2k_{layer} + k_p\right) - \chi\left(k_p - k_{layer}\right)\left(2k_{layer} + k_{eff}\right)}{\left(2k_{eff} + k_{layer}\right)\left(2k_{layer} + k_p\right) + 2\chi\left(k_p - k_{layer}\right)\left(k_{layer} - k_{eff}\right)} = 0$$

(3.9)

where:

$$\chi = \left(\frac{a_p}{a_p + \delta}\right)^3$$

where δ is the thickness of interfacial shell. Xie et al.[31] developed a new model by assuming that the thermal conductivity of interfacial layer has a linear distribution. The model is expressed as follows:

$$k_{eff} = \left(1 + 3\Theta\phi_T + \frac{3\Theta^2\phi_T^2}{1 - \Theta\phi_T}\right)k_{bf}$$

(3.10)

where:

$$\Theta = \frac{\left(\begin{array}{c}\left[\left(k_{layer} - k_{bf}\right)/\left(k_{layer} + 2k_{bf}\right)\right] \\ \left\{(1+\beta)^3 - \left[\left(k_p - k_{layer}\right)\left(k_{bf} + 2k_{layer}\right)/\left(k_p + 2k_{layer}\right)\left(k_{bf} - k_{layer}\right)\right]\right\}\end{array}\right)}{(1+\beta)^3 + 2\left[\left(k_{layer} - k_{bf}\right)/\left(k_{layer} + 2k_{bf}\right)\right]\left[\left(k_p - k_{layer}\right)/\left(k_p + 2k_{layer}\right)\right]}$$

$\phi_T = \phi(1+\beta)^3$ is the total volume fraction of the nanoparticle and the nanolayer

Leong et al.[32] derived a nanolayer-based new model similar as the effective medium approach.

$$k_{eff} = \frac{\left(k_p - k_{layer}\right)\phi k_{layer}\left(2\varepsilon_1^3 - \varepsilon^3 + 1\right) + \left(k_p + 2k_{layer}\right)\varepsilon_1^3\left[\phi\varepsilon^3\left(k_{layer} - k_{bf}\right) + k_{bf}\right]}{\varepsilon_1^3\left(k_p + 2k_{layer}\right) - \left(k_p - k_{layer}\right)\phi\left(\varepsilon_1^3 + \varepsilon^3 - 1\right)}$$

(3.11)

where:

$\varepsilon = 1 + \beta$

$\varepsilon_1 = 1 + \beta/2$

If $k_{\text{layer}} = k_{\text{bf}}$ and $\delta = 0$ ($\beta = 0$ and $\varepsilon = 1$), this model is consistent with the Maxwell model. Although Leong et al. showed that the proposed model is more consistent than the Yu and Choi model,[28] Doroodchi et al.[33] pointed out that the Yu and Choi model[28] is more reliable than the model of Leong et al.[32] to consider the nanolayer effect. Xue et al.[34] also showed that the effective k of nanofluids estimated by the Yu and Choi model[28] is not significantly greater than the prediction of the conventional Maxwell theory. This result indicates that the interfacial nanolayering is not sufficient to describe the enhanced thermal conductivity of nanofluids. Nanolayer-based models are summarized in Table 3.2.

Numerically, Xue et al.[34] studied the effect of the nanolayer on the thermal transport in nanofluids by means of the nonequilibrium molecular dynamics (MD) simulations. They demonstrated that the layering of the liquid atoms at the liquid-solid interface is minor effect on thermal

TABLE 3.2   Nanolayer-Based Models

Authors	Formulation
Yu and Choi[26]	$k_{\text{eff}} = \dfrac{\left[k_{\text{pe}} + 2k_{\text{bf}} + 2(k_{\text{pe}} - k_{\text{bf}})(1-\beta)^3 \phi \right]}{\left[k_{\text{pe}} + 2k_{\text{bf}} - (k_{\text{pe}} - k_{\text{bf}})(1+\beta)^3 \phi \right]} k_{\text{bf}}$
Xue[29]	$9\left[1 - (\phi/\lambda)\right]\left[\dfrac{(k_{\text{eff}} - k_{\text{bf}})}{(2k_{\text{eff}} + k_{\text{bf}})} \right]$ $+ (\phi/\kappa)\left\{ \left[\dfrac{k_{\text{eff}} - k_{c,x}}{\left[k_{\text{eff}} + B_{2,x}\left(k_{c,x} - k_{\text{eff}} \right) \right]} \right] + 4 \dfrac{\left[k_{\text{eff}} - k_{c,y} \right]}{\left[2k_{\text{eff}} + \left(1 - B_{2,x}\right)\left(k_{c,y} - k_{\text{eff}} \right) \right]} \right\} = 0$
Xue and Xu[30]	$\left[1 - (\varphi/\chi)\right]\left[\dfrac{(k_{\text{eff}} - k_{\text{bf}})}{(2k_{\text{eff}} + k_{\text{bf}})} \right]$ $+ (\varphi/\chi)\dfrac{\left[(k_{\text{eff}} - k_{\text{layer}})(2k_{\text{layer}} + k_{\text{p}}) - \chi(k_{\text{p}} - k_{\text{layer}})(2k_{\text{layer}} + k_{\text{eff}}) \right]}{\left[(2k_{\text{eff}} + k_{\text{layer}})(2k_{\text{layer}} + k_{\text{p}}) - 2\chi(k_{\text{p}} - k_{\text{layer}})(k_{\text{layer}} - k_{\text{eff}}) \right]} = 0$
Xie et al.[31]	$k_{\text{eff}} = \left[1 + 3\Theta\phi_{\text{T}} + \dfrac{\left(3\Theta^2\phi_{\text{T}}^2\right)}{\left(1 - \Theta\phi_{\text{T}}\right)} \right] k_{\text{bf}}$
Leong et al.[32]	$k_{\text{eff}} = \left\{ \begin{array}{l} \left(k_{\text{p}} - k_{\text{layer}}\right)\phi k_{\text{layer}}\left(2\varepsilon_1^3 - \varepsilon^3 + 1\right) \\[2ex] + \left(k_{\text{p}} + 2k_{\text{layer}}\right)\varepsilon_1^3 \dfrac{\left[\phi\varepsilon^3\left(k_{\text{layer}} - k_{\text{bf}}\right) + k_{\text{bf}} \right]}{\left[\varepsilon_1^3\left(k_{\text{p}} + 2k_{\text{layer}}\right) - \left(k_{\text{p}} - k_{\text{layer}}\right)\phi\left(\varepsilon_1^3 + \varepsilon^3 - 1\right) \right]} \end{array} \right\}$

transport properties of nanofluids. Because their results are obtained from the simple MD simulation using the monoatomic liquid system, this study may have limitation to apply in the complex nanofluids. However, Li et al.[35] numerically examined the effect of nanolayer using the equilibrium MD simulation and reported that a thin molecular layer at the fluid–solid interface can contribute to the enhanced thermal conductivity of nanofluids. The controversy between the researchers regarding the nanolayer mechanisms is still under debate. Therefore, researchers should find out a proper method to precisely obtain the thickness and the thermal conductivity of nanolayer to validate this mechanism.

3.2.1.3 Aggregation or Clustering of Nanoparticles

Keblinski et al.[25] proposed a concept that the aggregation of nanoparticles in nanofluids can enhance the effective k of nanofluids. As shown in Figure 3.5, clustered solid nanoparticles locally form the rapid heat conduction paths (percolation paths) and they also play a role to increase the

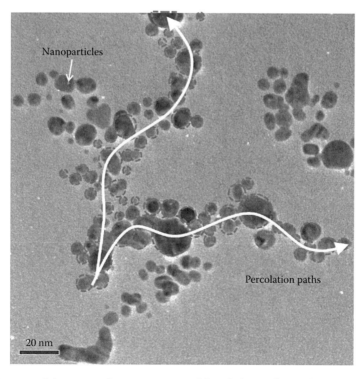

FIGURE 3.5 Schematic of aggregation model and thermal percolation paths.

effective volume fraction of nanofluids.[25] Wang et al.[36] derived an aggregation-based model depending on the EMT and fractal theory. The formula is given by

$$k_{\text{eff}} = \left\{ \frac{(1-\phi)+3\phi\int_0^\infty \dfrac{k_{\text{cl}}(a_{\text{cl}})n(a_{\text{cl}})}{\left[k_{\text{cl}}(a_{\text{cl}})+2k_{\text{bf}}\right]}da_{\text{cl}}}{(1-\phi)+3\phi\int_0^\infty \dfrac{k_{\text{bf}}(a_{\text{cl}})n(a_{\text{cl}})}{\left[k_{\text{cl}}(a_{\text{cl}})+2k_{\text{bf}}\right]}da_{\text{cl}}} \right\} k_{\text{bf}} \qquad (3.12)$$

where a_{cl}, $k_{\text{cl}}(a_{\text{cl}})$, and $n(a_{\text{cl}})$ are the equivalent radius of cluster, the effective thermal conductivity of nanoparticle clusters obtained by the Bruggeman model, and the radius distribution function, respectively.

Numerically, Prasher et al.[37] examined the effect of aggregation using Monte Carlo simulation and showed that the linearly clustered nanoparticles (called back bone) can highly contribute to increase the effective k of nanofluids. Evans et al.[38] extended Prasher et al.'s[37] study by considering the Kapitza resistance at the solid–liquid interface. In their study, the Kapitza resistance acts as a thermal barrier and makes to decrease the enhancement magnitude of thermal conductivity of nanofluids. Aggregation-based models are summarized in Table 3.3. More details about aggregation mechanism will be discussed in Section 3.2.3. Based on the previous analyses, well-aggregated or clustered nanoaparticles can be seen to increase the effective thermal conductivity of nanofluids. However, the level of aggregation is still unknown for individually manufactured nanofluids and there are no methods to determine the aggregation level of nanofluids. Moreover, a higher level of aggregation causes degradation of suspension stability of nanofluids.

TABLE 3.3   Aggregation-Based Models

Authors	Formulation
Wang et al.[36]	$k_{\text{eff}} = \left\{ \dfrac{\left[(1-\phi)+3\phi\int_0^\infty \left[k_{\text{cl}}(a_{\text{cl}})n(a_{\text{cl}})\right]/\left[k_{\text{cl}}(a_{\text{cl}})+2k_{\text{bf}}\right]da_{\text{cl}}\right]}{\left[(1-\phi)+3\phi\int_0^\infty \left[k_{\text{bf}}(a_{\text{cl}})n(a_{\text{cl}})\right]/\left[k_{\text{cl}}(a_{\text{cl}})+2k_{\text{bf}}\right]da_{\text{cl}}\right]} \right\} k_{\text{bf}}$

3.2.2 Dynamic Mechanisms and Models

The conventional EMTs are used to evaluate the effective thermal conductivities of suspensions with macro-sized particles. These theories assume that the including particles are homogeneously suspended in the base medium and there is no motion of particles. However, as described in Section 3.1, the EMT is not sufficient to explain the novel behaviors of thermal conductivity of nanofluids such as temperature and size dependence. As the nano-sized particles in the fluid medium are continuously moving with collision of fluid molecules under the thermodynamic energy (Brownian motion), several researchers proposed dynamic mechanisms based on the Brownian motion in order to explain the novel features of thermal conductivity of nanofluids. As a first dynamic concept, collision between nanoparticles suspended in nanofluids was proposed by Keblinski et al.[25] However, Wang et al.[39] and Keblinski et al.[25] showed that the effect of collisions between particles on the thermal conductivity enhancement of nanofluids is negligibly small. Alternatively, Jang and Choi[40] proposed the nanoconvection induced by the Brownian motion of nanoparticles as a second dynamic mechanism. They assumed that the Brownian motion of nano-sized particles agitates the surrounding fluid molecules and induces nanoscale convention. Figure 3.6 shows the concept of the nanoconvection induced by the Brownian motion. Jang and Choi[40] proposed a nanoconvection-based model for predicting the effective thermal conductivity of nanofluids. They divided the thermal conduction in nanofluids into four modes: (1) collision of base fluid molecules with each other, (2) thermal diffusion of nanoparticles in fluids, (3) collision of nanoparticles with each

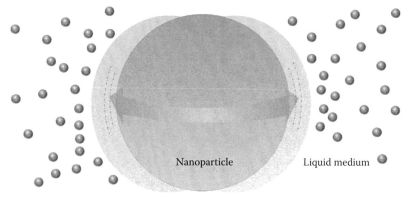

Brownian motion-induced nanoconvection

FIGURE 3.6   Schematic of Brownian motion-induced nanoconvection model.

other, and (4) Brownian motion-induced nanoconvection by the Brownian motion of nanoparticles suspended in nanofluids. Based on an order-of-magnitude analysis, they developed a new thermal conductivity model by superposition of modes 1, 2, and 4 as follows:

$$k_{\text{eff}} = k_{\text{bf}}(1-\phi) + \beta k_p \phi + C_1 \frac{d_{\text{bf}}}{d_p} k_{\text{bf}} \operatorname{Re}_{d_p}^2 \operatorname{Pr}\phi \tag{3.13}$$

where:

C_1 is a proportional constant
β is a constant for the Kapitza resistance per unit area
Re_{d_p} is the Reynolds number of nanoparticle

$$\operatorname{Re}_{d_p} = \frac{\bar{C}_{\text{R.M.}} d_p}{\nu}, \quad \bar{C}_{\text{R.M.}} = \frac{D_o}{l_{\text{bf}}}$$

where:

$\bar{C}_{\text{R.M.}}$, D_o, d_p, and l_{bf} are the random motion velocity of a nanoparticle, the diffusion coefficient, the particle diameter, and the liquid mean free path
ν is a constant for considering the kinematic viscosity of base fluid

Although the existence of nanoconvection induced by the Brownian motion and the definition of nanoparticle velocity has not been proved yet, their model is able to describe the temperature- and size-dependent behaviors of the effective k of nanofluids, unlike the EMT. Kumar et al.[41] derived a thermal conductivity model consisting of two modes. For a stationary particle model, Fourier's equation is used. A moving particle is modeled by the kinetic theory of gases and the Stokes–Einstein formula. The model is expressed by

$$k_{\text{eff}} = \left[1 + C\left(\frac{2k_{\text{B}}T}{\pi\mu d_p^2}\right) \frac{\phi a_{\text{bf}}}{k_{\text{bf}}(1-\phi)a_p}\right] k_{\text{bf}} \tag{3.14}$$

where:

C is a constant
k_{B} is the Boltzmann constant
μ is the dynamic viscosity
a_{bf} is the radius of the base fluid

This model can also predict the size-, temperature-, and volume fraction-dependent thermal conduction behaviors of nanofluids. Although this model well described previous experimental results of the dilute nanofluids, they used physically invalid assumptions such as very long mean free path of a nanoparticle[16], $O(10^{-2})$. Patel et al.[42] developed a cell model with two parallel heat conduction paths in nanofluids. The first path is the heat conduction via the liquid medium and the second path is the heat propagation through particle conduction and random movement-induced microconvection. Based on the concept of thermal resistance, they presented a new model under the same temperature-gradient condition as follows:

$$k_{\text{eff}} = \left[1 + \frac{k_p}{k_{\text{bf}}} \left(\frac{\pi}{6 + \left(\alpha_{\text{bf}} \pi \eta \, d_p / 2C k_B T \right)} \right) \left(\frac{6\phi}{\pi} \right)^{1/3} \right] k_{\text{bf}} \qquad (3.15)$$

where C, T, and α_{bf} are an empirical constant, the temperature, and the thermal diffusivity of the base fluid. As shown in Equation 3.15, the cell model can account for the effect of particle volume fraction, temperature, and particle size on the effective thermal conductivity of nanofluids. Nonlinear behavior of thermal conductivity of nanofluids at low particle concentrations can also be predicted. Koo and Kleinstreuer[43] presented the Brownian motion-based model with the combination of the Maxwell model. The model can be described as follows:

$$k_{\text{eff}} = \left[\frac{k_p + 2k_{\text{bf}} + 2\left(k_p - k_{\text{bf}} \right)\phi}{k_p + 2k_{\text{bf}} - \left(k_p - k_{\text{bf}} \right)\phi} \right] k_{\text{bf}}$$

$$+ \left[\left(5 \times 10^4 \right) \beta' \phi \rho_p c_{v,p} \sqrt{\frac{k_B T}{\rho_p d_p}} \, f(T, \phi, \text{etc.}) \right] \qquad (3.16)$$

where:

$c_{v,p}$ and f are the specific heat of the particle and a factorial function that depends on the temperature and the particle volume fraction.

ρ_p and β' indicate the density of the particle and the fraction of the liquid volume that moves with a nanoparticle

The first term on the right-hand side indicates the Maxwell model and the second term represents the dynamic model. In their model, some parameters are not easily obtained. For example, a factorial function f

should be determined by experiments and the average moving distance is obtained by MD simulations. Prasher et al.[44] derived a modified EMT model by considering the nanoparticle's random motion-induced fluctuation. Based on the order-of-magnitude analysis for several postulated mechanisms such as Brownian movement of nanoparticles, interparticle potential, and nanoconvection due to Brownian motion of nanoparticles, they have shown that the Brownian motion-induced nanoconvection is a dominant mechanism to describe the exceptional thermal conductivity enhancement of nanofluids. Considering the nanoconvection from multiple nanoparticles, their model is expressed as follows:

$$k_{\text{eff}} = \left(1 + A\,\mathrm{Re}^m\,\mathrm{Pr}^{0.333}\,\phi\right)\left\{\frac{k_p(1+2\kappa)+2k_{bf}+2\phi[k_p(1-\kappa)-k_{bf}]}{k_p(1+2\kappa)+2k_{bf}-\phi[k_p(1-\kappa)-k_{bf}]}\right\}k_{bf} \quad (3.17)$$

where:

$$\mathrm{Re} = \frac{1}{v}\sqrt{\frac{18k_B T}{\pi\rho_p d_p}}$$

A and m are empirical constants

When the empirical parameter A or the Reynolds number is 0, the model is reduced to the Hasselman and Johnson model.[23] Yang[45] presented a new model based on the kinetic theory under relaxation time approximations. Their model is written as follows:

$$k_{\text{eff}} = \left[\frac{k_p + 2k_{bf} + 2(k_p - k_{bf})\phi}{k_p + 2k_{bf} - (k_p - k_{bf})\phi}\right]k_{bf} + \left(157.5\phi C_f u_p^2 \tau\right) \quad (3.18)$$

where:
  C_f is the heat capacity per unit volume of the fluid
  τ *is* the relaxation time of particle

$$u_p = \sqrt{\frac{3k_B T}{m_p}}$$

where m_p and u_p are the mass of particle and the Brownian velocity of the particle, respectively. The first term on the right-hand side of the model represents the Maxwell model and the remaining term indicates

TABLE 3.4   Brownian Motion-Based Models

Authors	Formulation
Jang and Choi[40]	$k_{eff} = k_{bf}(1-\phi) + \beta k_p \phi + C_1 \left(d_{bf}/d_p\right) k_{bf} \, \text{Re}_{d_p}^2 \, \text{Pr} \, \phi$
Kumar et al.[41]	$k_{eff} = \left\{ 1 + C \left[\dfrac{(2k_B T)}{\left(\pi \mu d_p^2\right)}\right]\left[\dfrac{(\phi a_{bf})}{\left[k_{bf}(1-\phi)a_p\right]}\right] \right\} k_{bf}$
Patel et al.[42]	$k_{eff} = \left\{ 1 + \left(k_p/k_{bf}\right)\left[\pi/6 + (\alpha_{bf}\pi\eta d_p)/(2Ck_B T)\right]\left[(6\phi)/\pi\right]^{1/3} \right\} k_{bf}$
Koo and Kleinstreuer[43]	$k_{eff} = \left\{ \left[\dfrac{k_p + 2k_{bf} + 2\left(k_p - k_{bf}\right)\phi}{k_p + 2k_{bf} - \left(k_p - k_{bf}\right)\phi}\right] \right\} k_{bf}$ $+ \left[(5\times10^4)\beta'\phi\rho_p c_{v,p}\sqrt{(k_B T)/(\rho_p d_p)} f(T,\phi,\text{etc.}) \right]$
Prasher et al.[44]	$k_{eff} = (1 + A\,\text{Re}^m\,\text{Pr}^{0.333}\,\phi) \left\{ \left[\dfrac{k_p(1+2\kappa) + 2k_{bf} + 2\phi(k_p(1-\kappa) - k_{bf})}{k_p(1+2\kappa) + 2k_{bf} - \phi(k_p(1-\kappa) - k_{bf})}\right] \right\} k_{bf}$
Yang[45]	$k_{eff} = \left\{ \left[\dfrac{k_p + 2k_{bf} + 2\left(k_p - k_{bf}\right)\phi}{k_p + 2k_{bf} - \left(k_p - k_{bf}\right)\phi}\right] \right\} k_{bf} + \left(157.5\phi C_f u_p^2 \tau\right)$

a contribution of nanoscale convection caused by the random motion of nanoparticles. Yang[45] showed that the long relaxation time of the Brownian motion of nanoparticles significantly affects to enhance the heat transport properties of nanofluids. However, He[45] assumed the relaxation time, τ, to predict the effective thermal conductivity of nanofluids and its value has not been evaluated yet. Brownian motion-based models are summarized in Table 3.4. For numerical studies, Evans et al.[46] and Vladkov and Barrat[47] performed MD simulation of the Brownian motion of nanoparticles and showed that the effect of nanoconvection on the enhanced thermal conductivity of nanofluids is insignificant. However, Sarkar and Selvam[48] carried out the equilibrium MD simulation and showed that the thermal conductivity of nanofluids increases due to enhanced movement of liquid atoms adjacent to nanoparticles. Li and Peterson[49] numerically also showed that Browninan motion-induced nanoconvection and mixing are the key mechanisms for the heat transfer enhancement of nanofluids. Bhattacharya et al.[50] and Jain et al.[51] also numerically demonstrated by means of Brownian dynamics simulation that the random motion of nanoparticles can enhance the effective thermal conductivity of nanofluids. Although the dynamic mechanisms and models can describe the exceptional behavior of temperature-, size-, and volume fraction-dependent k of nanofluids, the debate of the dynamic mechanism is still ongoing.

3.2.3 Combined Models of Static and Dynamic Mechanisms

3.2.3.1 Combination of Nanolayer and Brownian Motion

In order to accurately explain the novel behavior of thermal conductivity of nanofluids, some researchers tried to combine both static and dynamic mechanisms. Ren et al.[52] proposed a combined thermal conductivity model for nanofluid by considering the interfacial nanolayering and the Brownian motion-induced convection. They assumed that the total heat flux consists of four fluxes through the nanoparticle, the nanolayer, the microconvection, and the base fluid. They also considered the linear variation of thermal conductivity of nanolayer. The derivation result is written as follows:

$$k_{\text{eff}} = \left[1 + F(\text{Pe}) + 3\Theta\phi_T + \frac{3\Theta^2\phi_T^2}{1 - \Theta\phi_T} \right] k_{\text{bf}} \qquad (3.19)$$

where Θ is the same as defined by Xie et al.[31] $F(\text{Pe})$ is given as

$$F(\text{Pe}) = 0.0556(\text{Pe}) + 0.1649(\text{Pe})^2 - 0.0391(\text{Pe})^3 + 0.0034(\text{Pe})^4$$

where:

$$\text{Pe} = \frac{\bar{u}L}{\alpha_{\text{bf}}} \phi_T^{0.75}, \quad \bar{u} = \sqrt{\frac{3k_B T}{m_c}}, \quad m_c = \frac{4}{3}\rho_p\pi a_p^3 \left\{ \frac{\rho_{\text{layer}}}{\rho_p} \left[\left(1 + \frac{\delta}{a_p}\right)^3 - 1 \right] + 1 \right\},$$

$$\text{and } L = (a_p + \delta)\sqrt[3]{\frac{4\pi}{3\phi_T}}$$

where:
  $\bar{u}$ is the mean velocity of the complex nanoparticle
  L is the specific length
  ρ_{layer} is the density of the nanolayer
  m_c is the mass of the complex nanoparticle

Murshed et al.[53] derived a combined thermal conductivity model consisting of nanolayer, Brownian motion, surface chemistry, and interaction potential. The model is shown as follows:

$$
\begin{aligned}
k_{\text{eff}} = \Bigg\{ & \frac{\phi\omega(k_p - \omega k_{bf})(2\varepsilon_1^3 - \varepsilon^3 + 1) + (k_p + 2\varepsilon k_{bf})\varepsilon_1^3 \left[\phi\varepsilon^3(\varepsilon-1)+1\right]}{\varepsilon_1^3(k_p + 2\omega k_{bf}) - (k_p - \varepsilon k_{bf})\phi(\varepsilon_1^3 + \varepsilon^3 - 1)} \Bigg\} k_{bf} \\
& + \left[\phi^2\varepsilon^6 k_{bf}\left(3\Lambda^2 + \frac{3\Lambda^2}{4} + \frac{9\Lambda^3}{16}\frac{k_{cp}+2k_{bf}}{2k_{cp}+3k_{bf}} + \frac{3\Lambda^4}{2^6} + \cdots\right)\right] \\
& + \left\{\frac{1}{2}\rho_{cp}c_{p,cp}L_s\left[\sqrt{\frac{3k_B T(1-1.5\varepsilon^3\phi)}{2\pi\rho_{cp}\varepsilon^3 a_p^3}} + \frac{G_T}{6\pi\mu\varepsilon a_p L_s}\right]\right\}
\end{aligned}
\tag{3.20}
$$

where:

$$k_{\text{layer}} = \omega k_{bf}$$

$$\Lambda = \frac{k_{cp} - k_{bf}}{k_{cp} + 2k_{bf}}$$

$$k_{cp} = \frac{2(k_p - k_{\text{layer}}) + e^3(k_p + 2k_{\text{layer}})}{(k_{\text{layer}} - k_p) + e^3(k_p + 2k_{\text{layer}})} k_{\text{layer}}$$

G_T denotes the total interparticle potential or Derjaguin–Landau–Verwey–Overbeck (DLVO) interaction potential

ω is an empirical parameter that depends on the orderliness of fluid molecules in the interface and the nature and surface chemistry of nanoparticles

L_s is the interparticle separation distance

The subscript cp indicates the complex particle defined as the particle with a thin nanolayer

When there are no particle–particle interaction and the interfacial thin layer and Brownian motion contribution, the proposed model is consistent with the Maxwell model. Their results indicated that the major contributions to the thermal conductivity enhancement of nanofluids come from the static mechanisms.

3.2.3.2 Combination of Aggregation and Brownian Motion

Xuan et al.[54] proposed a thermal model for the effective k of nanofluids considering Brownian motion and nanoparticle aggregation. The stochastic motion of nanoparticles in a liquid is modeled using the Brownian motion theory and the aggregation of particles is simulated by diffusion-limited aggregation model, which can account for the morphology of nanofluids and aggregation process of the suspended nanoparticles. The model is written as follows:

$$k_{\text{eff}} = \left[\frac{k_{\text{p}} + 2k_{\text{bf}} + 2(k_{\text{p}} - k_{\text{bf}})\phi}{k_{\text{p}} + 2k_{\text{bf}} - (k_{\text{p}} - k_{\text{bf}})\phi} \right] k_{\text{bf}} + \frac{\rho_{\text{p}}\phi c_{v,\text{p}}}{2} \sqrt{\frac{k_{\text{B}}T}{3\pi a_{\text{cl}}\mu}} \qquad (3.21)$$

where a_{cl} is the apparent radius of the nanoparticle cluster and is influenced by the fractal dimension of the cluster structure. The first term on the right-hand side of the model is the Maxwell model and the second term is the contribution of thermal conductivity of nanofluids due to Brownian motion of nanoparticle clustering. Prasher et al.[55] investigated aggregation kinetics with Brownian motion-induced microconvection, and they developed a combined thermal conductivity model of nanofluids. They assumed that the nanoparticles are initially well dispersed in a base liquid without any aggregation, and the level of aggregation increases as time goes on. Also, they divided the primary volume fraction of nanofluids into two volume fraction concepts.

$$\phi_{\text{p}} = \phi_{\text{int}}\phi_{\text{ag}} \qquad (3.22)$$

where ϕ_{p}, ϕ_{int}, and ϕ_{ag} are the volume fraction of the primary particles, the particles in the aggregates, and the aggregates in the entire fluid, respectively. When there is no agglomeration of nanoparticles, the volume fraction of particles in aggregates become 1 ($\phi_{\text{int}} = 1$) and $\phi_{\text{p}} = \phi_{\text{ag}}$, which indicates a well-dispersed system. On the contrary, for a fully aggregated system, the volume fraction of aggregates in a base fluid is 1 ($\phi_{\text{ag}} = 1$) and $\phi_{\text{p}} = \phi_{\text{int}}$. They also considered the interparticle interactions using the DLVO theory with several chemical parameters such as pH, ion concentration, Hamaker constant, and zeta potential. The proposed model is as follows:

$$k_{\text{eff}} = \left(1 + A\text{Re}^m \text{Pr}^{0.333}\,\phi\right)\left[\frac{k_{\text{ag}} + 2k_{\text{bf}} + 2(k_{\text{ag}} - k_{\text{bf}})\phi_{\text{ag}}}{k_{\text{ag}} + 2k_{\text{bf}} - (k_{\text{ag}} - k_{\text{bf}})\phi_{\text{ag}}} \right] k_{\text{bf}} \qquad (3.23)$$

where k_{ag} is the thermal conductivity of the aggregates and this value is obtained by the Bruggeman model.[18] If there are no interparticle interactions and Brownian motion-induced convection, the model reduces to the Maxwell model for a well-dispersed system. Although this model can explain the numerous aspects of thermal conductivity of nanofluids such as temperature, particle size, and pH dependence, this model requires a number of information such as fractal dimensions, aggregation time constant, and chemical parameters to determine the effective thermal conductivity of nanofluids.

3.2.3.3 Combination of Nanolayer and Aggregation

Feng et al.[56] proposed a combined model of the interfacial nanolayer and the cluster of aggregated nanoparticles. They assumed that the heat is transported through a unit cell, which consists of a nanoparticle, base liquid, and a nanolayer. The model is shown as follows:

$$k_{eff} = \left[(1-\phi_e) \frac{k_{pe} + 2k_{bf} + 2(k_{pe} - k_{bf})(1-\beta)^3 \phi}{k_{pe} + 2k_{bf} - (k_{pe} - k_{bf})(1+\beta)^3 \phi} \right] k_{bf}$$
$$+ \phi_e \left\{ \left(1 - \frac{3}{2}\phi_e \right) + \frac{3\phi_e}{\eta} \left[\frac{1}{\eta} \ln \frac{a_p + \delta}{(a_p + \delta)(1-\eta)} - 1 \right] \right\} k_{bf}$$

$$(3.24)$$

where ϕ_e is the equivalent volume fraction defined as $\phi_e = \phi(1+\beta)^3$ and $\eta = 1 - k_{bf}/k_{pe}$. The first term without $(1-\phi_e)$ term indicates the Yu and Choi model,[26] which represents the nanolayer effect for nonaggregated particles. The second term is derived from the thermal resistance approach for the unit column or cell of touching particles. As shown in Equation 3.24, this model depends on the thickness of nanolayer, the volume fraction, the size, and the thermal conductivity of nanoparticles. When $\phi_e = 2/3$, whole nanoparticles with nanolayer touch each other and this is an upper limit of equivalent volume fraction. If $\phi_e = 0$, their model is identical to the Yu and Choi model,[26] and this means that whole nanoparticles with interfacial nanolayer are well dispersed in the liquid medium without any aggregation. Table 3.5 summarizes the combined models for thermal conductivity of nanofluids.

TABLE 3.5 Combined Mechanism-Based Models

Mechanisms	Authors	Formulation
Nanolayer + Brownian motion	Ren et al.[52]	$k_{\mathrm{eff}} = \left\{ 1 + F(\mathrm{Pe}) + 3\Theta\phi_{\mathrm{T}} + \left[\dfrac{(3\Theta^2\phi_{\mathrm{T}}^2)}{(1-\Theta\phi_{\mathrm{T}})} \right] \right\} k_{\mathrm{bf}}$
	Murshed et al.[53]	$k_{\mathrm{eff}} = \left(\left\{ \dfrac{\phi\omega(k_{\mathrm{p}} - \omega k_{\mathrm{bf}})(2\varepsilon_i^3 - \varepsilon^3 + 1) + (k_{\mathrm{p}} + 2\varepsilon k_{\mathrm{bf}})\varepsilon_i^3\left[\phi\varepsilon^3(\varepsilon-1)+1\right]}{\varepsilon_i^3(k_{\mathrm{p}} + 2\omega k_{\mathrm{bf}}) - (k_{\mathrm{p}} - \varepsilon k_{\mathrm{bf}})\phi(\varepsilon_i^3 + \varepsilon^3 - 1)} \right\} k_{\mathrm{bf}} \right.$ $\left. + \phi^2\varepsilon^6 k_{\mathrm{bf}} \left[\dfrac{3\Lambda^2 + (3\Lambda^2)/4 + (9\Lambda^3)}{16(k_{\mathrm{cp}} + 2k_{\mathrm{bf}})/(2k_{\mathrm{cp}} + 3k_{\mathrm{bf}})} + \dfrac{(3\Lambda^4)}{(2^6)} + \cdots \right] \right] + \left[\dfrac{1}{2}\rho_{\mathrm{cp}}c_{p,\mathrm{cp}}L_s \left[\sqrt{\dfrac{3k_{\mathrm{B}}T(1-1.5\varepsilon^3\phi)}{(2\rho_{\mathrm{cp}}\varepsilon^3 a_{\mathrm{p}}^3)}} + \dfrac{(G_{\mathrm{T}})}{(6\pi\mu\varepsilon a_{\mathrm{p}}L_s)} \right] \right]$
Aggregation + Brownian motion	Xuan et al.[54]	$k_{\mathrm{eff}} = \left[\dfrac{\left[k_{\mathrm{p}} + 2k_{\mathrm{bf}} + 2\left(k_{\mathrm{p}} - k_{\mathrm{bf}}\right)\phi \right]}{\left[k_{\mathrm{p}} + 2k_{\mathrm{bf}} - \left(k_{\mathrm{p}} - k_{\mathrm{bf}}\right)\phi \right]} \right] k_{\mathrm{bf}} + \dfrac{(\rho_{\mathrm{p}}\phi c_{s,\mathrm{p}})}{2\sqrt{(k_{\mathrm{B}}T)/(3\pi a_{\mathrm{cl}}\mu)}}$
	Prasher et al.[55]	$k_{\mathrm{eff}} = \left(1 + A\,\mathrm{Re}^m\,\mathrm{Pr}^{0.333}\,\phi\right)\left\{ \dfrac{\left[k_{\mathrm{ag}} + 2k_{\mathrm{bf}} + 2\left(k_{\mathrm{ag}} - k_{\mathrm{bf}}\right)\phi_{\mathrm{ag}} \right]}{\left[k_{\mathrm{ag}} + 2k_{\mathrm{bf}} - \left(k_{\mathrm{ag}} - k_{\mathrm{bf}}\right)\phi_{\mathrm{ag}} \right]} \right\} k_{\mathrm{ag}}$
Nanolayer + aggregation	Feng et al.[56]	$k_{\mathrm{eff}} = \left\{ (1-\phi_{\mathrm{e}})\left[\dfrac{\left[k_{\mathrm{pe}} + 2k_{\mathrm{bf}} + 2(k_{\mathrm{pe}} - k_{\mathrm{bf}})(1-\beta)^3\phi \right]}{\left[k_{\mathrm{pe}} + 2k_{\mathrm{bf}} - (k_{\mathrm{pe}} - k_{\mathrm{bf}})(1+\beta)^3\phi \right]} \right] k_{\mathrm{bf}} + \phi_{\mathrm{e}} \left[\left[\left(1 - \dfrac{3}{2}\right)\phi_{\mathrm{e}} \right] + \dfrac{(3\phi_{\mathrm{e}})}{\eta} \left[\dfrac{1}{\eta}\ln\dfrac{a_{\mathrm{p}}+\delta}{(a_{\mathrm{p}}+\delta)(1-\eta)} - 1 \right] \right] k_{\mathrm{bf}} \right\}$

3.3 CONCLUSIONS

Because the conventional EMTs fail to sufficiently describe the novel aspects of thermal conduction in nanofluids, a number of new mechanisms and models have been proposed. This chapter summarized the classical models and the newly proposed models by dividing into three categories: the static, dynamic, and combined mechanisms. The classical effective medium theories, the nanolayer-based models, and the aggregation-based models are described as the static mechanisms that have the assumption of the stationary dispersed solid-liquid suspension system. However, focusing on the continuous moving of nanoparticles in a base fluid, Brownian motion-based models are developed as the dynamic mechanism. The combined models of both mechanisms are also introduced to accurately capture the unique features of thermal transport phenomenon in nanofluids. However, numerous prior information or assumptions are essential to calculate the proposed models for effective thermal conductivity of nanofluids. For example, the thickness and properties of nanolayer should be required for the nanolayer-based models. The existence of nanoconvection effect is also not experimentally evaluated at the nanoscale. Thus, the controversy of postulated mechanisms is not over and the validity of the new physics is still hotly under debate. In order to resolve this critical issue, more systematic experiments should be performed using well-characterized and well-controlled nanofluids. Moreover, the comprehensive understanding of the thermal transport phenomenon in nanofluids should be carried out in parallel. We thank the late Professor Stephen U. S. Choi for his helpful discussion during his lifetime.

REFERENCES

1. Maxwell, J. C. *A Treatise on Electricity and Magnetism*. (Oxford: Clarendon Press, 1873), 360–366.
2. Das, S. K., Putra, N., Thiesen, P., and Roetzel, W. "Temperature dependence of thermal conductivity enhancement for nanofluids." *ASME Journal of Heat Transfer* 125 no. 4, 2003: 567–574.
3. Li, C. H., and Peterson, G. P. "Experimental investigation of temperature and volume fraction variations on the effective thermal conductivity of nanoparticle suspensions (nanofluids)." *Journal of Applied Physics* 99 no. 8, 2006: 084314.
4. Chon, C. H., Kihm, K. D., Lee, S. P., and Choi, S. U. S. "Empirical correlation finding the role of temperature and particle size for nanofluid (Al_2O_3) thermal conductivity enhancement." *Applied Physics Letters* 87 no. 15, 2005: 153107.

5. Murshed, S. M. S., Leong, K. C., and Yang, C. "Investigations of thermal conductivity and viscosity of nanofluids." *International Journal of Thermal Sciences* 47 no. 5, 2008: 560–568.

6. Patel, H. E., Sundararajan, T., and Das, S. K. "An experimental investigation into the thermal conductivity enhancement in oxide and metallic nanofluids." *Journal of Nanoparticle Research* 12 no. 3, 2010: 1015–1031.

7. Chopkar, M., Das, P. K., and Manna, I. "Synthesis and characterization of nanofluid for advanced heat transfer applications." *Scripta Materialia* 55 no. 6, 2006: 549–552.

8. Hong, K. S., Hong, T.-K., and Yang, H.-S. "Thermal conductivity of Fe nanofluids depending on the cluster size of nanoparticles." *Applied Physics Letters* 88 no. 3, 2006: 031901.

9. Kim, S. H., Choi, S. R., and Kim, D. "Thermal conductivity of metal-oxide nanofluids: Particle size dependence and effect of laser irradiation." *ASME Journal of Heat Transfer* 129 no. 3, 2007: 298–307.

10. Lee, D., Kim, J.-W., and Kim, B. G., "A new parameter to control heat transport in nanofluids: Surface charge state of the particle in suspension." *The Journal of Physical Chemistry B* 110 no. 9, 2006: 4323–4328.

11. Dongsheng, Z., Li, X., Wang, N., Wang, X., Gao, J., and Li, H. "Dispersion behavior and thermal conductivity characteristics of Al_2O_3–H_2O nanofluids." *Current Applied Physics* 9 no. 1, 2009: 131–139.

12. Zhang, X., Gu, H., and Fujii, M. "Effective thermal conductivity and thermal diffusivity of nanofluids containing spherical and cylindrical nanoparticles." *Journal of Applied Physics* 100 no. 4, 2006: 044325.

13. Timofeeva, E. V., Gavrilov, A. N., McCloskey, J. M. et al. "Thermal conductivity and particle agglomeration in alumina nanofluids: Experiment and theory." *Physical Review E* 76 no. 6, 2007: 061203.

14. Chen, G., Yu, W., Singh, D., Cookson, D., and Routbort, J. "Application of SAXS to the study of particle-size-dependent thermal conductivity in silica nanofluids." *Journal of Nanoparticle Research* 10 no. 7, 2008: 1109–1114.

15. Beck, M. P., Yuan, Y., Warrier, P., and Teja, A. S. "The effect of particle size on the thermal conductivity of alumina nanofluids." *Journal of Nanoparticle Research* 11 no. 5, 2009: 1129–1136.

16. Lee, J.-H., Lee, S.-H., Choi, C. J., Jang, S. P., and Choi, S. U. S. "A review of thermal conductivity data, mechanisms and models for nanofluids." *International Journal of Micro-Nano Scale Transport* 1 no. 4, 2010: 269–322.

17. Maxwell-Garnett, J. C., "Colours in metal glasses and in metallic films." *Philosophical Transactions of the Royal Society of London, Series A* [Containing Papers of a Mathematical of Physical Character] 203, 1904: 385–420.

18. Bruggeman, D. A. G. "Berechnung Verschiedener Physikalischer Konstanten von Heterogenen Substanzen, I. Dielektrizitatskonstanten und Leitfahigkeiten der Mischkorper aus Isotropen Substanzen." *Annalen der Physik Leipzig* 24, 1935: 636–679.

19. Hamilton, R. L., and Crosser, O. K. "Thermal conductivity of heterogeneous two-component system." *Industrial and Engineering Chemistry Fundamentals* 1 no. 3, 1962: 187–191.

20. Hashin, Z., and Shtrikman, S. "A variational approach to the theory of the effective magnetic permeability of multiphase materials." *Journal of Applied Physics* 33 no. 10, 1962: 3125–3131.

21. Keblinski, P., Prasher, R., and Eapen, J. "Thermal conductance of nanofluids: Is the controversy over?." *Journal of Nanoparticle Research* 10 no. 7, 2008: 1089–1097.

22. Eapen, J., Rusconi, R., Piazza, R., and Yip, S. "The classical nature of thermal conduction in nanofluids." *ASME Journal of Heat Transfer* 132 no. 10, 2010: 102402.

23. Hasselman, D. P. H., and Johnson, L. F. "Effective thermal conductivity of composites with interfacial thermal barrier resistance." *Journal of Composite Materials* 21 no. 6, 1987: 508–515.

24. Nan, C.-W., Birringer, R., Clarke, D. R., and Gleiter, H. "Effective thermal conductivity of particulate composites with interfacial thermal resistance." *Journal of Applied Physics* 81 no. 10, 1997: 6692–6699.

25. Keblinski, P., Phillpot, S. R., Choi, S. U. S., and Eastman, J. A. "Mechanisms of heat flow in suspensions of nano-sized particles (nanofluids)." *International Journal of Heat and Mass Transfer* 45 no. 4, 2002: 855–863.

26. Yu, W., and Choi, S. U. S. "The role of interfacial layers in the enhanced thermal conductivity of nanofluids: A renovated Maxwell model." *Journal of Nanoparticle Research* 5 no. 1–2, 2003: 167–171.

27. Yu, C.-J., Richter, A. G., Datta, A., Durbin, M. K., and Dutta, P. "Molecular layering in a liquid on a solid substrate: An X-ray reflectivity study." *Physica B: Condensed Matter* 283 no. 1–3, 2000: 27–31.

28. Yu, W., and Choi, S. U. S. "The role of interfacial layers in the enhanced thermal conductivity of nanofluids: A renovated Hamilton–Crosser model." *Journal of Nanoparticle Research* 6 no. 4, 2004: 355–361.

29. Xue, Q.-Z. "Model for effective thermal conductivity of nanofluids." *Physics Letters A* 307 no. 5–6, 2003: 313–317.

30. Xue, Q., and Xu, W.-M. "A model of thermal conductivity of nanofluids with interfacial shells." *Materials Chemistry and Physics* 90 no. 2–3, 2005: 298–301.

31. Xie, H., Fujii, M., and Zhang, X. "Effect of interfacial nanolayer on the effective thermal conductivity of nanoparticle-fluid mixture." *International Journal of Heat and Mass Transfer* 48 no. 14, 2005: 2926–2932.

32. Leong, K. C., Yang, C., and Murshed, S. M. S. "A model for the thermal conductivity of nanofluids—The effect of interfacial layer." *Journal of Nanoparticle Research* 8 no. 2, 2006: 245–254.

33. Doroodchi, E., Evans, T. M., and Moghtaderi, B. "Comments on the effect of liquid layering on the thermal conductivity of nanofluids." *Journal of Nanoparticle Research* 11 no. 6, 2009: 1501–1507.

34. Xue, L., Keblinski, P., Phillpot, S. R., Choi, S. U. S., and Eastman, J. A. "Effect of liquid layering at the liquid–solid interface on thermal transport." *International Journal of Heat and Mass Transfer* 47 no. 19–20, 2004: 4277–4284.

35. Li, L., Zhang, Y., Ma, H., and Yang, M. "Molecular dynamics simulation of effect of liquid layering around the nanoparticle on the enhanced thermal conductivity of nanofluids." *Journal of Nanoparticle Research* 12 no. 3, 2010: 811–821.

36. Wang, B.-X., Zhou, L.-P., and Peng, X.-F. "A fractal model for predicting the effective thermal conductivity of liquid with suspension of nanoparticles." *International Journal of Heat and Mass Transfer* 46, no. 14, 2003: 2665–2672.

37. Prasher, R., Evans, W., Meakin, P., Fish, J., Phelan, P., and Keblinski, P. "Effect of aggregation on thermal conduction in colloidal nanofluids." *Applied Physics Letters* 89 no. 14, 2006: 143119.

38. Evans, W., Prasher, R., Fish, J., Meakin, P., Phelan, P., and Keblinski, P. "Effect of aggregation and interfacial thermal resistance on thermal conductivity of nanocomposites and colloidal nanofluids." *International Journal of Heat and Mass Transfer* 51 no. 5–6, 2008: 1431–1438.

39. Wang, X., Xu, X., and Choi, S. U. S. "Thermal conductivity of nanoparticle-fluid mixture." *Journal of Thermophysics and Heat Transfer* 13 no. 4, 1999: 474–480.

40. Jang, S. P., and Choi, S. U. S. "Role of Brownian motion in the enhanced thermal conductivity of nanofluids." *Applied Physics Letters* 84 no. 21, 2004: 4316–4318.

41. Kumar, D. H., Patel, H. E., Kumar, V., Rajeev, R., Sundararajan, T., Pradeep, T., and Das, S. K. "Model for heat conduction in nanofluids." *Physical Review Letters* 93 no. 14, 2004: 144301.

42. Patel, H. E., Sundararajan, T., and Das, S. K. "A cell model approach for thermal conductivity of nanofluids." *Journal of Nanoparticle Research* 10 no. 1, 2008: 87–97.

43. Koo, J., and Kleinstreuer, C., "A new thermal conductivity model for nanofluids." *Journal of Nanoparticle Research* 6 no. 6, 2004: 577–588.

44. Prasher, R., Bhattacharya, P., and Phelan, P. E. "Thermal conductivity of nanoscale colloidal solutions (nanofluids)." *Physical Review Letters* 94 no. 2, 2005: 025901.

45. Yang, B. "Thermal conductivity equations based on Brownian motion in suspensions of nanoparticles (nanofluids)." *ASME Journal of Heat Transfer* 130 no. 4, 2008: 042408.

46. Evans, W., Fish, J., and Keblinski, P. "Role of Brownian motion hydrodynamics on nanofluid thermal conductivity." *Applied Physics Letters* 88 no. 9, 2006: 093116.

47. Vladkov, M., and Barrat, J.-L. "Modeling transient absorption and thermal conductivity in a simple nanofluid." *Nano Letters* 6 no. 6, 2006: 1224–1228.

48. Sarkar, S., and Selvam, R. P. "Molecular dynamics simulation of effective thermal conductivity and study of enhanced thermal transport mechanism in nanofluids." *Journal of Applied Physics* 102 no. 7, 2007: 074302.

49. Li, C. H., and Peterson, G. P. "Mixing effect on the enhancement of the effective thermal conductivity of nanoparticle suspensions (nanofluids)." *International Journal of Heat and Mass Transfer* 50 no. 23–24, 2007: 4668–4677.

50. Bhattacharya, P., Saha, S. K., Yadav, A., Phelan, P. E., and Prasher, R. S. "Brownian dynamics simulation to determine the effective thermal conductivity of nanofluids." *Journal of Applied Physics* 95 no. 11, 2004: 6492–6494.

51. Jain, S., Patel, H. E., and Das, S. K. "Brownian dynamics simulation for the prediction of effective thermal conductivity of nanofluid." *Journal of Nanoparticle Research* 11 no. 4, 2009: 767–773.

52. Ren, Y., Xie, H., and Cai, A. "Effective thermal conductivity of nanofluids containing spherical nanoparticles." *Journal of Physics D: Applied Physics* 38 no. 21, 2005: 3958–3961.

53. Murshed, S. M. S., Leong, K. C., and Yang, C. "A combined model for the effective thermal conductivity of nanofluids." *Applied Thermal Engineering* 29 no. 11–12, 2009: 2477–2483.

54. Xuan, Y., Li, Q., and Hu, W. "Aggregation structure and thermal conductivity of nanofluids." *AIChE Journal* 49 no. 4, 2003: 1038–1043.

55. Prasher, R., Phelan, P. E., and Bhattacharya, P. "Effect of aggregation kinetics on the thermal conductivity of nanoscale colloidal solutions (nanofluid)." *Nano Letters* 6 no. 7, 2006: 1529–1534.

56. Feng, Y., Yu, B., Xu, P., and Zou, M. "The effective thermal conductivity of nanofluids based on the nanolayer and the aggregation of nanoparticles." *Journal of Physics D: Applied Physics* 40 no. 10, 2007: 3164–3171.

CHAPTER 4

Experimental Methods for the Characterization of Thermophysical Properties of Nanofluids

Sergio Bobbo and Laura Fedele

CONTENTS

4.1 INTRODUCTION

All the possible applications of nanofluids are strictly connected to the knowledge of their thermophysical properties as, for example, thermal conductivity, dynamic viscosity, and surface tension.

In the past few years, several theoretical or empirical models have been proposed to describe the thermophysical behavior of these suspensions, but none revealed to be rigorously reliable. Actually, the physics governing the real behavior of the nanofluids is far to be understood. For this reason, the thermophysical properties must be measured.

An exception could be given by density (ρ_{nf}) and specific heat capacity ($c_{p,nf}$) that can be calculated as weight average:

$$\rho_{nf} = (1-\varphi)\rho_f + \varphi\rho_{np} \qquad (4.1)$$

$$c_{p,nf} = (1-w)c_{p,f} + w\,c_{p,np} \qquad (4.2)$$

where:

ρ_f and ρ_{np} are the densities of base fluid and nanoparticles, respectively

ϕ is the volume fraction

$c_{p,f}$ and $c_{p,np}$ are the specific heat capacities of base fluid and nanoparticles, respectively

w is the mass fraction

However, as discussed below, few studies have also been published on the measurements of these two properties.

Several techniques could be used to characterize these fluids, but it should be worth noting that the intrinsic nature of them, given by a mixture of solid nanoparticles in liquids, must be taken into consideration to perform correct measurements.

In this chapter, the main thermophysical properties and the corresponding measurement methodologies are discussed.

4.2 THERMAL CONDUCTIVITY

Thermal conductivity λ is surely one of the most studied properties of nanofluids, starting from the idea that they should show enhanced thermal properties. It quantifies the capability of a material to conduct heat, without any motion, and it is defined as the heat Q transmitted in the time unit through a thickness l in the direction normal to a surface of area A, with a temperature gradient ΔT, under steady-state conditions:

$$\lambda = \frac{(Q/A)}{(\Delta T/l)} \tag{4.3}$$

For a correct measurement, all the studied suspensions must remain stationary during the experiments.

Although the literature is often contradictory, generally thermal conductivity increases with nanoparticles' concentration in nanofluids and with temperature.

Several techniques have been proposed to measure nanofluids' thermal conductivity (Paul et al. 2010). They can be mainly divided into two types: transient techniques and steady-state techniques. In the first case, the measurements are taken under a variation of temperature, due to a heating process, whereas in the second one, the temperature is maintained constant.

A third kind of technique, not often used and here only briefly described, is given by the thermal comparator method (Paul et al. 2010; Powell 1957).

The main outlines of principal techniques are described in Sections 4.2.1 through 4.2.3. For a more detailed description of the measurement methods and the basic equations for the thermal conductivity calculation for each nanofluid, the reader should refer to the cited literature.

4.2.1 Transient Techniques

In the nonsteady-state or transient techniques, the experimental measurements are performed during heating process. Generally, these methods are fast, even though the analysis of the recorded data is more difficult than a steady-state method.

4.2.1.1 Transient Hot-Wire

The first apparatus based on the transient hot-wire method was developed by Stâlhane and Pyk (1931) for solids and powders, and then modified for liquids by Eucken and Englert (1938).

This technique is based on the consideration that when a linear, infinitely, long, and vertical source of heat with infinite thermal conductivity and zero heat capacity is immersed in a fluid, if a heat flux per unit length (q) is imposed, all the energy will be conducted to the fluid. In this condition, based on a specific solution of Fourier's law, the relation between the temperature increase of the fluid and its thermal conductivity is given by (Roder 1981; Wakeham et al. 1991)

$$\Delta T = \frac{q}{4\pi\lambda} \ln\left(\frac{4\alpha t}{r^2 C}\right) \tag{4.4}$$

where:
  α is the thermal diffusivity
  t is the measurement time
  r is the wire radius
  C is a constant

This method has been widely employed for the measurements of nanofluid thermal conductivity, starting from Masuda et al. (1993), who first applied the transient hot-wire method to a suspension of nanoparticles in liquid. Several instruments based on this technique have been developed in the past few decades. They all consist of a probe made of a metallic (generally platinum) wire, serving both as heat source and thermometer, which must be

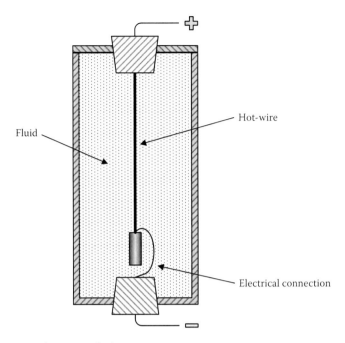

FIGURE 4.1   Schematic of a hot-wire apparatus.

introduced in the nanofluid, as roughly represented by Figure 4.1. The wire must be kept always under tension by a weight or firmly fixed to the top and bottom of the cell, possibly insulated to avoid nanoparticles deposition.

A constant current is supplied to the wire to raise its temperature and that of the nanofluid. This increment depends on the thermal conductivity of the sample. Interpolating all the experimental data in the line ΔT versus $\ln(t)$, given by Equation 4.4, its slope will give the reciprocal of λ for the sample.

In the literature, several modifications of this technique have been proposed, such as the liquid metal transient hot-wire (Omotani et al. 1982) or the transient short hot-wire (Xie et al. 2006).

4.2.1.2 Transient Plane Source

The transient plane source method was first proposed by Gustafsson (1967), and then developed and improved by Gustafsson (1991) and Gustavsson et al. (1994).

In this method, the probe, made by a metal strip sensor or a hot disk, often made of nickel, works as a continuous plane heat source and, at the same time, as temperature sensor. Its basic principle is similar to that of the hot-wire (He 2005), using Fourier's law of heat conduction, but with this method, measurements are faster, the experimental range of thermal conductivity is wider, and the sample preparation is easier (Paul et al. 2010). For the study of nanofluids, the sensor must be immersed in the suspension, which is contained in a vessel. The λ of the nanofluid is measured to determine the sensor resistance. Generally, the sensor is made of an electrically conducting material, which is formed in a thin foil assembled in an insulating layer. If an electric power is given to the probe, the resulting increase in temperature should be measured by the probe through the electric resistance. Always following Fourier's law of heat conduction, when convection does not occur, the increase in temperature can be calculated as

$$\Delta T(\psi) = \frac{Q}{\pi^{1.5} r \lambda} \Omega(\tau) \tag{4.5}$$

where:
$\psi = \sqrt{(t\alpha)/r^2}$
r is the probe radius
Q is the heat power supplied as electric power
Ω is a dimensionless specific function of ψ

Without natural convection, fitting the line given by Equation 4.5 to the experimental data, its slope gives $1/\lambda$.

4.2.1.3 Temperature Oscillation

The temperature oscillation method was first suggested by Roetzel et al. (1990), and further developed by Czarnetzki and Roetzel (1995).

This technique uses a test cylindrical cell, shown in Figure 4.2, in which the sample is introduced (Das et al. 2003). Generally, a Peltier element is used to change the temperature of the fluid, which is measured at the center and the opposite parallel surfaces of the cell by means of thermocouples. The fundamental principle of this method is that the response of the fluid, or nanofluid, to a temperature oscillation or to an imposed heat flux is given by its thermal conductivity. Solving the heat conduction law in an isotropic fluid with constant thermal conductivity,

$$\frac{1}{\alpha} \frac{\partial T}{\partial t} = \nabla^2 T \tag{4.6}$$

By measuring the phase and amplitude of the temperature oscillation between the two surfaces and the center of the cell, α can be estimated.

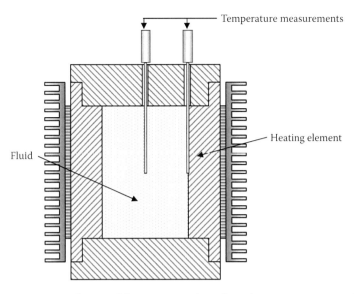

FIGURE 4.2   Basic scheme of the apparatus based on the temperature oscillation method.

Estimating density and specific heat capacity (e.g., Equations 4.1 and 4.2), λ can be calculated as

$$\lambda = \alpha \cdot \rho \cdot c_p \tag{4.7}$$

4.2.1.4 3ω

Like the transient hot-wire and transient plane source methods, in the 3ω method, first developed by Cahill (1990), a radial flow heat is generated from a single element that acts also as a thermometer. However, the measurement technique is based on the frequency of the temperature oscillation instead of the time-dependent temperature increase. The basic element is a metal wire through which a sinusoidal current (with frequency ω) passes, generating a heat wave, with frequency 2ω. Measuring the alternate voltage, containing both ω and 3ω elements, λ can be estimated. In fact, from the exact solution of the diffusion equation for a temperature oscillation at a distance d from an infinitely narrow line heat source on an infinite half-volume surface (Carslaw and Jaeger 1959), λ is given by

$$\Delta T = \frac{P}{l\pi\lambda} K_0 (bd) \tag{4.8}$$

where:

P/l is the power per unit length generated at frequency 2ω

K_0 is the zeroth-order modified Bessel function, with

$$\frac{1}{b} = \left(\frac{\alpha}{i2\omega} \right) \tag{4.9}$$

which is the diffusive thermal wavelength or the thermal penetration depth (Cahill 1990), with α as thermal diffusivity.

4.2.1.5 Laser Flash

The flash method, developed to determine thermal diffusivity, heat capacity, and thermal conductivity, was first described by Parker et al. (1961). With this technique, a laser beam, characterized by a high energy pulse with short duration, is used as a heat source, absorbed by the sample. In the case of liquids (Tada et al. 1978), in particular nanofluids (Lee et al. 2012), the sample is inserted between a metal disk and a holder. The laser beam is absorbed by the external surface of the metal disk, which increases the temperature rapidly, and then the heat flows into the liquid in one-dimension, as in Figure 4.3. The thermal conductivity is estimated from the temperature decrease on the surface of the metal disk, during the heat discharge in the fluid, by the solution of the diffusion equation. The fundamental assumptions are that the heat flow is one-dimensional, the layers around the liquid are considered as semi-infinite cylindrical rods, the contact resistance between the liquid and the metal disk is neglected, and the physical properties are independent on temperature, due to small temperature differences.

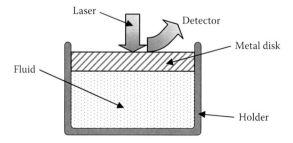

FIGURE 4.3   Basic scheme of the laser flash method.

4.2.1.6 Photoacoustic

Since the beginning, this method has been developed to measure thermal conductivity of thin film, measuring the acoustic wave amplitude generated near the sample surface by the periodic expansion and contraction of the gas layer due to the changes in surface temperature (Govorkov et al. 1997). When the sample is exposed to a modulated laser beam light, the surface temperature varies for the heating. For a certain light intensity, the temperature change at the surface is given by how much heat is dissipated through the film into the sample. This technique can also be applied to nanofluids (Raykar and Singh 2011). The sensor is based on open cell mode. A laser beam is used to excite the sample and generate the thermal waves, and then measured through a microphone as an acoustic signal. As a result of the laser radiation absorption, the temperature of the sample rises, the heat diffuses through the sample, and the pressure modifies. From the temperature dependence on time (t) in one dimension (x), the thermal diffusivity and then the thermal conductivity ($\lambda = \alpha \cdot \rho \cdot c_p$) can be estimated:

$$\frac{\partial T(x,t)}{\partial t} = \alpha \frac{\partial^2 T(x,t)}{\partial x^2} \qquad (4.10)$$

4.2.2 Steady-State Techniques

Steady-state methods are characterized by the constant temperature difference during the measurements. For this reason, the experimental data analysis is usually simpler than the transient techniques, because constant parameters describe the steady state.

4.2.2.1 Steady-State Parallel Plate

Challoner and Powell (1956) first built an experimental setup for the measurements of thermal conductivity by means of a steady-state parallel plate method and it has been further used for nanofluid measurements (i.e., Wang et al. 1999). The fluid sample is put between two parallel round plates generally made of copper at distance d in a volume, in which the liquid can adapt as a consequence of the thermal expansion. A heater electrically supplies a heat flux directed from the upper plate to the lower plate through the liquid sample. Temperatures of the two plates and the electric power are measured continuously, so, if the heat conduction remains one-dimensional, thermal conductivity can be calculated as

$$\lambda = \frac{Q}{\Delta T} \cdot \frac{d}{A} \qquad (4.11)$$

where:

A is the upper plate surface area

ΔT is the temperature difference between the two plates

However, the heat is not completely conducted from one plate to the other through the sample, so during these measurements it is necessary to evaluate all the heat losses (Wakeham and Assael 2014).

4.2.2.2 Cylindrical Cell

This method is based on the principle that a thin layer of the sample fluid should be limited between two coaxial cylinders of infinite length, with r_i and r_o as radii of the inner and the outer cylinder, respectively (Wakeham et al. 2007). If a heat flux is imposed on the sample from the inner to the external cylinder, it propagates through the liquid under steady-state condition. Measuring the difference ΔT between the temperatures of the internal and external cylinders and the heat flux Q, thermal conductivity can be calculated as

$$\lambda = \frac{Q}{\Delta T} \cdot \frac{\log(r_o/r_i)}{2\pi} \qquad (4.12)$$

The resulting λ corresponds to the thermal conductivity at $T = (T_i + T_o)/2$.

A particular apparatus based on this method is the ring gap apparatus, already used also for nanofluids, in which the sample volume is defined by the ring volume between an inner heated body and an outer body, as in Figure 4.4 (Buschmann 2012; Ehle et al. 2011; Glory et al. 2008). With this device, the gap width is generally short (around 1 mm) throughout the entire volume and it has been verified that convection is negligible during the measurements.

In an ideal ring gap apparatus, the measurement volume consists of a cylindrical ring gap and two hemispherical gaps, so thermal conductivity is calculated as

$$\lambda = \frac{Q}{\Delta T} \cdot \frac{1}{\left[l/\log(r_o/r_i) \right] + \left[2 r_i r_o / (r_o - r_i) \right]} \qquad (4.13)$$

Several effects disturb the measurements and must be taken into account: the thermal bridges formed by the spacers between the inner surfaces, the

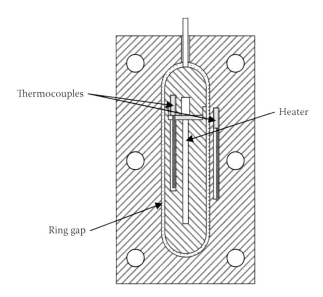

FIGURE 4.4   Ring gap apparatus scheme. (Adapted from Buschmann, M. H., *Int. J. Therm. Sci.*, 62, 19–28, 2012.)

deviations from the ideal gap geometry due to assembling inaccuracies, and the possible development of convection in the gap, among them. For this reason, calibration is necessary.

4.2.3 Thermal Comparator Technique

The idea at the base of this technology, here only briefly described, is that when two materials at two different temperatures are put in contact, an equilibrium temperature is reached at the contact point and it depends on the thermal conductivity of these materials. For this reason, the probe used in this technique has a contact point with the experimental sample. The temperature difference between the sample and the probe is measured by means of thermocouples. Starting from the sample of known thermal conductivity, a calibration curve can be drawn and, through it, the thermal conductivity of the unknown sample can be easily measured. More details can be found in the work of Paul et al. (2010).

4.3 DYNAMIC VISCOSITY

Considering a fluid ideally contained between two plates of area A at distance d in the y-direction, if a force F is applied to the upper plate, it moves continuously with velocity U as far as the force is operated. In this condition, the shear stress is defined as the ratio $\tau = F/A$, whereas the shear rate is equal to the velocity gradient $\gamma = \partial u/\partial y$, where $u = Uy/d$ represents the velocity profile within the fluid. Dynamic viscosity is defined as

$$\mu = \frac{\tau}{\gamma} \tag{4.14}$$

whereas kinematic viscosity is defined as

$$v = \frac{\mu}{\rho} \tag{4.15}$$

When a fluid is Newtonian, viscosity is independent on the shear rate, whereas for a non-Newtonian fluid, it varies with shear rate (Tilton 1997), following different specific laws depending on the fluid. Nanofluids can be either Newtonian or non-Newtonian, especially depending on the nanoparticle concentration. Generally, nanofluid viscosity increases with nanoparticle concentration and decreases with temperature. Although viscosity is an essential parameter in the analysis of the potentiality of nanofluids, it is still not enough studied (Mushed et al. 2008; Wang and Mujumdar 2007). Few apparatuses for the viscosity measurements are described in Sections 4.3.1 and 4.3.2. Within these instruments, it is worth to be mentioned that viscometers can measure the viscosity of Newtonian fluid, whereas rheometers should be used for measuring fluids whose viscosity changes with flow conditions.

The methods used for the nanofluids analysis can be divided into two groups: flow type and drag type (Behi 2012).

4.3.1 Flow Type

4.3.1.1 Capillary Viscometer

A capillary viscometer can be used only to measure the viscosity of Newtonian fluid, and it was also employed for nanofluids (Li et al. 2002). The fluid flows through a capillary tube and the flow rate is measured. The liquid can flow under gravity, in the gravity type viscometer, or under an external force. The time required for fluid to flow in a predefined section of the capillary is measured. Considering a tube of length l and radius r, the relation between the sample volume V and the time t required for it to flow is given by the Hagen–Poiseuille equation with a pressure difference ΔP:

$$\frac{V}{t} = \frac{\pi \Delta P}{8 \mu l} r^4 \tag{4.16}$$

For the gravity type viscometer, $\Delta P = \rho g z$, with density ρ, gravitational acceleration g, and fluid position z, and then

$$\mu = \frac{\pi g z r^4}{8lV}\rho t \qquad (4.17)$$

The capillary viscometers need to be calibrated with fluid of known viscosity. Then, based on Equation 4.17,

$$\mu = K\rho t \qquad (4.18)$$

Determining K through calibration, the viscosity can be measured from t.

To minimize the measurement errors, some corrections should be made, especially due to the kinetic energy and the end effect (Viswanath et al. 2007).

4.3.2 Drag Type

4.3.2.1 Rotational Viscometers and Rheometers

While a rheometer can work controlling the applied shear stress obtaining viscosity at different flow conditions, a viscometer allows to measure viscosity only at one flow condition. In the rotational rheometers, a known force is applied to rotate two surfaces, one with respect to the other, containing a viscous medium, in order to reach a fixed rotational speed. They are more difficult to operate than the capillary viscometers, but can also be used for non-Newtonian fluids and can work under steady-state conditions and different shear rates. The main rotational viscometers and rheometers are described in Sections 4.3.2.1.1 through 4.3.2.1.4.

4.3.2.1.1 Coaxial Cylinder Viscometer   In the coaxial cylinder viscometer, a cylinder is immersed in the sample, contained in an external second cylinder. One cylinder is rotated and the resistance to the rotation, due to the viscosity, is measured by the deflection of the spring hanging from the other cylinder. Krieger and Maron (1952) first developed a concentric cylinder viscometer able to measure non-Newtonian fluids, varying the ratio between the radii of the two coaxial cylinders.

4.3.2.1.2 Cone–Plate Viscometer   Originally, the cone–plate viscometer was constructed by Higginbotham (1950) and was further updated to reach better measurement accuracy and to measure Newtonian and non-Newtonian fluids. It is probably the most employed rotational viscometer, especially for the rheological analysis of non-Newtonian fluids.

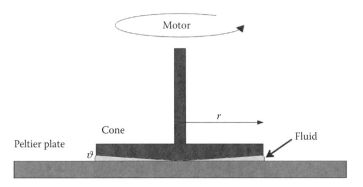

FIGURE 4.5 Geometry of the cone–plate viscometer.

Here, the sample is confined in the measuring gap between a cone and a plate, being the liquid layer as thin as possible to stabilize the temperature and reduce the temperature increase at high shear rate that can be changed continuously. The usual configuration is given in Figure 4.5.

As underlined by McKennell (1956), with this geometry all the problems faced with the coaxial cylinders, related to the impossibility of obtaining a constant shear rate or to the end effects due to the viscous traction on the internal cylinder ends, are not present. In fact, due to the small angle ϑ, the edge effects are negligible.

4.3.2.1.3 Plate–Plate Viscometer   Another rotational viscometer is the plate–plate viscometer, primarily developed by Mooney (1934). It consists of two parallel rotating disks, at a specified gap, containing the sample. The torque required to rotate the disk in the sample is proportional to the viscosity. Generally, this instrument is used for measuring high-viscosity fluids.

4.3.2.1.4 Cone–Cylinder Viscometer   The cone–cylinder viscometer was introduced by Mooney and Ewart (1934) to eliminate the end effects in the coaxial cylinder viscometer. Here, the two cylinders are modified with a conical extension. In this way, the shear rate in the sample will be the same at any point.

4.3.2.2 Falling Body Viscometer
In the falling body viscometer [as in Feng and Johnson (2012)], a body, generally a ball of known density and size, is left to fall through the sample, driven by gravity, in a vertical tube of known dimensions. Considering that, at a certain time, the falling body velocity reaches a constant value when the gravity is balanced by the viscous resistance, dynamic viscosity can be calculated by Stoke's law:

$$\mu = \frac{2gr_b^2(\rho_b - \rho_f)}{9u\pi} \tag{4.19}$$

The above equation has been subjected to several modifications to consider the errors due to the tube wall effects or the finite depth of the fluid sample.

4.4 CONVECTIVE HEAT TRANSFER COEFFICIENT

Nanofluids were first introduced for their potential application as heat transfer fluids, due to an expected enhanced heat transfer coefficient compared to the base fluids.

The heat transfer coefficient is defined as the ratio between the heat flux Q on a heated surface A and the temperature difference between the surface and the fluid that flows inside:

$$h = \frac{Q}{A \cdot \Delta T} \tag{4.20}$$

Relating to nanofluids, several authors studied their heat transfer coefficients in laminar or turbulent flow, generally in a circular horizontal tube.

The two principal measurement methods, based on different imposed boundary conditions, are described in Sections 4.4.1 and 4.4.2.

4.4.1 Constant Heat Flux

Generally, the experimental apparatus is composed of a pump, a pipeline, and a tube test section (Daungthongsuk and Wongwises 2007). This is the main part of the system, which is often made of a highly thermal conductive material (brass or copper), bounded by a heater, as an electric resistance wrapped around the pipe heated or directed by an electric current passing through the tube, to impose a constant wall heat flux. The fluid temperature profile, due to the constant heat flux, varies within the tube and, at any axial position, changes from the wall (T_w) to the center of the flow (T_f). Fluid velocity can be adjusted in order to measure h at different flow conditions.

T_w is usually measured by thermocouples fixed in the external wall of the pipe and T_f, at the inlet and the outlet of the test section, is generally measured by means of thermoresistances or thermocouples inserted into the pipe, in contact with the fluid. $T_{f,l}$ at some distance l from the inlet is calculated through the energy balance of the section considered.

Knowing the temperature difference between the wall and the fluid ($\Delta T_l = T_{w,l} - T_{f,l}$) at a certain position l along the tube section, a local heat transfer coefficient (h_l) can be defined from Equation 4.20 as

$$h_l = \frac{Q}{A \cdot \left(T_{w,l} - T_{f,l}\right)}$$

h_l changes as long as the region is not fully thermally developed. When the thermally developed region is reached, ΔT_l is constant along the tube (Janna 1999).

From the fluid thermal conductivity λ and the inner tube diameter D of the test section, the Nusselt number (Nu) can be calculated as

$$\mathrm{Nu}_l = \frac{h_l D}{\lambda} \tag{4.21}$$

4.4.2 Constant Wall Temperature

Normally, the experimental apparatus is similar to that already described in Section 4.4.1, but a different boundary condition for the determination of the convective heat transfer is imposed, that is, the constant temperature of the duct wall. This condition can be experimentally fixed exploiting the latent heat of a pure fluid condensing or evaporating at the external surface of the tube test section (e.g., Heris et al. 2006). A local convective heat transfer coefficient h_l can also be defined. In this case, ΔT_l decreases along the tube, $T_{w,l}$ remains constant and $T_{f,l}$ increases. However, also the ratio Q/A, where $Q = c_p \rho U A \Delta T$, with $\rho U A$ the mass flow rate, diminishes until h_l becomes constant when the fully thermally developed region is reached.

4.5 SURFACE TENSION

The surface tension Γ is the force acting on the liquid surface, defined as the ratio between the work W needed to impose an alteration ΔA to the surface area A of a liquid sample droplet and the surface alteration ΔA (Atkins 1995):

$$\Gamma = \frac{W}{\Delta A} \tag{4.22}$$

A liquid droplet tends to assume a spherical shape to minimize its surface area, this shape being the one with the lower ratio between the surface and the volume. All the forces acting on the droplet may affect its tendency to be spherical, for example, gravity.

This property is really important in the characterization of nano-fluids, because of, for example, its influence in the nanofluid wetting behavior, the boiling processes, and spray applications, but it is still rarely studied.

There are several methods for the surface tension measurements, depending on the considered parameter for the analysis (Masutani and Stenstrom 1984).

The methods that are mostly used for measuring nanofluids are described in Sections 4.5.1 through 4.5.8.

4.5.1 Pendant Drop

Considering a drop hanging from a vertical capillary tube, its surface tension can be calculated based on its physical characteristics. A camera device can continuously register the images of the drop from the tube, which can be analyzed by a proper software to acquire all the drop parameters (Tanvir and Qiao 2012), as the curvature radius at the drop apex, *r*. Considering that the only acting force is gravity, the surface tension can be calculated as

$$\Gamma = \frac{\Delta \rho \cdot g \cdot r^2}{\beta} \tag{4.23}$$

where:
  $\Delta \rho$ is the difference between the density of the drop and that of the adjacent medium
  β is a shape factor, considering the drop asymmetry (Hansen and Rødsrud 1991)

4.5.2 Ring

The ring method was first proposed by Du Noüy (1925). It has been rarely used for measuring nanofluids (Moosavi et al. 2010). The principle of measurement is based on the force *F* needed to lift a ring, with radius *r*, generally made of platinum and immersed in the liquid, from the liquid surface:

$$\Gamma = \frac{F}{4 \pi r} \tag{4.24}$$

Using this method, the ring must lie on a plane that must be aligned with the surface of the liquid.

4.5.3 Sessile Drop

In the sessile drop method, a liquid droplet is placed on a surface and the surface tension is determined from the contact angle between the drop and the surface.

Generally, a high contact angle means low chemical affinity between the liquid and the solid surface and then low degree of wetting (Sefiane and Bennacer 2009).

4.5.4 Capillary Rise

In the capillary rise method, a small tube is immersed in the liquid sample. Due to capillarity, the liquid adheres to the tube walls and its height H in the column inside the tube depends on the liquid surface tension Γ, the density ρ, and the tube radius r (Golubovic et al. 2009):

$$H = 2\frac{\Gamma \cos \vartheta}{\rho g r} \tag{4.25}$$

and then

$$\Gamma = \frac{H \rho g r}{2 \cos \vartheta}$$

where ϑ is the contact angle between the liquid inside the tube and the tube wall.

4.5.5 Bubble Pressure

The bubble pressure tensiometer is based on the measurement of the pressure needed to produce bubbles, that is, air bubbles, through a capillary tube immersed in a liquid (Godson et al. 2010). The obtained maximum value of pressure p_{max} is related to the formation of a bubble with the same diameter r of the capillary. At this point, the surface tension can be calculated by the Laplace law as

$$\Gamma = \frac{p_{max} \cdot r}{2} \tag{4.26}$$

4.5.6 Wilhelmy Plate

As in the ring method, in the Wilhelmy plate method, the force F needed to lift a solid body, in this case a thin plate, from the liquid sample surface is the measured parameter for the surface tension estimation. The thin plate is maintained in a vertical position by a tensiometer or a micro-balance. When the plate is placed in contact with the liquid sample, it is dragged from the liquid for its surface tension, causing an increase in the force needed to maintain suspended the plate. Then, if l is the perimeter of the plate,

$$\Gamma = \frac{F}{l} \tag{4.27}$$

when the contact angle between the liquid sample and the plate is assumed to be 0 (Thiessen and Man 1999).

4.5.7 Spinning Drop

The spinning drop method is used to measure the interfacial tension between two immiscible fluids, as performed by Hendraningrat et al. (2013) for crude oil and nanofluids. Generally, the fluid with higher density is put into a glass tube and then a drop of the other fluid is placed in the first fluid. The tube is put under rotation and the drop deforms and elongates due to the centrifugal forces. When the interfacial tension balances the centrifugal forces, the deformation ends and the surface tension between the two fluids can be calculated by analyzing the drop structure.

Measuring the radius r of the cylindrical drop, the difference between densities of the two fluids $\Delta\rho$, and the rotational speed ω, surface tension can be calculated as (Drelich et al. 2002)

$$\Gamma = \frac{1}{4} r^3 \Delta\rho\omega^2 \tag{4.28}$$

4.5.8 Stalagmometric

The basic principle of the stalagmometric method is the measurement of the mass of a droplet falling from a vertical capillary tube. In fact, the sample droplet comes out from the tube when its weight is equal to the force due to its surface tension. For this reason, the surface tension can be calculated as

$$\Gamma = \frac{mg}{2\pi r} \tag{4.29}$$

where:
  m is the mass of the droplet
  r is the capillary tube radius

The capillary dimension can be calibrated by measuring a well-known fluid, water. Considering the water surface tension Γ_{H_2O} and its mass m_{H_2O}, from Equation 4.29 the radius r is

$$r = \frac{m_{H_2O} g}{\Gamma_{H_2O} 2\pi} \tag{4.30}$$

and then (Kosmala 2012)

$$\Gamma = \Gamma_{H_2O} \frac{m}{m_{H_2O}} \tag{4.31}$$

4.6 LIQUID DENSITY AND HEAT CAPACITY

Heat transfer coefficient is generally obtained through dimensionless relations expressing the Nusselt number (Nu) as a function of other dimensionless number, the most important of them being Reynolds (Re) and Prandtl (Pr) numbers. The calculation of these numbers requires the knowledge of several thermophysical properties, basically thermal conductivity, viscosity, density, and specific heat capacity. As already described, thermal conductivity and viscosity of nanofluids, being not directly proportional to volume fraction of nanoparticles, need to be measured. Vice versa, density and specific heat capacity are generally considered in the literature as linearly dependent on the volume fraction and then are not measured, but simply calculated through the Equations 4.1 and 4.2 (Pak and Cho 1998). However, some measurements are reported in the literature for both properties.

4.6.1 Liquid Density

Only few papers in the literature report measurements for nanofluids density and a comparison between experimental results and the ideal behavior described by Equation 4.1.

Though the difference is generally small, in some cases it is not negligible and a sort of excess specific volume v^E can be calculated through the expression (Pastoriza-Gallego et al. 2009, 2011):

$$v^E = \frac{1}{\rho_{nf}} - \sum_{j=1}^{n} \frac{w_j}{\rho_j} \tag{4.32}$$

This excess volume depends on the dimension of nanoparticles and the concentration, whereas it is much less influenced by the temperature and pressure.

4.6.1.1 Vibrating Tube Densimeter

Generally, density is measured by means of vibrating tube densimeter (Pastoriza-Gallego et al. 2009, 2011; Vajjha et al. 2009). In this apparatus, a hollow vibrating U-tube is filled with the sample and the oscillation period π is proportional to the sample density as

$$\pi^2 = A \cdot \rho + B \tag{4.33}$$

where A and B are constants that must be determined through calibration of the densimeter, measuring a well-known fluid and vacuum or atleast two fluids and correlating its density to the measured π.

4.6.1.2 Analytical Balance

Hwang et al. (2009) and Feng and Johnson (2012) used an analytical balance to determine the nanofluid density. They simply measured the liquid mass of given volumes (generally between 1 and 4 mL), sampled by means of graduated pipette, by calculating density basically dividing mass m by volume V:

$$\rho = \frac{m}{V} \tag{4.34}$$

4.6.2 Specific Heat Capacity

In the case of specific heat capacity, an alternative model to Equation 4.2, sometimes used in the literature, is the one proposed by Xuan and Roetzel (2000):

$$\left(\rho c_p\right)_{nf} = \varphi\left(\rho c_p\right)_{np} + \left(1 - \varphi\right)\left(\rho c_p\right)_f \tag{4.35}$$

It is worth noting that, as shown by Duangthongsuk and Wongwises (2010), the two models for specific heat capacity give quite different results depending on the volume fraction of nanoparticles; however, the influence of this difference on calculation of the Nusselt number and then of the heat transfer coefficient is weak.

In the literature, only few papers have been published on the specific heat capacity measurements, mainly based on the differential scanning calorimetry (e.g., Hwang et al. 2009; O'Hanley et al. 2012).

4.6.2.1 Differential Scanning Calorimetry

The nanofluid sample is contained in a little pot and a heating regime is imposed through an applied electric voltage. The equipment is composed of two calorimetric channels, in which three measurements are made under a temperature variation ΔT, with velocity $u = dT/dt$. In the first measurement, two empty cells occupy the two channels and the voltage difference S, due to the ΔT between the two channels, is measured as zero reference of the instrument. In the second and third measurements, one cell channel contains an empty cell, whereas in the other the nanofluid sample, first, and a reference sample of known specific heat, second, are introduced. The voltage differences S between the two channels are measured as a function of temperature and velocity of temperature increasing or decreasing in all the three measurement modes. All the samples are subjected to the same variation of temperature T at the same velocity u.

Using these measurements, specific heat capacity can be calculated as follows:

$$c_{p,nf} = \frac{\left(S_{nf} - S_0\right)}{\left(S_{ref} - S_0\right)} \frac{m_{ref}}{m_{nf}} c_{p,ref} \tag{4.36}$$

where the subscripts nf, 0, and ref refer to cells with nanofluid, empty, or filled with the reference fluid.

LIST OF SYMBOLS

A	surface area (m²)
c_p	heat capacity [J/(kg K)]
D	diameter (m)
d	distance (m)
F	force (N)
f	frequency (1/s)
g	gravitational acceleration (m/s²)
h	heat transfer coefficient (W/m²·K)
H	height (m)
i	nanofluid component
l	thickness or length (m)
m	mass (kg)
n	number of components in a nanofluid
Nu	Nusselt number
P	power (W)
p	pressure (MPa)
q	heat flux per unit length (W/m)
Q	heat power (W)

r	radius (m)
S	voltage signal (V)
T	temperature (K)
t	time (s)
u, U	velocity (m/s or K/s)
v	specific volume (m³/kg)
V	volume (m³)
w	mass fraction
w	work (J)
y	distance in the y-direction (m)
z	position (m)

Greek symbols

Δ	difference
Γ	surface tension (J/m²)
ϑ	angle (rad)
α	thermal diffusivity (m²/s)
β	shape factor
γ	shear rate (1/s)
φ	volume fraction
λ	thermal conductivity (W/m·K)
μ	dynamic viscosity (Pa.s)
π	oscillation period (s)
ρ	density (kg/m³)
τ	shear stress (N/m²)
ω	rotational speed (rad/s) or frequency (1/s)
ν	Kinematic viscosity (m²/s)

Superscript

E	Excess property

Subscripts

0	empty reference
b	ball
f	fluid or base fluid
H_2O	water
i	inner
J	component
l	local

max   maximum
nf   nanofluid
np   nanoparticles
o   outer
ref   reference known fluid
w   surface wall

REFERENCES

Thermal conductivity

Buschmann, M. H. 2012. Thermal conductivity and heat transfer of ceramic nanofluids. *Int. J. Therm. Sci.* 62:19–28.

Cahill, D. G. 1990. Thermal conductivity measurement from 30 to 750 K: The 3ω method. *Rev. Sci. Instrum.* 61:802–808.

Carslaw, H. S., and Jaeger, J. C. 1959. *Conduction of Heat in Solids.* Oxford: Oxford University Press.

Challoner, A. R., and Powell, R. W. 1956. Thermal conductivity of liquids: New determinations for seven liquids and appraisal of existing values. *Proc. R. Soc. Lond. Ser. A* 238:90–106.

Czarnetzki, W., and Roetzel, W. 1995. Temperature oscillation techniques for simultaneous measurement of thermal diffusivity and conductivity. *Int. J. Thermophys.* 16:413–422.

Das S. K., Putra, N., Thiesen, P., and Roetzel, W. 2003. Temperature dependence of thermal conductivity enhancement for nanofluids. *J. Heat Transf.* 125:567–574.

Ehle, A., Feja, S., and Buschmann, M. H. 2011. Temperature dependency of ceramic nanofluids shows classical behavior. *J. Therm. Heat Transf.* 25:378–385.

Eucken, A., and Englert, H. 1938. Die experimentelle Bestimmung des Wärmeleitvermögens einiger verfestigter Gase und Flüssigkeiten. *Z. Ges. Kalte-Ind.* 45:109–118.

Glory, J., Bonetti, M., Helezen, M., Mayne-L'Hermite, M., and Reynaud, C. 2008. Thermal and electrical conductivities of water-based nanofluids prepared with long multiwalled carbon nanotubes. *J. Appl. Phys.* 103:94309.

Govorkov, S., Ruderman, W., Horn, M. W., Goodman, R. B., and Rothschild, M. 1997. A new method for measuring thermal conductivity of thin films. *Rev. Sci. Instrum.* 68:3828–3834.

Gustafsson, S. E. 1967. A non-steady-state method of measuring the thermal conductivity of transparent liquids. *Z. Naturf.* 22a:1005–1011.

Gustafsson, S. E. 1991. Transient plane source techniques for thermal conductivity and thermal diffusivity measurements of solid materials. *Rev. Sci. Instrum.* 62:797–804.

Gustavsson, M., Karawacki, E., and Gustafsson, S. E. 1994. Thermal conductivity, thermal diffusivity and specific heat of thin samples from transient measurement with hot disk sensors. *Rev. Sci. Instrum.* 65:3856–3859.

He, Y. 2005. Rapid thermal conductivity measurement with a hot disk sensor: Part 1. Theoretical considerations. *Thermochim. Acta* 436:122–129.

Lee, S. W., Park, S. D., Kang, S., Shin, S. H., Kim, J. H., and Bang I. C. 2012. Feasibility study on molten gallium with suspended nanoparticles for nuclear coolant applications. *Nucl. Eng. Des.* 247:147–159.

Masuda, H., Ebata, A., Teramae, K., and Hishinuma, N. 1993. Alternation of thermal conductivity and viscosity of liquid by dispersing ultra-fine particles (dispersion of γ-Al_2O_3, SiO_2 and TiO_2 ultra-fine particles). *Netsu Bussei* 4:227–233.

Omotani, T., Nagasaka, Y., and Nagashima, A. 1982. Measurement of the thermal conductivity of KNO_3-$NaNO_3$ mixtures using a transient hot-wire method with a liquid metal in a capillary probe. *Int. J. Thermophys.* 3:17–26.

Parker, W. J., Jenkins, R. J., Butler, C. P., and Abbott, G. L. 1961. Flash method of determining thermal diffusivity, heat capacity, and thermal conductivity. *J. Appl. Phys.* 32:1679–1684.

Paul, G., Chopkar, M., Manna, I., and Das, P. K. 2010. Techniques for measuring the thermal conductivity of nanofluids: A review. *Renew. Sust. Energ. Rev.* 14:1913–1924.

Powell, R. W. 1957. Experiments using a simple thermal comparator for measurement of thermal conductivity, surface roughness and thickness of foils or of surface deposits. *J. Sci. Instrum.* 34:485–92.

Raykar, V. S., and Singh, A. K. 2011. Photoacoustic method for measurement of thermal effusivity of Fe_3O_4 nanofluid. *J. Therm.* Article ID 464368.

Roder, H. M. 1981. A transient hot wire thermal conductivity apparatus for fluids. *J. Res. Natl. Bur. Stand.* 86:457–493.

Roetzel, W., Prinzen, S., and Xuan, Y. 1990. Measurement of thermal diffusivity using temperature oscillations thermal conductivity. In Cremers, C. Y., and Fine, H. A. Eds. *Thermal Conductivity 21, Proceedings of the 21st Conference on Thermal Conductivity.* New York: Plenum Press.

Stålhane, B., and Pyk, S. 1931. Ny Metod för Bestämning Av Värmelednings-Koefficienter. *Teknisk Tidskr.* 61:389–398.

Tada, Y., Harada, M., Tanigaki, M., and Eguchi, W. 1978. Laser flash method for measuring thermal conductivity of liquids—Application to low thermal conductivity liquids. *Rev. Sci. Instrum.* 49:1305–1314.

Wakeham, W. A., and Assael, M. J. 2014. Thermal conductivity measurement. In Webster, J. G., and Eren, H. Eds. *Measurement, Instrumentation, and Sensors Handbook*, 2nd ed. Spatial, Mechanical, Thermal, and Radiation Measurement, Boca Raton, FL: CRC Press, pp. 66-3–66-4.

Wakeham, W. A., Assael M. J., Marmur, A., De Coninck, J., Blake, T. D., and Theron S. A., Zussman, E. 2007. Material properties: Measurement and data. In Tropea, C., Yarin A., and Foss, J. F. Eds. *Handbook of Experimental Fluid Mechanics.* Berlin/Heidelberg, Germany: Springer-Verlag, pp. 139–140.

Wakeham, W. A., Nagashima, A., and Sengers, J. V. 1991. *Measurement of the Transport Properties of Fluids.* Oxford: Blackwell.

Wang, X., Xu, X., and Choi, S. U. S. 1999. Thermal conductivity of nanoparticle–fluid mixture. *J. Therm. Phys. Heat Transf.* 13:474–480.

Xie, H., Gu, H., Fujii, M., and Zhang, X. 2006. Short hot wire technique for measuring thermal diffusivity of various materials. *Meas. Sci. Technol.* 17:208–214.

Dynamic viscosity

Behi, M. 2012. Investigation on thermal conductivity, viscosity and stability of nanofluids. Master of Science Thesis EGI-2012, Royal Institute of Technology (KTH), School of Industrial Engineering and Management, Department of Energy Technology, Division of Applied Thermodynamics and Refrigeration, Stockholm, Sweden.

Feng, X., and Johnson, D. W. 2012. Mass transfer in SiO_2 nanofluids: A case against purported nanoparticle convection effects. *Int. J. Heat Mass Transf.* 55:3447–3453.

Higginbotham, R. S. 1950. A cone and plate viscometer. *J. Sci. Instrum.* 27:139.

Krieger, I. M., and Maron, S. H. 1952. Direct determination of the flow curves of non-Newtonian fluids. *J. Appl. Phys.* 23:147–149.

Li, J., Li, Z., and Wang, B. 2002. Experimental viscosity measurements for copper oxide nanoparticle suspensions. *Tsinghua Sci. Technol.* 7:199–201.

McKennell, R. 1956. Cone-plate viscometer. *Anal. Chem.* 28:1710–1714.

Mooney, M. 1934. A shearing disk plastometer for unvulcanized rubber. *Ind. Eng. Chem. Anal. Ed.* 6:147–151.

Mooney, M., and Ewart, R. H. 1934. The conicylindrical viscometer. *J. Appl. Phys.* 5:350–354.

Murshed, S. M. S., Leong, K. C., and Yang C. 2008. Thermophysical and electrokinetic properties of nanofluids—A critical review. *Appl. Therm. Eng.* 28:2109–2125.

Tilton, J. N. 1997. Fluid dynamics. In Perry, R. H., Green, D. W., and Maloney, J. O. Eds. *Perry's Chemical Engineers' Handbook*, 7th ed. Australia: McGraw-Hill, pp. 6-4–6-5.

Viswanath, D. S., Ghosh, T. K., Prasad, D. H. L., Dutt, N. V. K., and Rani K. Y. 2007. *Viscosity of Liquids—Theory, Estimation, Experiment, and Data*. Dordrecht, the Netherlands: Springer.

Wang, W. Q., and Mujumdar, A. S. 2007. Heat transfer characteristics of nanofluids: A review. *Int. J. Therm. Sci.* 46:1–19.

Convective Heat transfer coefficient

Daungthongsuk, W., and Wongwises, S. 2007. A critical review of convective heat transfer of nanofluids. *Renew. Sust. Energ. Rev.* 11:797–817.

Heris, S. Z., Etemad, S. G., and E. M. Nasr. 2006. Experimental investigation of oxide nanofluids laminar flow convective heat transfer. *Int. Commun. Heat Mass* 33:529–535.

Janna, W. S. 1999. *Engineering Heat Transfer*. Boca Raton, FL: CRC Press, pp. 295–301.

Surface tension

Atkins, P. W. 1995. *Physical Chemistry*, 5th ed. Oxford: Oxford University Press, pp. 962–963.

Drelich, J., Fang, Ch., and White, C. L. 2002. Measurement of interfacial tension in fluid-fluid system. In: Hubbard, A. T. Ed. *Encyclopedia of Surface and Colloid Science*. New York: Marcel Dekker, p. 3152.

Du Noüy, P. L. 1925. An interfacial tensiometer for universal use. *J. Gen. Physiol.* 7: 625–632.

Godson, L., Raj, V., Raja, B., and Lal, D. M. 2010. Measurement of viscosity and surface tension of silver-deionized water nanofluid. *Proceedings of the 37th National and 4th International Conference on Fluid Mechanics and Fluid Power*, December 16–18, IIT Madras, Chennai, India.

Golubovic, M. N., Madhawa Hettiarachchi, H. D., Worek, W. M., and Minkowycz, W. J. 2009. Nanofluids and critical heat flux, experimental and analytical study. *Appl. Therm. Eng.* 29:1281–1288.

Hansen, F. K., and Rødsrud, G. 1991. Surface tension by pendant drop: I. A fast standard instrument using computer image analysis. *J. Colloid. Interface Sci.* 141:1–9.

Hendraningrat, L., Li, S., and Torsæter, T. 2013. A core flood investigation of nanofluid enhanced oil recovery. *J. Petrol. Sci. Eng.* 111:128–138.

Kosmala, A., 2012. Development of high loading Ag nanoparticle inks for inkjet printing and Ag nanowire dispersions for conducting and transparent coatings. PhD Thesis, Cranfield University, Bedford.

Masutani, G., and Stenstrom, M. K. 1984. A review of surface tension measuring techniques, surfactants and their implication for oxygen transfer in wastewater treatment plants, Water Resources Program, School of Engineering and Applied Science, University of California, Los Angeles, CA.

Moosavi, M., Goharshadi, E. K., and Youssefi, A. 2010. Fabrication, characterization, and measurement of some physicochemical properties of ZnO nanofluids. *Int. J. Heat Fluid Fl.* 31:599–605.

Sefiane, K., and Bennacer, R. 2009. Nanofluids droplets evaporation kinetics and wetting dynamics on rough heated substrates. *Adv. Colloid. Interface Sci.* 147–148:263–271.

Tanvir, S., and Qiao, L. 2012. Surface tension of nanofluid-type fuels containing suspended nanomaterials. *Nanoscale Res. Lett.* 7:226.

Thiessen, D. B., and Man, K. F. 1999. Surface tension measurement. In Webster, J. G., Ed. *The Measurement, Instrumentation, and Sensors Handbook*. Boca Raton, FL: CRC Press, pp. 31-7–31-8.

Liquid density and heat capacity

Duangthongsuk, W., and Wongwises, S. 2010. Comparison of the effects of measured and computed thermophysical properties of nanofluids on heat transfer performance. *Exp. Therm. Fluid Sci.* 34:616–624.

Feng, X., and Johnson, D. W. 2012. Mass transfer in SiO_2 nanofluids: A case against purported nanoparticle convection effects. *Int. J. Heat Mass Transf.* 55:3447–3453.

Hwang, K. S., Jang, S. P., and Choi, S. U. S. 2009. Flow and convective heat transfer characteristics of water-based Al_2O_3 nanofluids in fully developed laminar flow regime. *Int. J. Heat Mass Transf.* 52:193–199.

O'Hanley, H., Buongiorno, J., McKrell, T., and Hu, L.-W. 2012. Measurement and model validation of nanofluid specific heat capacity with differential scanning calorimetry. *Adv. Mech. Eng.* Article ID 181079.

Pak, B. C., and Cho, Y. I. 1998. Hydrodynamic and heat transfer study of dispersed fluids with submicron metallic oxide particles. *Exp. Heat Transf.* 11:151–170.

Pastoriza-Gallego, M. J., Casanova, C., Páramo, R., Barbés, B., Legido, J. L., and Piñeiro M. M. 2009. A study on stability and thermophysical properties (density and viscosity) of Al_2O_3 in water nanofluid. *J. Appl. Phys.* 106: 64301.

Pastoriza-Gallego, M. J., Lugo, L., Legido, J. L., and Piñeiro M. M. 2011. Enhancement of thermal conductivity and volumetric behavior of Fe_xO_y nanofluids. *J. Appl. Phys.* 110: 14309.

Vajjha, R. S., Das, D. K., and Mahagaonkar, B. M. 2009. Density measurement of different nanofluids and their comparison with theory. *Pet. Sci. Tech.* 27:612–624.

Xuan, Y., and Roetzel, W. 2000. Conceptions for heat transfer correlation of nanofluids. *Int. J. Heat Mass Transf.* 43:3701–3707.

CHAPTER 5

Nanofluid Forced Convection

Gilles Roy

CONTENTS

5.1 INTRODUCTION

Industrial processes involving heating and cooling of fluids flowing inside the conduits of all sorts are widespread and represent some of the most common and important processes found in engineering today (Kreith and Bohn 1997). Indeed, in thermal engineering, forced convection is probably the most effective and widely used means to transfer heat. Applications include various types of heat exchangers, heating and cooling units, impinging jets, and a multitude of other flow-induced heat transfer situations. As heat transfer is of course directly related to the fluid's thermophysical properties, the possibilities of increasing, in particular, a fluid's thermal conductivity is quite appealing. Increased demands for improved cooling and heat dissipation technologies as well as the growing need for more compact and energy-efficient thermal management systems are continuously challenging engineers to come up with innovative solutions. The high level of interest for nanofluids in recent years is directly related to its potential of becoming a next-generation heat transfer medium. Conventional coolants such as water, ethylene glycol, and oils typically have lackluster heat transfer performances. Engineers have therefore considered adding various types of solid particles in suspension in these coolants in order to provide heat transfer enhancement. Initial attempts using particle sizes in the micron range were less than successful due to considerable adverse effects. Recent advances in manufacturing technologies have made the production of particles in the nanometer scale possible. Initial studies revealed that the use of suspensions containing nanoscale particles as heat transfer mediums were very promising. In essence, high-performance coolants could provide more compact, energy-efficient heat management systems and provide more efficient localized cooling. Examples of benefits could include smaller/lighter heat transfer systems, reductions of heat transfer fluid inventories, reductions in emissions, and so on. Engineers are therefore hopeful that these special solid–liquid mixtures will offer opportunities for development and innovation in a wide range of technological sectors. Although effective thermal conductivity enhancement is very promising, it is certainly not the only factor weighing in on a nanofluid's potential as a heat transfer medium. Other important factors include other thermophysical properties such as viscosity and specific heat, as well as flow-related parameters and considerations such as particle clustering and migration, nanofluid stability, and, of course, flow conditions. Furthermore, these factors are often interdependent (Figure 5.1).

As one can see, from this somewhat simplistic perspective, the interrelated factors influencing forced convection heat transfer performance can be quite complex. It becomes quite clear that an increase in fluid thermal conductivity is not a guarantee of increased performance in a forced convection application.

This chapter will attempt to present a snapshot of the current theory, research, potential uses, and challenges of forced convection applications using nanofluids as heat transfer mediums. Further reading on the subject

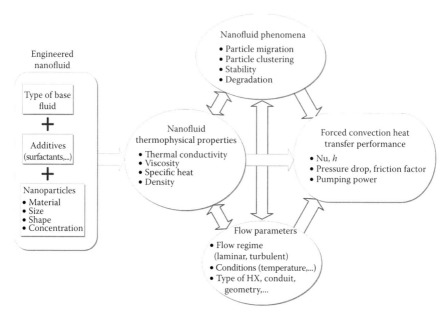

FIGURE 5.1   Factors influencing nanofluid forced convection heat transfer performance. (Adapted from Timofeeva, E. V., *Two Phase Flow, Phase Change and Numerical Modeling*, InTech, Rijeka, Croatia, 2011.)

can also be found elsewhere in this book, as distinct chapters on nanofluid flow and heat transfer in heat exchangers (Chapter 7) and microchannels (Chapter 8) present these subjects in more detail.

5.2 FORCED CONVECTION FUNDAMENTALS

It is generally recognized that heat can be transferred via three distinct modes, that is, conduction, convection, and radiation. Convective heat transfer is the typical mode found in fluids. In reality, convection is the combination of two distinct energy transfer mechanisms, specifically heat diffusion (or conduction) and advection (due to bulk fluid flow). Fluid motion results from either a density gradient (natural buoyancy forces) or a pressure difference. In the first case, the process is typically called "natural convection." In the case of a fluid flow caused by a pressure gradient, the fluid motion normally results from external surface forces produced through pumping and blowing, or in another mechanically assisted manner. This is most often called "forced convection." Of course, heat transmission in a fluid can be a combination of both natural and forced convection, which is referred to as "mixed convection." In any event, forced convection is a mode of heat transmission that is present in countless engineering processes, including various types of heat exchangers, fan-assisted cooling, impinging jets, and so on.

5.2.1 Convection Heat Transfer Coefficient

Generally speaking, the rate of heat transfer by convection between a surface and a fluid can be calculated from Equation 5.1, initially proposed by Isaac Newton (often called Newton's law of cooling):

$$\dot{Q}_c = \overline{h}_c A \left(T_w - T_{f,\infty} \right) = \overline{h}_c A \Delta T \qquad (5.1)$$

where:
$\dot{Q}_c$ is the rate of heat transfer (W)
$\overline{h}_c$ is the average convective heat transfer coefficient (W/m²K)
A is the heat transfer surface area (m²)
T_w and $T_{f,\infty}$ are, respectively, the temperature at the surface of the solid and the fluid temperature at a reference location away from the surface (K)

By rewriting Equation 5.1, the average heat transfer coefficient can be expressed as

$$\overline{h}_c = \frac{\dot{Q}_c}{A \left(T_w - T_f \right)} = \frac{q_c''}{T_w - T_{f,\infty}} \qquad (5.2)$$

where q_c'' is the heat flux at the solid–fluid interface (W/m²). One should note that this equation represents the average heat transfer coefficient. Normally, the local value is determined and the average value is then obtained through integration over the entire surface. The convective heat transfer coefficient, unlike thermal conductivity, is not a physical property of the materials involved in the heat transfer process. Rather, the value of the convective heat transfer coefficient depends on a number of parameters such as flow velocity, temperature difference, and thermophysical properties of the fluid. As the values of these parameters normally are not constant over a surface or body, the numerical value of the heat transfer coefficient will therefore also vary along the surface. In engineering applications, forced convection is typically used to increase the rate of heat exchange between a body and a fluid. As fluid temperature and velocity gradients are important factors in determining the heat transfer coefficient, an understanding of the underlying physics of fluid flow is essential in order to evaluate the heat transfer mechanisms in a fluid. The basic governing equations for fluid flow consist of a set of coupled partial derivative equations. These will be discussed briefly in Section 5.2.2.

5.2.2 Governing Equations: Conservation of Mass, Momentum, and Energy for Laminar Flow

To derive the conservation equations for laminar flow, one usually uses a control volume approach in which the rates of mass, momentum, and energy flowing into and out of the control volume are accounted for. The conservation of momentum arises from applying Newton's second law to fluid motion, whereas the conservation of energy equation represents the first law of thermodynamics. Details of the development of these equations can be found in any standard fluid mechanics textbook. Assuming that the fluid is incompressible and Newtonian and that both compression work and viscous dissipation are assumed negligible in the energy equation, the general conservation equations for steady flow can be written in vector form as follows (Warsi 1999):

- Conservation of mass

$$\mathrm{div}\left(\rho \vec{V}\right)=0 \tag{5.3}$$

- Conservation of momentum

$$\mathrm{div}\left(\rho \vec{V}\vec{V}\right)=-\mathrm{grad}\ p+\mu\nabla^2\vec{V} \tag{5.4}$$

- Conservation of energy

$$\mathrm{div}\left(\rho \vec{V}C_p T\right)=\mathrm{div}\left(k\mathrm{grad}\ T\right) \tag{5.5}$$

where:
   ρ, μ, C_p, and k are, respectively, the fluid density, viscosity, specific heat, and thermal conductivity of the considered fluid
   T, p, and $\vec{V}$ are, respectively, the temperature, pressure and velocity vector

These equations are, except for highly simplified cases, impossible to solve analytically.

A few examples of simple cases with analytical solutions include steady plane flows such as Couette–Poiseuille flows and steady axisymmetric flow such as the Hagen–Poiseuille flow in circular pipes (Schlichting 1979). Because of the complexity of determining velocity and temperature gradients analytically in more complex flow fields, other methods are required to evaluate the convective heat transfer coefficient.

5.2.3 Evaluation of the Convective Heat Transfer Coefficient

As previously mentioned, the convective heat transfer coefficient is a function of the hydrodynamic and thermal fields, including the associated velocity and temperature gradients near the fluid–wall interface. The heat transfer coefficient can be determined by five general methods (Kreith and Bohn 1997), which are described in Sections 5.2.3.1 through 5.2.3.5.

5.2.3.1 Dimensional Analysis Combined with Experiments

As exact and/or analytical solutions to most fluid flow and heat transfer problems are either extremely complicated or even impossible to achieve, engineers have made use of empirical results obtained from experiments to characterize certain physical situations. By itself, experimental data are often difficult to analyze and use in situations that are not exactly the same as the conditions in which the data were extracted. Dimensional analysis provides the engineer with a tool to develop qualitative relationships between different physical variables and quantities. These quantities can be, for example, related to fluid properties, flow conditions, and geometry. Examples of important parameters obtained by dimensional analysis are the Reynolds, Prandtl, and Nusselt numbers. The approach, by itself, does not provide solvable equations describing certain phenomena. However, provided that the parameters describing a physical situation are adequately identified, dimensional analysis will permit the development of useful dimensionless parameters, which, when combined with experimental data, can form very useful correlations that will permit the determination of other parameters such as the heat transfer coefficient.

5.2.3.2 Exact Mathematical Solutions of the Boundary Layer Equations

Exact mathematical solutions for the equations describing fluid motion and heat transfer in fluid flows are possible for a very limited number of relatively simple laminar flow systems such as those, for example, over a flat plate. In order to obtain these solutions, certain assumptions are required to significantly simplify the governing equations. Solutions typically yield velocity and temperature profiles in the boundary layer, thus permitting the determination of the heat transfer coefficient.

5.2.3.3 Approximate Analysis of the Boundary Layer Equations by Integral Methods

With the use of integral methods, one can avoid the onerous mathematical descriptions of the flow inside the boundary layer and, as such, avert solving the resulting set of partial differential equations. These methods will yield fairly accurate results for friction coefficients, heat transfer coefficients, and, ultimately, Nusselt numbers for laminar or turbulent flows. The basic procedure consists of using a simple equation (such as a polynomial function) to describe the velocity and temperature distributions inside the boundary layer. Considering an elemental control volume, the integral formulations for the conservation of momentum and energy are obtained. By substituting the polynomial functions with appropriately evaluated constants into the integral momentum and energy equations, one can evaluate parameters such as the friction and heat transfer coefficients.

5.2.3.4 Analogy between Heat and Momentum Transfer

Turbulent flows are found in most practical engineering applications involving forced convection. As rigorous mathematical representations of turbulent transfer processes are currently impossible, the analogy between heat and momentum transfer is therefore an interesting approach and can be useful for analyzing heat transfer processes in certain cases. The concepts were initially developed by Osborne Reynolds (i.e., the *Reynolds analogy*) and Ludwig Prandtl. Essentially, the method establishes an approximate relation between skin friction (momentum flux) and heat transfer (heat flux) to the wall.

5.2.3.5 Numerical Analysis

Numerical methods are widely used in fluid flow and heat transfer studies. Computational fluid dynamics (or CFD) is now an established discipline. As previously discussed, the governing equations for fluid flow are, except for very simple cases, impossible to solve analytically. Numerical methods can solve in an approximate form these exact equations of motion by solving their discretized versions (see, e.g., Patankar [1980]). The field variables (velocity, pressure, and temperature) are calculated at discrete points in space (i.e., in the flow domain of interest) and time. The accuracy of the solution is highly dependent on the number of nodes considered and the various discretization schemes used in the considered algorithm. Turbulence modeling is readily possible with numerical methods. Various numerical methods and algorithms have been developed over the years, the most popular families being finite difference methods, finite element methods, and finite volume methods (see, e.g., Pletcher, Tannehill, and Anderson [2013]). With the advent of the exponential growth in high-performance computing, highly complex problems can now be tackled by CFD.

5.2.4 Dimensionless Groups of Importance in Forced Convection

This section presents the most common dimensionless groups that are often encountered in forced convection applications such as thermal management systems.

- *Reynolds number (Re)*: The Reynolds number (Equation 5.6) is based on fluid properties (density ρ and viscosity μ), a velocity "U_∞", and a characteristic length. It is the ratio of inertial forces to viscous forces for a given flow situation. As such, it represents the relative importance of these types of forces and is commonly used to characterize different flow regimes for fluid flows. For internal flows, the Reynolds number is typically based on the hydraulic diameter D_h of the conduit and the mean flow velocity "$\bar{U}$".

$$\mathrm{Re} = \frac{\rho U_\infty L}{\mu} = \frac{U_\infty L}{\nu_f}; \qquad \mathrm{Re} = \frac{\rho \bar{U} D_h}{\mu} \qquad (5.6)$$

- *Prandtl number (Pr)*: The Prandtl number (Equation 5.7) represents the ratio of molecular momentum diffusivity (or kinematic viscosity ν) to thermal diffusivity α. Both are molecular transport properties. Essentially, the Prandtl number relates the temperature distribution to the velocity distribution. In practice, this means that for two geometrically similar configurations, similar temperature distributions will be found if both systems have the same Reynolds and Prandtl numbers.

$$\mathrm{Pr} = \frac{C_p \mu}{k} = \frac{\nu}{\alpha} \qquad (5.7)$$

- *Nusselt number (Nu)*: The Nusselt number (Equation 5.8) is also known as the dimensionless heat transfer coefficient and is, as such, an essential parameter in forced convection applications. It represents the ratio of convection heat transfer to conduction in a fluid layer of thickness L. Indeed, in equation 5.8, "$\overline{h_c}$" represents the average convective heat transfer coefficient and "k_f" is the fluid thermal conductivity. In conduit flow, the characteristic length is typically the hydraulic diameter. In forced convection applications, the Nusselt number is usually expressed as a function of both Reynolds and Prandtl numbers, $Nu = f(Re, Pr)$. A quick review of literature will reveal that empirical correlations of this type for a wide range of flows with heat transfer are available. One such example is the Dittus–Boelter correlation for pipe flow.

$$Nu = \frac{\overline{h_c} L}{k_f} \tag{5.8}$$

- *Peclet number (Pe or Pé)*: The Peclet number is the product of Reynolds and Prandtl numbers ($Pe = Re \cdot Pr$). By developing the Reynolds and Prandtl numbers in the Peclet number, one will see that it represents the ratio of the heat transport by convection to the heat transport by conduction.

- *Grashof number (Gr)*: The Grashof number (Equation 5.9) represents the ratio of the buoyancy forces to the viscous forces present in a fluid flow situation with heat transfer. It is often used in studies with natural or mixed convection. In equation 5.9, "g" is the acceleration due to gravity, "L" is a characteristic length, "β" is the coefficient of expansion of the fluid, "$T_s - T_\infty$" is the temperature difference and "v" is the fluid kinematic viscosity.

$$Gr = \frac{g\beta(T_s - T_\infty)L^3}{v^2} \tag{5.9}$$

- *Friction factor (f)*: The friction factor (or *Darcy friction factor*) (Equation 5.10) represents the dimensionless pressure drop for internal flows through conduits. This coefficient is a function of factors such as the Reynolds number and, in the case of turbulent flow, the relative roughness of the pipe. In equation 5.10, Δp is the pressure drop, L and D are respectively the tube length and diameter, ρ is the fluid density and U_m is the mean velocity.

$$f = \frac{\Delta p}{(L/D)(\rho U_m^2 / 2)} \tag{5.10}$$

5.3 A BRIEF INTRODUCTION TO PARTICULATE FLOWS

Engineers have been interested in particulate flows for well over a century. The wide range of applications consisting of suspended particles in a fluid flow has led to numerous studies on the subject. One can think of engineered products such as aerosols, paints, papers, cements, slurries, and fluidized beds. In general terms, particulate flows are two-phase fluid flows, in which one phase is referred to as the continuous or carrier phase and the other phase consists of small, immiscible particles and is named the "dispersed" phase. Particle density is usually different than that of the carrier phase. The concentration of particles is often referred to as "particle loading" and this IS usually expressed as a volume fraction or a mass fraction. Generally speaking, if the particle loading is small, it is often assumed that the dynamics of the particle phase is influenced by the fluid; however, the effects of the particles on the fluid are negligible. As such, particle properties such as velocity and temperature are influenced by the carrier phase, but the reverse is not the case. This is often referred to as *one-way coupling*. If the particle loading becomes more important, both constituents influence each other and the resulting flow is then considered to be *two-way coupled*. An example of a two-way coupled situation is where turbulence modifies particle behavior, which, in turn, modifies the turbulence (Gouesbet and Berlemont 1999).

5.3.1 Particulate Flow Special Characteristics

Particulate flows have special characteristics such as hydrodynamic and colloidal forces that may or may not be present in nanofluids but are nevertheless important factors that should be considered as some are thought to have effects on heat transfer enhancement. The following nonexhaustive list of special features is found in particulate flows. Due to limited space, the reader is invited to consult other references for detailed descriptions of these factors (see, e.g., Chandrasekar and Suresh [2009] and Das et al. [2008]):

- *Fluid–particle interaction:* The dynamics of particles in suspension in a base fluid can be quite complex, especially in turbulent flow. Of course, important factors include particle size (Kolmogorov scale) and density as well as particle concentration. Particle size and density difference between the particles and the carrier fluid will have a great effect on particle inertia and will, of course, dictate to what extent the particles follow the variations in the continuous phase. Furthermore, the presence of particles may enhance or attenuate turbulence in a particulate flow (Tanaka and Eaton 2008).

- *Particle–particle and particle–wall collision:* Fundamental knowledge of how particles behave when entering in collision with walls or with other particles is also an important area of current research. Indeed, such topics as the influence of lubrication fluid forces, London–van der Waals attraction forces, and electrical double-layer repulsive forces on particle motion and surface deformation are of particular interest. Knowledge of the mechanisms on how particles are attracted, deform, and rebound is of interest, for example, for predicting whether particles adhere together of whether they rebound following an impact.

- *Brownian motion:* Brownian motion is essentially the random movement of particles suspended in a liquid or gas under the action of collisions with surrounding molecules in the fluid. It is well established that the intensity of these chaotic motions increases with temperature. A link with the temperature effect on thermal conductivity is therefore natural. Brownian motion is considered by many as a particularly important factor in the thermal enhancement of nanofluids (others, however, disagree).

- *Thermophoresis, Soret, and Dufour effects:* Thermophoresis and the Soret effect are phenomena that are associated with temperature gradients, whereas the Dufour effect is an energy (or heat) flux created by a chemical potential gradient. Basically, thermophoresis is observed when solid particles suspended in a fluid experience a force in the direction opposite to the imposed temperature gradient. Likewise, the Soret effect in a liquid mixture is associated with the separation of both components under the influence of a temperature gradient, thus creating a concentration gradient. The Dufour effect is, in essence, the reciprocal phenomenon to the Soret effect.

- *Shear lift force:* Particles placed in suspension in shear flows experience lift forces perpendicular to the direction of the flow. Such forces in fluid flows have been studied by quite a few authors (see, e.g., Auton [1987] and Saffman [1965]). Of interest, the study of flow fields around small particles moving and rotating in shear flows close to walls by considering drag and lift forces on the particles can lead to interesting insight into the formation of particle bridges and clustering (Poesio et al. 2006).

5.3.2 Lagrangian versus Eulerian Approaches

Particulate flow modeling is typically done by considering one of two families of approaches (or frames of reference), either Lagrangian or Eulerian.

In the Eulerian approach (static frame), the dispersed phase is treated as a continuum. As a result, the problem is treated as if there were two distinct but interacting "fluids" present (or *two-fluid model*) and, as such, the appropriate continuum equations for the fluid and particles phases are solved. The term "Eulerian–Eulerian approach" is often used for cases where the fluid and the particles are both treated in the Eulerian frame. This approach is most often used when higher particle volume fractions are encountered. However, in a Lagrangian approach (frame moving with particle), the particulate phase is treated as single particles for which the trajectories are predicted inside the continuous phase. These predictions result from applying external forces on the particles and then solving for acceleration, velocity, and position. The Eulerian–Lagrangian approach is often used when the particle is tracked in the Lagrangian frame and the fluid in the Eulerian frame. This approach is typically only used when a low particle volume fraction is considered as computational times become quite important. Detailed discussions on these methods can be found elsewhere (see, e.g., Durst, Milojevic, and Schönung [1984] and Gouesbet and Berlemont [1999]).

5.3.3 Heat Transfer in Particulate Flows

Particulate flows with heat transfer are found in quite a few applications. For example, in the case of fluidized beds, a better understanding of the parameters that control the interaction mechanisms between both phases may permit engineers to increase the efficiency of these processes (Avila and Cervantes 1995). Most certainly, these interaction mechanisms between phases are complex and are functions of several parameters such as thermophysical properties of both phases, velocity, turbulence intensity, particle size, and particle loading. Considerable amounts of early research on particulate flows considered solid dispersed particles in gases (see, e.g., Avila and Cervantes [1995], Boothroyd and Haque [1970], Michaelides [1986], and Murray [1994]). General conclusions from these studies were as follows:

- The wall heat transfer coefficient and Nusselt number may either increase or decrease, depending on certain parameters such as particle loading, particle size, and flow conditions.

- The presence of solid particles in a gas flow seems to interfere substantially with the turbulence generation in the flow.

- The presence of solids increases the frictional pressure drop by a greater percentage than the increase in wall heat transfer coefficient.

More recently, relatively new CFD techniques such as direct numerical simulation have been used extensively for the determination of aerodynamic and thermal fields of suspension flows consisting of gas and solid particles (Dan and Wachs 2010; Tavassoli et al. 2013).

Studies on heat transfer characteristics of suspensions consisting of solid particles in liquids have also been considered. In their study of effective thermal conductivities in disperse two-phase (solid-in-liquid) mixtures, Sohn and Chen found that in shear flows, the mixtures exhibited a significant increase in effective thermal conductivity at higher Peclet numbers (Ped > 300) as a result of microconvection (Sohn and Chen 1981). For mixtures, they used 2.9 mm beads in a mixture of silicone oil and kerosene and 0.3 mm polystyrene particles in a mixture of silicone oil and Freon-113. Ahuja (1975a, 1975b, 1982) studied the effects of particle suspensions in laminar flow. In his body of work, he considered 50–100 μm diameter polystyrene spheres in aqueous NaCl or glycerine fluid phases. According to his experimental results, heat transport was seen to increase by as much as 200% with the particles. Important parameters that were found to influence heat transfer performance were particle size, particle loading, tube diameter, fluid phase physical properties, and so on. When considering the suspensions in a heat exchanger, the Nusselt number increased by 82% for a 100 μm diameter particle suspension with a volume fraction of 4.64%.

Generally speaking, although suspensions consisting of millimeter- or micrometer-sized particles have interesting heat transfer enhancements, their use as coolants in practical engineering applications never really materialized as they were considered undesirable due to rapid sedimentation, pipeline erosion, channel clogging, and important pressure drops. With the advent of new manufacturing technologies in the late 1980s and the early 1990s, the production of nanoscale particles became technically possible and the interest in solid-liquid mixtures for cooling was rekindled.

5.4 NANOFLUID FORCED CONVECTION MODELING

From a theoretical standpoint, suspensions consisting of liquids and dispersed ultrafine particles (nanoparticles) represent a relatively new and technologically interesting sector. Developing new theories and models that would help predict nanofluid flow and heat transfer behaviors in engineering applications are complex challenges for researchers. Indeed, according to some, it appears quite difficult to formulate theories that could reasonably predict the flow and heat transfer of nanofluids by considering it as a multicomponent fluid (Drew and Passman 1999). However, because a nanofluid is by nature a two-phase fluid, one could expect that it will most certainly possess some common features with conventional solid–fluid mixture behaviors and theories briefly discussed in Section 5.3. Due to the minute size of the particles, certain interesting phenomena are believed to present in nanofluids but are not found in conventional mixtures containing larger sized particles. These include thermal dispersion, intermolecular energy exchange, and liquid layering on the solid–liquid interface as well as phonon effects on the heat transport inside the particle itself. Such phenomena have been under considerable investigation for the past 20 years and more insight into these features can be found in Chapters 1 and 3. As a general consensus does not seem to currently exist on what is the best approach to model nanofluid flows with heat transfer, a general discussion on the current approaches and their limitations is presented in Section 5.4.1.

5.4.1 Single-Phase versus Two-Phase Approaches

As previously discussed, some believe that, because of the extremely small particle sizes, traditional two-phase theories developed for larger sized particle suspensions are not appropriate for nanofluids. However, it seems logical that they will at least possess some common features with conventional two-phase theories. Therefore, generally speaking, two approaches to nanofluid forced convection modeling have been used and sometimes compared by various research groups. Essentially, some have treated nanofluids as single-phase fluids with "effective" thermophysical properties, whereas others have considered them as two component mixtures using various two-phase modeling approaches. As most nanofluids considered for practical heat transfer applications are typically composed of ultrafine particles, it is conceivable that these may be easily fluidized. Consequently, by assuming negligible motion slip between the particles and the continuous phase and that thermal equilibrium conditions prevail between the two components, the nanofluid may then be considered as a conventional single-phase fluid, with effective thermophysical properties being the function of the properties of both constituents and their respective concentrations (Pak and Cho 1998; Xuan and Roetzel 2000). If one views such an assumption as feasible, the classic theories developed for conventional single-phase fluids can then be applied to nanofluids as well. Thus, all the equations of conservation (mass, momentum, and energy) governing single-phase fluids can be directly extended and employed for nanofluids using effective thermophysical properties. This approach has been used quite extensively in numerical research studies to date. Indeed, most of the early work on nanofluid forced convection used this approach. The most challenging aspect of using a single-phase fluid approach is specifying adequate effective thermophysical properties. The determination and use of these effective properties will be discussed in Section 5.5.

In the case of two-phase modeling, as discussed previously, there are generally two approaches that can be used. Essentially, for mixtures containing low particle loadings, the Lagrangian–Eulerian approach is typically used. In this approach, the mixture is modeled using an Eulerian frame for the base fluid and a Lagrangian frame for the particles. For higher volume fractions of particles, the approach most often used is the Eulerian–Eulerian. In the particular case of nanofluids, even for a small volume fraction, the number of particles will be extremely large due to their minute sizes. Therefore, the use of a Lagrangian–Eulerian approach becomes unpractical in a computational sense. As a result, Eulerian–Eulerian approaches are most-often used for nanofluid two-phase flow modeling. Several Eulerian–Eulerian models exist, including the "volume of fluid" (VOF), the mixture model, and the Eulerian model. These are typically incorporated in popular commercially available CFD packages such as FLUENT. The VOF model was designed for two or more immiscible fluids, in which a single set of momentum equations is shared by all fluids (or phases). The volume fraction of each component is tracked over the entire computational domain by solving a continuity equation for the secondary phase (the sum of volume fractions of all phases equals unity). The mixture model was specifically designed for two or more phases. This model is adequate for low volume fraction particulate flows. It solves the general conservation equations for the mixture as well as a volume fraction equation for the secondary phases. It then prescribes relative velocities to describe the dispersed phases. Finally, the Eulerian model is a more complex multiphase model. Although the pressure is shared by all phases, it solves a separate set of momentum, continuity, and energy equations for each phase. Momentum exchange between the phases is dependent on the type of mixture being modeled. More information on these two-phase modeling approaches used in nanofluid forced convection problems can be obtained in various references, including Akbari, Galanis, and Behzadmehr (2011) and Corcione, Habib, and Quintino (2013). Several research groups have made comparisons between single-phase and two-phase approaches. One of the first numerical studies to consider two-phase modeling for nanofluids was conducted by Behzadmehr, Saffar-Avval, and

Galanis (2007). In their work, they considered turbulent forced convection of a water—1% volume copper nanofluid using a two-phase mixture model. For comparison purposes, they also considered the same problem using a single-phase approach with constant weighted average properties of the nanofluid. The results obtained by both modeling techniques were then compared to available experimental data and the authors concluded that the mixture model was more precise than the single-phase model. Bianco et al. (2009) used a "discrete particle model" for two-phase modeling to investigate laminar forced convection flow of a water–Al_2O_3 nanofluid in a circular tube. A comparison of results obtained using single-phase modeling found that results were quite similar, with a maximum difference of 11% in the average heat transfer coefficient between the two approaches. The authors found that the results compared even better at higher particle volume fractions (4%). Turbulent forced convection of a water–Al_2O_3 nanofluid in a horizontal tube was considered by Lotfi, Saboohi, and Rashidi (2010). For their analysis, they considered two different two-phase models (Eulerian and mixture) as well as a single-phase approach. Results were also compared with experimental data by Wen and Ding (2004). They found that the mixture model was more precise than the other two models. In the case of laminar mixed convection, for cases with low particle volume fraction, a two-phase approach was found to give better agreement with experimental data (Akbari, Galanis, and Behzadmehr 2011). However, the rate of increase of the heat transfer coefficient with particle volume fraction predicted by the two-phase models was not consistent with experimental data. In their comprehensive study of nanofluid turbulent forced convection, Akbari, Galanis, and Behzadmehr (2012) concluded that single-phase modeling can lead to not only satisfactory but even better results compared to two-phase models. They compared the single-phase approach with three different two-phase models (VOF, mixture, and Eulerian) for turbulent forced convection of water-based nanofluids in a long horizontal tube with experimental results reported by Syam Sundar and Sharma (2010) (Al_2O_3 nanoparticles) and Xuan and Li (2003) (Cu nanopoarticles). In their conclusions, they stated that although the three two-phase models yielded essentially the same results, they found that they considerably overestimated the thermal fields compared to available experimental data. The differences were also found to increase with particle volume fraction. Results obtained with the single-phase approach were, however, compared rather favorably with experimental data. Sample results from this study are presented in Figure 5.2.

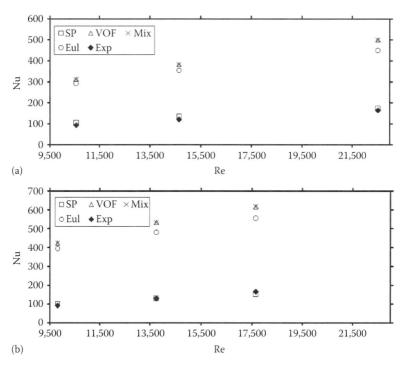

FIGURE 5.2 Comparison of calculated Nusselt numbers by various single- and two-phase models with experimental data from Xuan and Li for a water–Cu particle nanofluid: (a) φ = 1%; (b) φ = 2%. Eul, Eulerian; Mix, mixture. SP, Single-phase; Exp, Experimental; VOF, Volume of fluid. (Reproduced from *Int. J. Heat Fluid Fl.*, 37, Akbari, M. et al., Comparative assessment of single and two-phase models for numerical studies of nanofluid turbulent forced convection. 136–146, Copyright 2012, with permission from Elsevier Masson SAS.)

Hejazian and Moraveji (2013) considered turbulent forced convection tube flow with nanofluids consisting of TiO_2 nanoparticles. In the study, they compared the single-phase approach with a two-phase mixture model and compared the results with available experimental data obtained by Sajadi and Kazemi (2011). Results show that both approaches compared favorably with experimental data, although the authors stated that the two-phase approach yielded better results.

5.4.2 Concluding Remarks with Respect to Nanofluid Forced Convection Modeling

As one will easily assess from the preceding discussions, no general consensus has yet been made on the preferred modeling scheme for nanofluid forced convection problems. Generally speaking, two-phase models seem to give better results in certain cases, but in other situations, they were found to differ considerably from available experimental data. Other studies confirmed that the simplest approach based on the single-phase assumption can lead to satisfactory results, provided adequate nanofluid properties are specified. This highly important question for nanofluid forced convection modeling is therefore discussed next.

5.5 MODELING NANOFLUID EFFECTIVE PROPERTIES

Obviously, the choice of nanofluid effective properties will have considerable effects on the obtained results for any forced convection problem. Indeed, contradicting results by various research groups can, at least to a certain extent, be blamed on the dispersion of experimental data on such properties as nanofluid effective thermal conductivity and viscosity. Furthermore, the multitude of theories and correlations that have been used and developed for these thermophysical properties will undoubtedly yield results, predictions, and conclusions that can be strikingly different. This issue has been discussed and illustrated quite convincingly by some authors (see, e.g., Mansour, Galanis, and Nguyen [2007]). Although nanofluid effective properties have been discussed in Chapters 1 and 3, a brief discussion on the approaches that various authors have or are presently using to model nanofluid effective properties in their studies of nanofluids in forced convection applications is deemed important. Properties such as density and heat capacity are most commonly based on the basic mixing theory for a two-component mixture, whereas nanofluid effective thermal conductivity and viscosity are more complex to accurately model. Various approaches have been used by different authors; these include using relationships developed over the past century for conventional solid–liquid mixtures, simple correlations developed strictly from experimental data obtained for specific types of nanofluids (e.g., as a function of particle volume fraction only), or correlations encompassing various important parameters (e.g., temperature, particle size, particle volume fraction, etc.). As more knowledge is gained on the behavior of nanofluids, more adequate correlations encompassing various factors will be developed.

5.5.1 Density

Nanofluid density is typically obtained by measuring the volume and weight of the mixture. The particle volume fraction ϕ can be estimated knowing the densities of both constituents (Pak and Cho 1998):

$$\rho_{nf} = (1-\phi)\rho_{bf} + \phi\rho_p \qquad (5.11)$$

By rearranging the above equation, one can therefore determine the volume fraction ϕ of the mixture (Equation 5.12). Nanoparticle densities are usually values obtained in the available literature. This approach has been used by most nanofluid researchers to date.

$$\phi = \frac{\rho_{nf} - \rho_{bf}}{\rho_p - \rho_{bf}} \tag{5.12}$$

5.5.2 Specific Heat

In some cases, considering nanofluid convective heat transfer flows, it was suggested that nanofluid specific heat be calculated using the following equation based on the simple mixing theory (see, e.g., Hwang et al. [2007], Jang and Choi [2007], Maiga et al. [2005], and Pak and Cho [1998]):

$$C_{p,nf} = (1 - \phi)C_{p,bf} + \phi C_{p,p} \tag{5.13}$$

Others, however, have used an alternate approach based on heat capacity concept (can be deduced from the first law of thermodynamics). Some have stated that this approach is more accurate (see, e.g., Buongiorno [2006]; Manca, Nardini, and Ricci [2012]; and Xuan and Roetzel [2000]):

$$(\rho C_p)_{nf} = (1 - \phi)(\rho C_p)_{bf} + \phi(\rho C_p)_p \tag{5.14}$$

$$C_{p,nf} = \frac{\phi(\rho C_p)_p + (1 - \phi)(\rho C_p)_{bf}}{\phi \rho_p + (1 - \phi)\rho_{bf}} \tag{5.15}$$

Both these formulations will, of course, produce different results for specific heat. Available experimental results for nanofluid specific heat seem to confirm that the heat capacity concept does indeed yield more accurate predictions (Barbes et al. 2013; Zhou and Ni 2008).

5.5.3 Thermal Conductivity

Most researchers seem to agree that various theories and correlations developed for conventional solid–fluid flows are not appropriate for nanofluids. A multitude of theories on the mechanisms responsible for the (according to some) anomalous increases in heat transfer capabilities of nanofluids and their corresponding relationships and correlations have been developed over the past 20 years. However, there does not seem to be a consensus on what is the most appropriate or accurate approach. For example, in their literature review, Wang and Mujumdar (2008a) identified over 20 different thermal conductivity models that are/could potentially be used for nanofluids. Undoubtedly, this number has considerably increased since then. Interestingly, some studies have shown that relationships originally derived for conventional liquid–solid particle mixtures appear to be reasonably adequate for liquid–nanoparticle mixtures, especially for low particle volume concentrations. Indeed, Maxwell's original equation for particulate suspensions (Maxwell and Niven 1881) as well as subsequent extensions and developments, including the Hamilton–Crosser correlation (Hamilton and Crosser 1962), has been used by various authors for numerical modeling of confined flows using nanofluids:

$$\frac{k_{\text{eff}}}{k_{\text{f}}} = \frac{k_{\text{p}} + (n-1)k_{\text{f}} - (n-1)\phi(k_{\text{f}} - k_{\text{p}})}{k_{\text{p}} + (n-1)k_{\text{f}} + \phi(k_{\text{f}} - k_{\text{p}})} \tag{5.16}$$

where n is the empirical shape factor given by $n = 3/\psi$. For spherical particles ($\psi = 1$), the Hamilton–Crosser equation reduces to the following equation (see also Wasp [1977]):

$$\frac{k_{\text{eff}}}{k_{\text{f}}} = \frac{k_{\text{p}} + 2k_{\text{f}} - 2\phi(k_{\text{f}} - k_{\text{p}})}{k_{\text{p}} + 2k_{\text{f}} + \phi(k_{\text{f}} - k_{\text{p}})} \tag{5.17}$$

In various studies on convective nanofluid flows, this equation is often expressed, for example, for water–Al_2O_3 nanofluids:

$$k_{\text{nf}} = k_{\text{bf}}\left(4.97\phi^2 + 2.72\phi + 1\right) \tag{5.18}$$

Several authors found that results obtained with this approach can give results in relative agreement with their experimental results, at least in certain conditions (Barbes et al. 2013; Mintsa et al. 2009; Xuan and Li 2000). The Hamilton–Crosser equation and derivatives have been used quite extensively, especially for early work on forced convection problems, in quite a few studies (see, e.g., Akbarinia and Behzadmehr [2007] and Khanafer, Vafai, and Lightstone [2003]). Visibly, however, the most notable drawbacks of the Hamilton–Crosser model (and others of the same type) are that important physical parameters such as temperature and particle size are not considered as well as other important phenomena. As such, other authors used various expressions developed over the past few years based on various experimental data sets. For example, some authors used experimental data obtained in-house in order to better compare numerical results with the corresponding experimental data. One such case is presented in the study on confined nanofluid radial flow by Gherasim et al. (2011), in which the data obtained by Mintsa et al. (2009) was used for nanofluid effective thermal conductivity.

Some of the most recent numerical research projects have used multivariable correlations, which can be used to predict nanofluid effective thermal conductivity. These include those developed by various groups (see, e.g., Chon et al. [2005] and Koo and Kleinstreuer [2004]). The Chon et al. correlation has been used by Nguyen, Roy et al. (2007) in their study of heat transfer enhancement using nanofluids in an electronic liquid cooling system and by Abu-Nada (2009) in his study on heat transfer enhancement in natural convection. The correlation developed by Corcione (2011a) is similar in structure to the one developed by Chon et al. and has also been used by some authors as it is based on a comprehensive set of experimental data obtained by several research groups:

$$\frac{k_{eff}}{k_f} = 1 + 4.4 Re_p^{0.4} \; Pr_f^{0.66} \left(\frac{T}{T_{fr}}\right)^{10} \left(\frac{k_s}{k_f}\right)^{0.03} \phi^{0.66} \tag{5.19}$$

where:

Re_p is the Reynolds number of the nanoparticle
Pr_f is the Prandtl number of the base fluid
T is the temperature of the nanofluid
T_{fr} is the freezing point of the base liquid
k_s is the thermal conductivity of the solid nanoparticles
ϕ is the nanoparticle volume fraction

Re_p and Pr_f are defined as follows:

$$Re_p = \frac{2\rho_{bf} k_b T}{\pi \mu_{bf}^2 d_p}; \quad Pr_f = \frac{C_{p,bf} \mu_{bf}}{k_{bf}} \tag{5.20}$$

where k_b is the Boltzmann constant ($k_b = 1.38066 \times 10^{-23}$ J/K). Typically, the base fluid properties in these equations are temperature dependent. This approach has been used in the studies of Corcione (2011b) and Roy et al. (2012).

5.5.4 Dynamic Viscosity

Several rheological studies have determined that at least certain types of nanofluids behave as Newtonian fluids for varying ranges of particle volume fractions (see, e.g., Halelfadl et al. [2013]; Heris, Etemad, and Esfahany [2006]; and Kulkarni, Das, and Patil [2007]). Generally speaking, a fluid is considered "Newtonian" if its viscosity is independent of shear rate. In probably the first investigation interested in this question, Pak and Cho (1998) considered 13 nm Al_2O_3 and 27 nm TiO_2 particles suspended in water. From their study, they determined that the viscosities of the dispersed fluids with Al_2O_3 and TiO_2 particles at a 10% volume concentration were, respectively, approximately 200 and 3 times greater than that of water. Furthermore, they found that mixtures containing Al_2O_3 and TiO_2 particles showed shear-thinning behavior at or above the volume concentration of, respectively, 3% and 10%. Similarly, Das et al. (2003) found that in their mixtures containing Al_2O_3 nanoparticles dispersed in water, the suspensions showed Newtonian behavior up to 4%. In the investigations carried out by Kulkarni et al. (2007), results revealed that for nanofluids consisting of copper oxide nanoparticles dispersed in a 60:40 propylene glycol–water mixture, the fluids exhibited a Newtonian behavior for the considered range of $0 \leq \phi \leq 6\%$.

The effective dynamic viscosity of nanofluids can be approximated by various means. On one hand, it can be calculated using several existing theoretical formulas that have been derived for two-phase mixtures. One such relationship was proposed by Brinkman based on Einstein's equation for viscous fluids containing diluted, small, rigid, and spherical particles in suspension (Brinkman 1952):

$$\frac{\mu_{nf}}{\mu_{bf}} = \frac{1}{(1-\phi)^{2.5}} \tag{5.21}$$

This formulation was commonly used in the early work on nanofluids (see, e.g., Chein and Huang [2005] and Khanafer, Vafai, and Lightstone [2003]). In his theoretical analysis, Batchelor (1977) considered the effect of Brownian motion of particles on the bulk stress of an approximately isotropic suspension of rigid and spherical particles by developing the following relationship:

$$\frac{\mu_{nf}}{\mu_{bf}} = 1 + 2.5\phi + 6.5\phi^2 \tag{5.22}$$

In general, both these formulas have been found to be somewhat adequate in some cases for very low particle concentrations (i.e., $\phi \leq 1\%$), although they seem to considerably underestimate this property compared to experimental data for cases with more important particle concentrations (Kulkarni, Das, and Patil 2007; Murshed, Leong, and Yang 2008; Pak and Cho 1998). Other developments for the viscosity of fluids containing suspended solid particles can be found in the works of Graham (1981) and Lundgren (1972). It should be noted that these correlations are for the general case of conventional fluids with suspended solid particles and were not specifically developed for nanofluids. The applicability of these formulas is therefore questionable. In order to get around the lack of developed theory that could accurately represent nanofluid effective viscosity for numerical forced convection studies, Maïga et al. (2004) performed a least-squares fitting of available experimental data. The data were collected from early studies on nanofluids (Choi 1995; Lee et al. 1999; Masuda et al. 1993). For a water–Al_2O_3 nanofluid, the corresponding result is given as follows:

$$\frac{\mu_{nf}}{\mu_{bf}} = 123\phi^2 + 7.3\phi + 1 \tag{5.23}$$

As the above equation does not consider other important effects such as temperature, Palm, Roy, and Nguyen (2006) performed curve fitting of the temperature-dependent data published by Putra, Roetzel, and Das (2003). The results presented by them illustrated the importance of considering fluid temperature in numerical simulations. As the temperature increases, so does the thermal conductivity. However, the effective viscosity decreases with temperature. Others have investigated rheological properties of nanofluids "in-house" and have used the collected data in their numerical models (see, e.g., the forced convection and rheological studies of Namburu, Kulkarni, Dandekar et al. [2007]; Namburu, Kulkarni, Misra et al. [2007]; and Namburu et al. [2009]). In the numerical work by Gherasim et al. (2011) on nanofluid heat transfer and fluid flow in a confined radial flow cooling system, the authors used viscosity data obtained in-house (data published in Nguyen, Desgranges et al. 2007). The same nanofluids were used in their experimental work on radial flow systems (Gherasim et al. 2009) and therefore more accurate numerical–experimental comparisons were made possible. As in the case of the effective thermal conductivity of nanofluids, some of the most recent numerical studies have used multivariable correlations to predict nanofluid effective viscosity. The correlation developed by Corcione (2011b) is one such equation, Equation 5.24. This correlation is also based on a comprehensive set of experimental data by various authors.

$$\frac{\mu_{eff}}{\mu_f} = \frac{1}{1 - 34.87(d_p/d_f)^{-0.3}\phi^{1.03}} \tag{5.24}$$

where d_f is the equivalent diameter of a base fluid molecule:

$$d_f = 0.1\left(\frac{6M}{N\pi\rho_{f0}}\right) \tag{5.25}$$

where:
M and N are, respectively, the molecular weight of the base fluid and the Avogadro number (6.022×10^{-23} mol^{-1})
ρ_{f0} is the mass density of the base fluid calculated at $T_0 = 293$ K

5.5.5 General Concluding Remarks Regarding Nanofluid Effective Properties

The determination of adequate effective thermophysical properties for a considered nanofluid remains one of the critical issues when modeling nanofluid forced convection problems. Incredible dispersion in experimental results as well as a plethora of correlations and theoretical expressions developed for multiple types of nanofluids obtained by various research groups has resulted in numerous approaches and, ultimately, different results obtained for nanofluid forced convection problems. It therefore becomes imperative for researchers involved in this interesting field of work to adequately justify the assumptions and models used in their forced convection studies.

5.6 NANOFLUID FORCED CONVECTION APPLICATIONS—A REVIEW

A review of literature on nanofluid research will illustrate that the bulk of the initial work was on the determination of thermophysical properties and on the theories behind the important enhancements in thermal conductivity. The interest in this initial fundamental research was most certainly because of the potential that these specially engineered fluids had with regard to practical applications as high-performance coolants. With promising findings, the number of studies on forced convection with nanoparticle suspensions has grown exponentially over the past 5–10 years. Forced convection applications using nanofluids currently considered by engineers and researchers include various types of flow configurations. Of particular importance, nanofluid flow and heat transfer in circular tubes, heat exchangers, and microchannels are arguably the types that have generated the most attention and are therefore considered in distinct chapters in this book. Thus, the focus of this section will be on other types of applications.

5.6.1 The Genesis of Nanofluid Forced Convection

Among the earliest published results on nanofluid forced convection, the work by Pak and Cho (1998) considered water-based nanofluids with Al_2O_3 and TiO particles in turbulent flow in a circular pipe. Results showed that the Nusselt number for the dispersed fluids increased with particle volume concentration as well as with Reynolds number. However, under a condition of constant average velocity, they noted that the heat transfer coefficient of the nanofluid was 12% smaller than that of pure water. They therefore recommended that a better selection of particles having higher thermal conductivity and larger sizes be considered. For the types of nanofluids and ranges considered ($0 < \phi < 3\%$, $10^4 < Re < 10^5$, and $6.54 < Pr < 12.33$), Pak and Cho (1998) propose the following correlation for the Nusselt number. One can note the resemblance to the Dittus–Boelter equation for conventional fluids:

$$Nu = 0.021Re^{0.8}Pr^{0.5} \tag{5.26}$$

The work by Xuan and Li (2003) also provided empirical correlations for computing Nusselt numbers in laminar and turbulent tube flows using water-based nanofluids consisting of Cu and TiO_2 nanoparticles (Equations 5.27 and 5.28). These correlations are valid for concentrations of up to 2%. Results show an increase in heat transfer performance with an increase in particle volume fraction for the same Reynolds numbers.

For laminar flow,

$$Nu_{nf} = 0.4328\left(1.0 + 11.285\phi^{0.745}Pe_d^{0.218}\right)Re_{nf}^{0.333}Pr_{nf}^{0.4} \tag{5.27}$$

For turbulent flow,

$$Nu_{nf} = 0.0059\left(1.0 + 7.6286\phi^{0.6886}Pe_d^{0.001}\right)Re_{nf}^{0.9238}Pr_{nf}^{0.4} \tag{5.28}$$

Investigations into the convective heat transfer for a laminar water–Al_2O_3 nanofluid flowing inside a copper tube were conducted by Wen and Ding

(2004). Again, considerable heat transfer enhancements were found. Furthermore, these enhancements were significantly more important in the entrance region. Wen and Ding believe that nanoparticle migration is the most probable reason for this localized enhancement. They also considered the effect of nanoparticle suspensions on the heat transfer inside micro- or minichannels (Wen and Ding 2005). In the work, particle migration resulted in a significant nonuniformity in particle concentration inside the minichannel cross section. They found that their developed model considering particle migration yielded higher Nusselt numbers compared to cases considering a constant thermal conductivity. Others also considered the use of nanofluids in microchannels (Jang and Choi 2006). Their results show that, for a fixed pumping power, the cooling performance of a microchannel heat sink with water-based nanofluids containing diamond (1 vol.%, 2 nm) is enhanced by 10% compared to the results obtained with pure water.

Early numerical work on the application of nanofluids in tube flow has shown that considerable heat transfer enhancements are possible in both applications (Maiga et al. 2005, 2006). The nanofluids that were considered consisted of Al_2O_3 nanoparticles in water or ethylene glycol. In these early studies, the authors used the single-phase approach using the Hamilton–Crosser equation for effective thermal conductivity and a curve-fitting equation based on the available experimental data for effective viscosity. In the specific case of tube flow, the Maiga et al., presented the following correlations:

For laminar flow, constant wall heat flux,

$$Nu_{nf} = 0.086 Re_{nf}^{0.55} Pr_{nf}^{0.5} \tag{5.29}$$

For laminar flow, constant wall temperature,

$$Nu_{nf} = 0.28 Re_{nf}^{0.35} Pr_{nf}^{0.36} \tag{5.30}$$

For turbulent flow,

$$Nu_{nf} = 0.085 Re_{nf}^{0.71} Pr_{nf}^{0.35} \tag{5.31}$$

Other uses of nanofluids in confined flow situations are also considered in early studies, in which graphite nanoparticles in water inside a horizontal tube heat exchanger are considered (Yang et al. 2005), as well as laminar flow in micro heat sinks (Koo and Kleinstreuer 2005). In their study of heat transfer and flow behavior of aqueous suspensions of titanate nanotubes, Chen et al. (2008) found that, despite a rather small thermal conduction enhancement, the enhancement of the convective heat transfer coefficient is much more important. This seems to indicate that the enhancement in convective heat transfer seemingly cannot be attributed solely on enhancement in thermal conductivity. This was also observed by other authors, such as Wen and Ding (2004) and Xuan and Li (2003). The enhancement in convective heat transfer was found to be dependent on nanotube concentration, the Reynolds number, and the axial position. They also noticed that the highest enhancement was observed at the entrance region. Furthermore, in comparison with nanofluids containing spherical titanium nanoparticles under similar conditions, the enhancement of both thermal conductivity and convective heat transfer coefficient of the titanate nanotube nanofluids is considerably higher, thus indicating the importance of particle geometry on heat transfer enhancement. Cases considering carbon nanotube (CNT)-based nanofluids in confined flow heat transfer applications also started to appear (see, e.g., Ko et al. [2007]). In their experimental study, they have considered the pressure drop of aqueous suspensions of CNTs in a horizontal tube. The nanotube suspensions were stabilized using two different methods. The first consisted of using a surfactant, whereas in the other case stabilization was achieved by acid treatment. It was found that the nanofluids prepared by the acid treatment have much smaller viscosities than the ones made by surfactants. Thus, under laminar flow conditions, the friction factors of nanofluids with surfactants were much larger. Under turbulent flow conditions, however, friction factors of both types of stabilization techniques yielded similar values. This was explained by the shear-thinning nature of CNT nanofluids. They also noticed that the laminar flow regime was extended beyond pure water values.

As previously mentioned, as tube flow, heat exchanger applications, and microchannel applications will be covered in detail in other chapters of this book, the following sections will cover other types of flow configurations with an emphasis on specific potential engineering applications. A number of reviews on nanofluid convection problems can be found in recent literature, (see, e.g., Sarkar [2011], Wang and Mujumdar [2008a], Wen et al. [2009], and Yu et al. [2012]).

5.6.2 Radial Flow and Confined/Unconfined Impinging Jet Cooling Systems

Several research groups have considered using nanofluids in confined impinging jet and/or radial flow cooling devices. These types of flows have numerous practical applications for situations where high localized cooling capabilities are required. Numerical and experimental investigations into laminar and turbulent flow cases of heat transfer and fluid flow of nanofluids in these types of applications have been considered by the nanofluid research group at the Université de Moncton (see Gherasim et al. [2009, 2011]; Maiga et al. [2005]; Palm, Roy, and Nguyen [2006]; Roy, Nguyen, and Lajoie [2004]; and Roy et al. [2012]). Results presented by these authors indicate that aqueous-based Al_2O_3 and CuO nanofluids have the potential of becoming interesting alternatives to traditional liquids such as water. In general, increases in the Nusselt number with particle volume fraction and the Reynolds number were noticed. All numerical simulations were conducted using the single-phase approach. However, as better experimental results and correlations became available, correlations became available, different approaches for nanofluid effective properties were used. For turbulence modeling, the Reynolds-averaged Navier–Stokes equations (RANS) with $\kappa-\omega$ shear stress transport (SST) turbulence model were used with comprehensive relationships for temperature-dependent nanofluid effective thermal conductivities and viscosities. A sample of results for the turbulent flow of water–Al_2O_3 nanofluids is presented in Figure 5.3. As can clearly be seen, the average Nusselt number increases with Re and ϕ. Roy et al. (2012) also compared various types of nanofluids (CuO, Al_2O_3, and TiO particles in water). For the same volume fraction, the TiO nanofluid yielded slightly better results. Also of interest, the results presented by Feng and Kleinstreuer (2010) for a similar geometry seem to indicate that nanofluids reduce the radial cooling system's total entropy generation rate with minimal effects on the required pumping power. They concluded that the Nusselt number increases with higher nanoparticle volume fraction, smaller nanoparticle diameter, reduced disk spacing, and larger inlet Reynolds number. Numerical results presented by Yang and Lai (2010, 2011) show the same general conclusions from similar studies but also add those for fixed values of pumping power; nanofluids do not always offer improved heat transfer rates. They also found that the heat transfer enhancement of nanofluids increases with heat flux. On a related note, Manca et al. (2011) presented a numerical study of the case of a confined slot impinging jet with nanofluids. As in the radial flow cases previously discussed, they found that for an increase in particle concentrations, nanofluids produce related increases in the fluid bulk temperature, the Nusselt number, the heat transfer coefficient, and the pumping power (Manca et al. 2011).

Nguyen et al. (2009) considered the case of a confined and submerged impinging jet heat transfer of a water–Al_2O_3 nanofluid for particle loadings

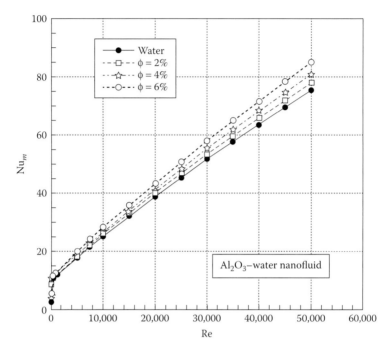

FIGURE 5.3 Average Nusselt number as a function of Reynolds number and particle volume fraction for a radial flow configuration using water–Al$_2$O$_3$ nanofluids. (Reproduced from *Int. J. Therm. Sci.*, 58, Roy, G. et al., Heat transfer performance and hydrodynamic behavior of turbulent nanofluid radial flows, 120–129. Copyright 2012, with permission from Elsevier Masson SAS.)

ranging from 0 to 6 vol.%. Both laminar and turbulent regimes were studied for Reynolds numbers of up to 88,000. They found that the heat transfer coefficient for the considered nanofluid flow can either increase or decrease compared to water depending on certain combinations of particle volume fraction and nozzle-to-surface distance.

A numerical study of pulsating rectangular jets with nanofluids was presented by Selimefendigil and Oztop (2014). They solved the unsteady Navier–Stokes and energy equations with a commercial finite volume-based CFD code. Heat transfer enhancements of up to 18.8% were found for a 6% volume fraction Al$_2$O$_3$–water nanofluid for a Reynolds number of 200.

5.6.3 Electronic Component Cooling

With the technological advances in the design and manufacturing of micro-electronic devices and power electronics systems, important size reductions in components with corresponding increases in power densities are found. Furthermore, CPU speeds are increasing at dramatic rates and, as a result, are requiring more power. Usually, computer performance enhancement directly translates into larger heat outputs. Inefficient heat dissipation will undoubtedly yield higher operating temperatures that could most certainly damage various components. These increases in heat production have traditionally been dealt with by more efficient heat sinks and larger fans. Increases in heat removal performance with the use of such techniques are rapidly becoming insufficient. Indeed, thermal management is now considered a major limitation for development in the electronics industry. Therefore, engineers have found themselves with interesting challenges and the search for more efficient cooling has led to various alternatives. Liquid cooling devices as well as heat pipes and microchannels have recently made their way into personal computers and other electronic devices. In the particular case of liquid cooling, a number of various liquid block designs have also been proposed, including slots, "cross-flow," and confined impinging jet/radial flow configurations. It is quite easily conceivable, however, that those devices may soon become insufficient for adequate cooling in this rapidly evolving technological sector. Nanofluids could therefore become an alternative to traditional coolants used in these applications.

A numerical investigation into the potential use of water-based alumina nanofluids in an electronic component cooling system was presented by Roy, Nguyen, and Comeau (2006). They considered a simplified computational domain based on a liquid cooling device for computers. For a nanoparticle volume fraction of 5%, an increase of approximately 30% in the average wall heat transfer coefficient was found. The heat transfer coefficient was influenced by the channel height and the Reynolds number. Local values were also found to vary considerably with the behavior of the hydrodynamic field (i.e., flow separation areas). With the same type of liquid cooling device in mind, Nguyen, Roy et al. (2007) studied heat transfer enhancement experimentally. Using a commercially available "water block" cooling system placed upon a heated surface, they built a closed fluid flow circuit. Nanofluids that were considered consisted of Al_2O_3 nanoparticles of two average sizes (36 and 47 nm) dispersed in water. For a particle volume fraction of 6.8%, the heat transfer coefficient was increased by as much as 40% compared to that of the base fluid. It was observed that increases in particle concentrations produced clear decreases in the surface temperature. For the same mass flow rate, a nanofluid with 36 nm particle size provided higher convective heat transfer coefficients than those obtained with 47 nm particles. Other authors have also since considered using nanofluids in similar experimental setups using commercially available water blocks. Roberts and Walker (2010) considered such a system using alumina nanoparticles in water. Nanoparticles of various sizes were considered. Results showed that an enhancement in convective heat transfer of approximately 20% was possible. Figure 5.4 presents results from this study. As can be seen, the temperature gain in the water block can be increased by several degrees with only a slight addition of nanoparticles (0.5%–1.5%). Roberts and Walker also stated that the results are promising as they performed well in a commercial system with results comparable to those obtained in a well-controlled experimental system.

Other types of devices that could use nanofluids as heat transfer mediums for electronic device cooling include minichannels and heat pipes (see, e.g., Putra and Iskandar [2011]).

5.6.4 Automotive/Transportation Industry

Along with electronics cooling, the automotive industry is probably one of the sectors that nanofluids have good potential of seeing practical use. Indeed, as ethylene glycol (typically mixed with water) and engine oil are

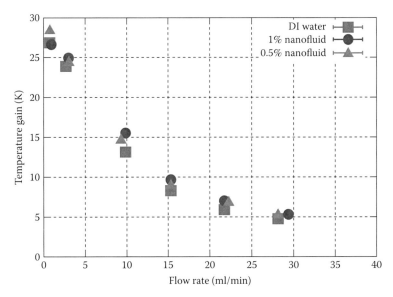

FIGURE 5.4 Temperature gain through a commercial water block for pure DI water and deionized (DI) water with 20–30 nm alumina nanoparticles. (Reproduced from *Appl. Therm. Eng.* 30, Roberts, N. A. and Walker, D. G. Convective performance of nanofluids in commercial electronics cooling systems, 2499–2504. Copyright 2010, with permission from Elsevier Masson SAS.)

the universal heat transfer fluids for the automotive industry, it is quite conceivable that the addition of nanoparticles to these traditional fluids could considerably improve the performance of vehicle thermal management systems (Wang and Mujumdar 2008b). It is well known that ethylene glycol and engine oil are relatively poor heat transfer mediums. For example, better fluid performance could lead to smaller, more efficient radiators, which in turn could increase fuel economy. Other areas in modern vehicle systems require efficient cooling as well. These include, for example, transmissions and generators. Interest in using nanofluids as coolants in motor vehicle thermal management systems originated in the early years of nanofluid development (see, e.g., Choi et al. [2001]).

Nanofluids have been considered for use as coolants in automatic transmissions (Tseng, Lin, and Huang 2005). The authors dispersed CuO and Al_2O_3 nanoparticles in transmission oil, with the intention of evaluating their heat transfer performance inside a four-wheel drive vehicle transmission with a rotary blade coupling. Results showed that the CuO transmission oil nanofluids produced the lowest blade temperatures. Kulkarni, Vajjha et al. (2008) studied the application of Al_2O_3 nanofluids in a diesel electric generator as jacket water coolants. They concluded that the use of this type of nanofluid decreases the cogeneration efficiency of the diesel generator as nanoparticle concentration increases because of the decrease in specific heat. However, the heat exchanger efficiency increases with particle volume fraction because of the higher heat transfer coefficients of nanofluids.

Ollivier et al. (2006) presented a study on the detection of knock occurrence in a spark ignition gas engine from a heat transfer analysis standpoint. They considered unsteady heat transfer through the cylinder and inside the coolant flow. The amplification of wall heat transfer due to knock is taken into account by a self-developed program. By calculating and analyzing the transient component of the thermal signal, it is possible to predict the knock occurrence. They found that, with the use of a nanofluid as a coolant, temperature variations are enhanced when knock occurs. This is explained by the higher diffusivity of the nanofluid. This would therefore favor the detection of knock occurrence from a heat transfer analysis.

In a numerical study, Al_2O_3 and CuO nanoparticle suspensions in an ethylene glycol and water mixture were considered for use as the heat transfer fluid circulating in the flat tubes of an automobile radiator (Vajjha, Das, and Namburu 2010). Considerable improvement in convective heat transfer coefficients was found with the nanofluids compared to the base fluid without the nanoparticles. Although the pressure losses increased with particle volume fraction, Vajjha, Das, and Namburu found that the pumping power was reduced due to the reduced volumetric flow needed for the same amount of heat extracted. They also presented new correlations for Nusselt numbers. Sarkar and Tarodiya (2013) analyzed the performance of a louvered fin tube automotive radiator using various types of nanofluids as coolants. The base fluid consisted of a 80% water and 20% ethylene glycol mixture, with Cu, SiC, Al_2O_3, and TiO_2 nanoparticles being evaluated. They found that the use of nanofluids as coolants in this type of radiator improved the effectiveness and cooling capacity of the radiator with a reduction in pumping power. They found that the SiC particles provided the best performance with an increase of over 15% in maximum cooling capacity.

5.6.5 Solar Energy

Potential applications of nanofluids in solar collectors (including solar water heaters, solar cells, solar stills, and thermal energy storage systems) have been considered by quite a few researchers. Indeed, it is believed that nanofluids could permit enhancements in energy efficiencies and performances of such systems. As in investigations into other types of potential applications, multiple fluid/particle mixtures have been considered for various types of collectors and results vary greatly from research team to research team. For example, solar collector efficiencies were found to increase anywhere between 5% and 28% (see, e.g., Tyagi, Phelan, and Prasher [2007] and Yousefi et al. [2012]). One of the advantages of nanofluids in these types of applications is that they have been reported to have excellent capabilities of absorbing incoming sunlight (in some cases, over 95%) (Taylor et al. 2011). Some also have stated that the use of nanofluids in such systems will help reduce CO_2 emissions (Khullar and Tyagi 2012; Otanicar and Golden 2009). For more information on potential applications of nanofluids in solar collectors, one can consult the recent review by Mahian et al. (2013).

5.6.6 Refrigeration Systems

Several authors have studied the use of various types of nanoparticles placed in suspension in several types of common refrigerants and/or lubricant oils in hopes of enhancing efficiency and reliability of refrigerators and chillers. For example, Bi, Shi, and Zhang (2008) considered the use of mineral oil–TiO_2 nanoparticle mixtures as lubricants instead of polyolester (POE) oil in an HFC134a refrigerator. Results indicated the combination of HFC134a and mineral oil with the nanoparticles worked normally and safely in the refrigerator while enhancing its performance by reducing energy consumption by 26%. In another study, CuO nanoparticles were placed in an R113 refrigerant in order to predict the resulting heat transfer performance (Peng et al. 2009). Experimental results presented by the authors illustrated that the convective heat transfer coefficient of the resulting mixture was larger than that of the pure refrigerant. The maximum enhancement of heat transfer coefficient was found to be 29.7%.

5.6.7 Other Potential Applications

Other potentially interesting forced convection applications for nanofluids as high-performance coolants include the nuclear energy industry (e.g., as reactor coolants) (Buongiorno et al. 2008), heating, ventilation and air-conditioning (HVAC) systems (Kulkarni, Das, and Vajjha 2009), and space and defense applications where power densities are high and the component size is very important (Yu and Xie 2012).

5.6.8 General Concluding Remarks Regarding Nanofluid Forced Convection Applications

This section presents an overview of potential forced convection applications of nanofluids. Obviously, it is practically impossible to cover all aspects of this fascinating subject as nanofluid forced convection studies have grown exponentially over the past few years. One thing is certain; nanofluids have generated considerable interest in a wide variety of areas of thermal management systems. Although nanofluids have the potential of becoming next-generation coolants, certain drawbacks will have to be addressed. Some of these issues will be considered in Sections 5.7 and 5.8.

5.7 NANOFLUID EFFICACY AND PERFORMANCE

The addition of nanoparticles into a base fluid will not only enhance heat transfer but also increase shear stresses and pressure losses. This is an important factor to consider while choosing an appropriate coolant for a specific application. Although some have indicated that the presence of nanoparticles placed in suspension in traditional coolants does not have significant negative effects (see, e.g., Chein and Huang [2005]), others have found otherwise. As previously stated, considerable increases in nanofluid effective viscosities were found by various authors. This can only translate into increases in pressure losses and pumping power. Furthermore, Li and Kleinstreuer (2008) found that the use of CuO nanofluids in microchannels increases the pressure drop (or pumping power), thus decreasing somewhat the beneficial effects. In the case of turbulent flow of water-based nanofluids with CuO, Al_2O_3, and SiO_2 particles, it was found that nanofluids with higher particle volume fraction not only have greater heat transfer enhancement but also have higher pressure drops (Namburu et al. 2009). Indeed, they found that a 6% volume fraction CuO nanofluid typically has wall shear stresses that are almost twice as great as the base fluid. They suggest that a careful choice of the concentration of nanoparticles should be made in order to balance heat transfer enhancement and pressure drop penalty. In their study on nanofluid viscosity, Prasher et al. (2006) suggested that in order to make the nanofluid thermal performance worse than that of the base fluid, the viscosity has to be increased by more than a factor of 4 (relative to the increase in thermal conductivity). Others considered pressure losses in convective heat transfer study of SiO_2 nanoparticles in an ethylene glycol–water mixture (Kulkarni, Namburu et al. 2008). They determined that pressure losses are a function of concentration due to the increases in viscosity. Recently, the case of turbulent nanofluid flow inside a ribbed channel has been considered (Manca, Nardini, and Ricci 2012). Compared to cases using water and smooth channels, Nusselt number values, heat transfer rates, and pumping power all increased with particle volume fraction by factors of 2.4, 2.66, and 31 times, respectively. Parameters such as pumping power are typically used to make efficacy comparisons (see, e.g., Agrawal and Varma [1991]):

$$\text{PP} = \dot{W} = \dot{V}\Delta p_t = \frac{\dot{m}\Delta p_t}{\rho} \qquad (5.32)$$

where Δp_t is the total pressure difference between the inlet and outlet sections, $\dot{V}$ is the volumetric flow rate, $\dot{m}$ is the mass flow rate and ρ is the fluid density. In order to appreciate the overall performance of a coolant, one can use different approaches to try to quantify and compare alternatives. For example, one can determine the pumping power required to evacuate a given heat load for various types of coolants. A more efficient coolant should be able to handle the same heat duty using less fluid and with the same or less pumping power. In an investigation into the potential use of nanofluids as coolants, it was concluded that in plate heat exchangers containing relatively large quantity of coolants, the substitution of conventional heat transfer fluids by nanofluids is not beneficial (Figure 5.5) (Pantzali, Mouza, and Paras 2009). For the same heat load,

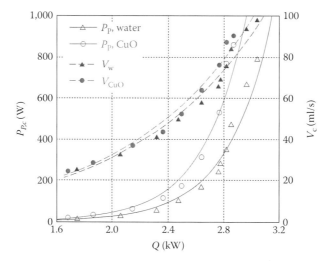

FIGURE 5.5 Pumping power and volumetric flow rate vs. heat flow rate for coolants. (Reproduced from *Chem. Eng. Sci.*, 64, Pantzali, M. N., Investigating the efficacy of nanofluids as coolants in plate heat exchangers (PHE), 3290–3300. Copyright 2009, with permission from Elsevier Masson SAS.)

CuO nanofluid requires larger volumetric flow rates and more important pumping powers compared to pure water.

Another criterion that has been used to evaluate nanofluid overall energy performance is the performance evaluation criterion (PEC) (Ferrouillat et al. 2011). This criterion is based on the ratio of heat transferred to the required pumping power in the test section:

$$\text{PEC} = \frac{\dot{m}c_p\left(T_{out} - T_{in}\right)}{\dot{V}\Delta p} \tag{5.33}$$

where $\dot{m}$ is the mass flow rate, c_p is the fluid specifc heat, T is the temperature, $\dot{V}$ is the volumetric flow rate and Δp is the pressure difference between the outlet and the inlet

In their experimental study, Ferrouillat et al. found that for the application considered (SiO_2–water nanofluids in horizontal tubes), the PEC values for all tested nanofluids were below those for water (Figure 5.6). This would seem to indicate that when compared to water, the nanofluids have unfavorable energy balances. The same type of analysis was performed for the case of a radial flow system using Al_2O_3, CuO, and TiO nanofluids. The same conclusions were found for all cases considered (Roy et al. 2012).

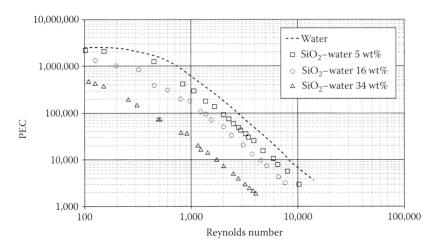

FIGURE 5.6 Evolution of the performance evaluation criterion vs. Reynolds number for water and various SiO₂–water nanofluids. (Reproduced from *Int. J. Heat Fluid Fl.*, 424–439, Ferrouillat, S. et al., Hydraulic and heat transfer study of SiO₂/water nanofluids in horizontal tubes with imposed wall temperature boundary conditions, 32. Copyright 2013, with permission from Elsevier Masson SAS.)

In any event, care should be taken when presenting and interpreting results pertaining to nanofluid efficacy. As Pantzali, Mouza, and Paras (2009) stated, different ways of presenting results can lead to different conclusions. This has been exemplified in the study made by Gherasim, Galanis, and Nguyen (2013) on the potential usage of nanofluids in plate heat exchangers. Indeed, when presenting certain results using both the "PP versus Q" and "PEC versus Re" approaches, they noticed an apparent contradiction with the results. They found that certain nanofluids provided better heat removal rates with less pumping power; however, when presenting their results on a PEC versus Re graph, all nanofluids had inferior performances compared to pure water. These apparent discrepancies between results analysis can be explained by the fact that the approach compares PEC values in function of the Reynolds number. However, using comparisons based on the Reynolds number can be misleading. Indeed, the considerable differences in coolant properties will result in corresponding differences in flow rates for the same Reynolds numbers, thus making clear comparisons difficult. Similarly, in a study of thermal performances of two different types of aqueous nanofluids (Al₂O₃ and CNT nanoparticles), it was found that although alumina and CNTs showed better thermal performances compared to water, the increases in pumping power were considered very important. However, when the comparison was based on a practical benefit parameter, the overall gains of 22% and 150% were, respectively, found for the Al_2O_3 and CNT nanofluids, thereby yielding better overall performance compared to pure water (Mare et al. 2011).

5.8 OTHER CHALLENGES AND LIMITATIONS OF NANOFLUIDS WITH REGARD TO ENGINEERING FORCED CONVECTION APPLICATIONS

Although a considerable amount of research has been done on nanofluids over the past decade, the conclusions on their behavior, characteristics, and performances remain somewhat controversial. Clearly, the lack of consistency in experimental values found in the current literature is problematic and quite probably reflects inadequate sample preparation. Hence, it is imperative to conduct more investigations in order to properly quantify the effects of particle size and shapes, clustering of particles, adequate dispersion of particles and sedimentation, surfactant effects, nanofluid temperature, adequate experimental techniques and procedures, and so on. Some of these issues have been addressed in part by a few authors. The most important are discussed in Sections 5.8.1 through 5.8.4.

5.8.1 Nanofluid Stability

Nanoparticles are often hydrophobic and therefore cannot typically be dispersed in most heat transfer fluids such as water or ethylene glycol without surface treatments and/or dispersants or surfactants (Chen and Xie 2010; Xie et al. 2003). Furthermore, without these special treatments, the nanoparticles would most certainly agglomerate and/or settle, thereby creating other problems such as channel clogging and a reduction in thermal conductivity of the mixture (Yu and Xie 2012). Surfactants or dispersion agents are therefore commonly used in nanofluids. Although they are beneficial for stabilizing the suspension, they may also create certain problems for heat transfer mediums. Li, Zhu, and Wang (2007) considered the dispersion behavior of aqueous copper nano-suspensions for various pH values and different dispersant types and concentrations. Dispersant agents (or surfactants) included hexadecyl trimethyl ammonium bromide, sodium dodecylbenzenesulfonate (SDBS), and polyoxyethylene nonylphenyl ether (TX-10). Results clearly show that surfactants lead to the enhancement of the stability of Cu suspensions. The authors proposed optimized concentrations for various types of surfactants tested (i.e., yielding better dispersion results). Similarly, Li et al. (2008) studied the effects of pH and chemical surfactant (SDBS) on thermal conductivity enhancement for Cu–H_2O nanofluids. Results show that the thermal conductivity of the nanofluids is highly dependent on the pH values and SDBS surfactant concentration. Both factors exhibit optimal values for increased thermal conductivities (for the considered cases, a pH in the range of 8.5–9.5 and SDBS mass fraction of 0.02 wt% yielded the best thermal conductivity values). In their work on the use of dispersants in transformer oil, Choi, Yoo, and Oh (2008) noted that an excessive quantity of surfactant had a negative effect on viscosity, thermal properties, and chemical stability of the nanofluid. Stability of mixtures remains an issue for which more investigation is required. Long-term stability, stability in practical applications, and the stability after thousands of thermal cycles are questions that will need to be addressed more carefully (Yu and Xie 2012). Time effects on nanofluid properties and behaviors are still not well understood.

As a side effect of improper mixture stabilization, the effects of clustering on the thermal conductivity of nanofluids is certainly of interest. This problem was considered by Karthikeyan, Philip, and Raj (2008) using suspensions consisting of CuO nanoparticles of 8 nm average diameter. The authors proceeded to study the thermal properties of the suspensions. For a 1% loading (volume), a 54% enhancement was found in thermal conductivity. With time, however, the thermal conductivity decreased. With microscopic observation, this was attributed to nanoparticle clustering. According to the authors, nanoparticle size, cluster size, and volume fraction of particles have a significant influence on thermal conductivity of the mixture.

5.8.2 High Cost of Nanofluids

So far, nanofluid manufacturing has been limited to laboratory-scale production. High-cost will, for at least the near future, limit potential widespread applications. Until manufacturing processes allow the (mass) production of nanoparticles and associated suspensions, costs will undoubtedly remain high.

5.8.3 Degradation of Fluid Transfer Components

Erosion and/or corrosion of fluid transfer components remains an issue for nanofluids. Furthermore, systems using nanofluids will most likely require more maintenance than those using traditional coolants (Roberts and Walker 2010).

5.8.4 Health, Safety, and Environmental Issues Related to the Manufacturing and Usage of Nanofluids and Nanoparticles

As in most other technological and biomedical areas where potential benefits of nanoparticles are generating a great deal of interest, the health and environmental risks associated with the manufacturing and widespread use of nanofluids are still unknown and are of concern. Nanoparticles can most certainly enter the human body via the lungs and quite possibly the skin (Hoet, Bruske-Hohlfeld, and Salata 2004). Identified potential health hazards of nanoparticles include, among others, inflammation, respiratory diseases, and carcinogenic effects. It is clear that health and environmental risks will have to be quantified and evaluated in order to develop safe nanoparticle and nanofluid handling and application procedures and parameters. In fact, controlled application of new emerging nanoparticles will require knowledge-based development, production, and application, including sustainable risk assessment prior to their widespread use (Kreyling, Semmler-Behnke, and Moller 2006). Indeed, risk assessment is a considerable challenge due to the lack of sufficient data and information; some of the related uncertainties include the following (Kandlikar et al. 2007):

- How particle characteristics affect toxicity (size, shape, chemical composition, etc.)

- The fate and transport of nanoparticles through the environment

- The routes of exposure (inhalation, dermal, ingestion, ocular) and the metrics by which exposure are measured

- The mechanisms of translocation to different parts of the body

- The mechanisms of toxicity and disease

Several studies indicate that nanoparticles are considerably more hazardous to human health than larger particles due to the fact that they can more easily migrate to other parts of the body (translocate) and end up in the bloodstream and neuronal pathways into the central nervous system and the brain. What is particularly alarming is the potential for nanoparticles to move into certain other sensitive areas such as the bone marrow, the lymph nodes, the spleen, and the heart (Kandlikar et al. 2007). On a side note, as stated by Kandlikar et al., the ability that nanoparticles have to reach different regions of the body, seemingly unaffected by the body's natural blocking mechanisms, is, ironically, the reason that nanoparticles potentially have useful medical applications.

Standards for workplace safety with regard to nanoparticles are currently emerging, and therefore engineers will have to compose with the restrictions and procedures not only associated with the preparation and handling of nanoparticles and nanofluids but also in the usage and discarding of these special fluids. With the prospect of widespread nanofluid use in thermal management systems, engineers will undoubtedly be challenged into providing environmentally friendly and health-safe products and systems. This may include the use of "green" and/or biodegradable nanoparticles.

As can be seen, although nanofluids have the potential of becoming the next-generation coolants, several key challenges will undoubtedly need to be addressed before they can be considered for widespread industrial applications.

REFERENCES

Abu-Nada, E. 2009. "Effects of variable viscosity and thermal conductivity of Al_2O_3-water nanofluid on heat transfer enhancement in natural convection." *International Journal of Heat and Fluid Flow* no. 30 (4):679–690.

Agrawal, K. N., and H. K. Varma. 1991. "Experimental-study of heat-transfer augmentation versus pumping power in a horizontal R12 evaporator." *International Journal of Refrigeration—Revue Internationale Du Froid* no. 14 (5):273–281.

Ahuja, A. S. 1975a. "Augmentation of heat transport in laminar flow of polystyrene suspensions—1. Experiments and results." *Journal of Applied Physics* no. 46 (8):3408–3416. doi:10.1063/1.322107.

Ahuja, A. S. 1975b. "Augmentation of heat transport in laminar flow of polystyrene suspensions—2. Analysis of the data." *Journal of Applied Physics* no. 46 (8):3417–3425.

Ahuja, A. S. 1982. "Thermal design of a heat exchanger employing laminar flow of particle suspensions." *International Journal of Heat and Mass Transfer* no. 25 (5):725–728.

Akbari, M., N. Galanis, and A. Behzadmehr. 2011. "Comparative analysis of single and two-phase models for CFD studies of nanofluid heat transfer." *International Journal of Thermal Sciences* no. 50 (8):1343–1354.

Akbari, M., N. Galanis, and A. Behzadmehr. 2012. "Comparative assessment of single and two-phase models for numerical studies of nanofluid turbulent forced convection." *International Journal of Heat and Fluid Flow* no. 37:136–146.

Akbarinia, A., and A. Behzadmehr. 2007. "Numerical study of laminar mixed convection of a nanofluid in horizontal curved tubes." *Applied Thermal Engineering* no. 27 (8–9):1327–1337.

Auton, T. R. 1987. "The lift force on a spherical body in a rotational flow." *Journal of Fluid Mechanics* no. 183:199–218.

Avila, R., and J. Cervantes. 1995. "Analysis of the heat transfer coefficient in a turbulent particle pipe flow." *International Journal of Heat and Mass Transfer* no. 38 (11):1923–1932.

Barbes, B., R. Paramo, E. Blanco, M. J. Pastoriza-Gallego, M. M. Pineiro, J. L. Legido, and C. Casanova. 2013. "Thermal conductivity and specific heat capacity measurements of Al_2O_3 nanofluids." *Journal of Thermal Analysis and Calorimetry* no. 111 (2):1615–1625.

Batchelor, G. K. 1977. "The effect of Brownian motion on the bulk stress in a suspension of spherical particles." *Journal of Fluid Mechanics* no. 83 (1):97–117.

Behzadmehr, A., M. Saffar-Avval, and N. Galanis. 2007. "Prediction of turbulent forced convection of a nanofluid in a tube with uniform heat flux using a two phase approach." *International Journal of Heat and Fluid Flow* no. 28 (2):211–219.

Bi, S. S., L. Shi, and L. L. Zhang. 2008. "Application of nanoparticles in domestic refrigerators." *Applied Thermal Engineering* no. 28 (14–15):1834–1843.

Bianco, V., F. Chiacchio, O. Manca, and S. Nardini. 2009. "Numerical investigation of nanofluids forced convection in circular tubes." *Applied Thermal Engineering* no. 29 (17–18):3632–3642.

Boothroyd, R. G., and H. Haque. 1970. "Fully developed heat transfer to a gaseous suspension of particles flowing turbulently in ducts of different size." *Journal of Mechanical Engineering Science* no. 12 (3):191–200.

Brinkman, H. C. 1952. "The viscosity of concentrated suspensions and solutions." *Journal of Chemical Physics* no. 20:571–581.

Buongiorno, J. 2006. "Convective transport in nanofluids." *Journal of Heat Transfer—Transactions of the ASME* no. 128 (3):240–250.

Buongiorno, J., L. W. Hu, S. J. Kim, R. Hannink, B. Truong, and E. Forrest. 2008. "Nanofluids for enhanced economics and safety of nuclear reactors: An evaluation of the potential features, issues, and research gaps." *Nuclear Technology* no. 162 (1):80–91.

Chandrasekar, M., and S. Suresh. 2009. "A review on the mechanisms of heat transport in nanofluids." *Heat Transfer Engineering* no. 30 (14):1136–1150.

Chein, R. Y., and G. M. Huang. 2005. "Analysis of microchannel heat sink performance using nanofluids." *Applied Thermal Engineering* no. 25 (17–18):3104–3114.

Chen, H. S., Y. Wei, Y. R. He, W. Ding, L. L. Zhang, C. Q. Tan, A. A. Lapkin, and D. V. Bavykin. 2008. "Heat transfer and flow behaviour of aqueous suspensions of titanate nanotubes (nanofluids)." *Powder Technology* no. 183 (1):63–72.

Chen, L., and H. Xie. 2010. "Surfactant-free nanofluids containing double- and single-walled carbon nanotubes functionalized by a wet-mechanochemical reaction." *Thermochimica Acta* no. 497:67–71.

Choi, C., H. S. Yoo, and J. M. Oh. 2008. "Preparation and heat transfer properties of nanoparticle-in-transformer oil dispersions as advanced energy-efficient coolants." *Current Applied Physics* no. 8 (6):710–712.

Choi, S. U. S. 1995. Enhancing thermal conductivity of fluids with nanoparticles. *Proceedings of the 1995 ASME International Mechanical Engineering Congress and Exposition*, November 12–17, 1995, San Francisco, CA. New York: American Society of Mechanical Engineers.

Choi, S. U. S., W. Yu, J. R. Hull, Z. G. Zhang, and F. E. Lockwood. 2001. Nanofluids for vehicle thermal management. *Proceedings of the Vehicle Thermal Management Systems Conference and Exhibition*, May 14–17, 2001, Nashville, TN. Warrendale, PA: Society of Automotive Engineers (SAE International).

Chon, C. H., K. D. Kihm, S. P. Lee, and S. U. S. Choi. 2005. "Empirical correlation finding the role of temperature and particle size for nanofluid (Al_2O_3) thermal conductivity enhancement." *Applied Physics Letters* no. 87 (15), Article ID 153107, 3pp.

Corcione, M. 2011a. "Empirical correlating equations for predicting the effective thermal conductivity and dynamic viscosity of nanofluids." *Energy Conversion and Management* no. 52:789–793.

Corcione, M. 2011b. "Rayleigh-Benard convection heat transfer in nanoparticle suspensions." *International Journal of Heat and Fluid Flow* no. 32 (1):65–77.

Corcione, M., E. Habib, and A. Quintino. 2013. "A two-phase numerical study of buoyancy-driven convection of alumina-water nanofluids in differentially-heated horizontal annuli." *International Journal of Heat and Mass Transfer* no. 65:327–338.

Dan, C., and A. Wachs. 2010. "Direct numerical simulation of particulate flow with heat transfer." *International Journal of Heat and Fluid Flow* no. 31 (6):1050–1057.

Das, S. K., S. U. S. Choi, W. Yu, and T. Pradeep. 2008. *Nanofluids: Science and Technology*. Hoboken, NJ: John Wiley & Sons.

Das, S. K., N. Putra, P. Thiesen, and W. Roetzel. 2003. "Temperature dependence of thermal conductivity enhancement for nanofluids." *Journal of Heat Transfer—Transactions of the ASME* no. 125 (4):567–574.

Drew, D. A., and S. L. Passman. 1999. *Theory of Multicomponent Fluids*. Berlin, Germany: Springer.

Durst, F., D. Milojevic, and B. Schönung. 1984. "Eulerian and Lagrangian predictions of particulate two-phase flows: A numerical study." *Applied Mathematical Modelling* no. 8:101–115.

Feng, Y., and C. Kleinstreuer. 2010. "Nanofluid convective heat transfer in a parallel-disk system." *International Journal of Heat and Mass Transfer* no. 53 (21–22):4619–4628.

Ferrouillat, S., A. Bontemps, J.-P. Ribeiro, J.-A. Gruss, and O. Soriano. 2011. "Hydraulic and heat transfer study of SiO_2/water nanofluids in horizontal tubes with imposed wall temperature boundary conditions." *International Journal of Heat and Fluid Flow* no. 32 (2):424–439.

Gherasim, I., N. Galanis, and C. T. Nguyen. 2013. "Numerical study of nanofluid flow and heat transfer in a plate heat exchanger." *Computational Thermal Sciences* no. 5 (4):317–332.

Gherasim, I., G. Roy, C. T. Nguyen, and D. Vo-Ngoc. 2009. "Experimental investigation of nanofluids in confined laminar radial flows." *International Journal of Thermal Sciences* no. 48 (8):1486–1493.

Gherasim, I., G. Roy, C. T. Nguyen, and D. Vo-Ngoc. 2011. "Heat transfer enhancement and pumping power in confined radial flows using nanoparticle suspensions (nanofluids)." *International Journal of Thermal Sciences* no. 50:369–377.

Gouesbet, G., and A. Berlemont. 1999. "Eulerian and Lagrangian approaches for predicting the behaviour of discrete particles in turbulent flows." *Progress in Energy and Combustion Science* no. 25 (2):133–159.

Graham, A. L. 1981. "On the viscosity of suspensions of solid spheres." *Applied Scientific Research* no. 37 (3–4):275–286.

Halelfadl, S., P. Estelle, B. Aladag, N. Doner, and T. Mare. 2013. "Viscosity of carbon nanotubes water-based nanofluids: Influence of concentration and temperature." *International Journal of Thermal Sciences* no. 71:111–117.

Hamilton, R. L., and O. K. Crosser. 1962. "Thermal conductivity of heterogeneous two-component systems." *I&EC Fundamentals* no. 1 (3):187–191.

Hejazian, M., and M. K. Moraveji. 2013. "A comparative analysis of single and two-phase models of turbulent convective heat transfer in a tube for TiO$_2$ nanofluid with CFD." *Numerical Heat Transfer, Part A: Applications* no. 63 (10):795–806.

Heris, S. Z., G. Etemad, and M. N. Esfahany. 2006. "Experimental investigation of oxide nanofluids laminar flow convective heat transfer." *International Communications in Heat and Mass Transfer* no. 33:529–535.

Hoet, P. H. M., I. Bruske-Hohlfeld, and O. V. Salata. 2004. "Nanoparticles—Known and unknown health risks." *Journal of Nanobiotechnology* no. 2:12. doi:10.1186/1477-3155-2-12.

Hwang, Y., J. K. Lee, C. H. Lee, Y. M. Jung, S. I. Cheong, C. G. Lee, B. C. Ku, and S. P. Jang. 2007. "Stability and thermal conductivity characteristics of nanofluids." *Thermochimica Acta* no. 455 (1–2):70–74.

Jang, S. P., and S. U. S. Choi. 2006. "Cooling performance of a microchannel heat sink with nanofluids." *Applied Thermal Engineering* no. 26 (17–18):2457–2463.

Jang, S. P., and S. U. S. Choi. 2007. "Effects of various parameters on nanofluid thermal conductivity." *Journal of Heat Transfer* no. 129 (5):617–623.

Kandlikar, M., G. Ramachandran, A. Maynard, B. Murdock, and W. A. Toscano. 2007. "Health risk assessment for nanoparticles: A case for using expert judgment." *Journal of Nanoparticle Research* no. 9 (1):137–156.

Karthikeyan, N. R., J. Philip, and B. Raj. 2008. "Effect of clustering on the thermal conductivity of nanofluids." *Materials Chemistry and Physics* no. 109 (1):50–55.

Khanafer, K., K. Vafai, and M. Lightstone. 2003. "Buoyancy-driven heat transfer enhancement in a two-dimensional enclosure utilizing nanofluids." *International Journal of Heat and Mass Transfer* no. 46 (19):3639–3653.

Khullar, V., and H. Tyagi. 2012. "A study on environmental impact of nanofluid-based concentrating solar water heating system." *International Journal of Environmental Studies* no. 69:220–232.

Ko, G. H., K. Heo, K. Lee, D. S. Kim, C. Kim, Y. Sohn, and M. Choi. 2007. "An experimental study on the pressure drop of nanofluids containing carbon nanotubes in a horizontal tube." *International Journal of Heat and Mass Transfer* no. 50 (23–24):4749–4753.

Koo, J., and C. Kleinstreuer. 2004. "A new thermal conductivity model for nanofluids." *Journal of Nanoparticle Research* no. 6 (6):577–588.

Koo, J., and C. Kleinstreuer. 2005. "Laminar nanofluid flow in microheat-sinks." *International Journal of Heat and Mass Transfer* no. 48 (13):2652–2661.

Kreith, F., and M. S. Bohn. 1997. *Principles of Heat Transfer*, 5th ed. Boston, MA: PWS Publishing Company.

Kreyling, W. G., M. Semmler-Behnke, and W. Moller. 2006. "Health implications of nanoparticles." *Journal of Nanoparticle Research* no. 8 (5):543–562.

Kulkarni, D. P., D. K. Das, and S. L. Patil. 2007. "Effect of temperature on rheological properties of copper oxide nanoparticles dispersed in propylene glycol and water mixture." *Journal of Nanoscience and Nanotechnology* no. 7 (7):2318–2322.

Kulkarni, D. P., D. K. Das, and R. S. Vajjha. 2009. "Application of nanofluids in heating buildings and reducing pollution." *Applied Energy* no. 86 (12):2566–2573.

Kulkarni, D. P., P. K. Namburu, H. Ed Bargar, and D. K. Das. 2008. "Convective heat transfer and fluid dynamic characteristics of SiO_2–ethylene glycol/water nanofluid." *Heat Transfer Engineering* no. 29 (12):1027–1035.

Kulkarni, D. P., R. S. Vajjha, D. K. Das, and D. Oliva. 2008. "Application of aluminum oxide nanofluids in diesel electric generator as jacket water coolant." *Applied Thermal Engineering* no. 28 (14–15):1774–1781.

Lee, S., S. U. S. Choi, S. Li, and J. A. Eastman. 1999. "Measuring thermal conductivity of fluids containing oxide nanoparticles." *Journal of Heat Transfer— Transactions of the ASME* no. 121 (2):280–289.

Li, J., and C. Kleinstreuer. 2008. "Thermal performance of nanofluid flow in microchannels." *International Journal of Heat and Fluid Flow* no. 29 (4):1221–1232.

Li, X., D. Zhu, and X. Wang. 2007. "Evaluation on dispersion behavior of the aqueous copper nano-suspensions." *Journal of Colloid and Interface Science* no. 310 (2):456–463.

Li, X. F., D. S. Zhu, X. J. Wang, N. Wang, J. W. Gao, and H. Li. 2008. "Thermal conductivity enhancement dependent pH and chemical surfactant for Cu-H_2O nanofluids." *Thermochimica Acta* no. 469 (1–2):98–103.

Lotfi, R., Y. Saboohi, and A. M. Rashidi. 2010. "Numerical study of forced convective heat transfer of nanofluids: Comparison of different approaches." *International Communications in Heat and Mass Transfer* no. 37 (1):74–78.

Lundgren, T. S. 1972. "Slow flow through stationary random beds and suspensions of spheres." *Journal of Fluid Mechanics* no. 51 (part 2):273–299.

Mahian, O., A. Kianifar, S. A. Kalogirou, I. Pop, and S. Wongwises. 2013. "A review of the applications of nanofluids in solar energy." *International Journal of Heat and Mass Transfer* no. 57 (2):582–594.

Maiga, S. E. B., C. T. Nguyen, N. Galanis, and G. Roy. 2004. "Heat transfer behaviours of nanofluids in a uniformly heated tube." *Superlattices and Microstructures* no. 35 (3–6):543–557.

Maiga, S. E. B., C. T. Nguyen, N. Galanis, G. Roy, T. Mare, and M. Coqueux. 2006. "Heat transfer enhancement in turbulent tube flow using $Al2O_3$ nanoparticle suspension." *International Journal of Numerical Methods for Heat and Fluid Flow* no. 16 (3):275–292.

Maiga, S. E. B., S. J. Palm, C. T. Nguyen, G. Roy, and N. Galanis. 2005. "Heat transfer enhancement by using nanofluids in forced convection flows." *International Journal of Heat and Fluid Flow* no. 26 (4 SPEC. ISS.):530–546.

Manca, O., P. Mesolella, S. Nardini, and D. Ricci. 2011. "Numerical study of a confined slot impinging jet with nanofluids." *Nanoscale Research Letters* no. 6:1–16.

Manca, O., S. Nardini, and D. Ricci. 2012. "A numerical study of nanofluid forced convection in ribbed channels." *Applied Thermal Engineering* no. 37:280–292.

Mansour, R. B., N. Galanis, and C. T. Nguyen. 2007. "Effect of uncertainties in physical properties on forced convection heat transfer with nanofluids." *Applied Thermal Engineering* no. 27 (1):240–249.

Mare, T., S. Halelfadl, O. Sow, P. Estelle, S. Duret, and F. Bazantay. 2011. "Comparison of the thermal performances of two nanofluids at low temperature in a plate heat exchanger." *Experimental Thermal and Fluid Science* no. 35 (8):1535–1543.

Masuda, H., A. Ebata, K. Teramae, and N. Hishinuma. 1993. "Alteration of thermal conductivity and viscosity of liquid by dispersing ultra-fine particles (dispersion of g-Al$_2$O$_3$, SiO$_2$ and TiO$_2$ ultra-fine particles)." *Netsu Bussei* no. 4 (4):227–233.

Maxwell, J. C., and W. D. Niven. 1881. *A Treatise on Electricity and Magnetism, Clarendon Press Series*. Oxford: Clarendon Press.

Michaelides, E. E. 1986. "Heat transfer in particulate flows." *International Journal of Heat and Mass Transfer* no. 29 (2):265–273.

Mintsa, H. A., G. Roy, C. T. Nguyen, and D. Doucet. 2009. "New temperature dependent thermal conductivity data for water-based nanofluids." *International Journal of Thermal Sciences* no. 48 (2):363–371.

Murray, D. B. 1994. "Local enhancement of heat transfer in a particulate cross flow—I. Heat transfer mechanisms." *International Journal of Multiphase Flow* no. 20 (3):493–504.

Murshed, S. M. S., K. C. Leong, and C. Yang. 2008. "Investigations of thermal conductivity and viscosity of nanofluids." *International Journal of Thermal Sciences* no. 47 (5):560–568.

Namburu, P. K., D. K. Das, K. A. Tanguturi, and R. S. Vajjha. 2009. "Numerical study of turbulent flow and heat transfer characteristics of nanofluids considering variable properties." *International Journal of Thermal Sciences* no. 48 (2):290–302.

Namburu, P. K., D. P. Kulkarni, A. Dandekar, and D. K. Das. 2007. "Experimental investigation of viscosity and specific heat of silicon dioxide nanofluids." *Micro and Nano Letters* no. 2 (3):67–71.

Namburu, P. K., D. P. Kulkarni, D. Misra, and D. K. Das. 2007. "Viscosity of copper oxide nanoparticles dispersed in ethylene glycol and water mixture." *Experimental Thermal and Fluid Science* no. 32:397–402.

Nguyen, C. T., F. Desgranges, G. Roy, N. Galanis, T. Mare, S. Boucher, and H. Angue Mintsa. 2007. "Temperature and particle-size dependent viscosity data for water-based nanofluids—Hysteresis phenomenon." *International Journal of Heat and Fluid Flow* no. 28 (6):1492–1506.

Nguyen, C. T., N. Galanis, G. Polidori, S. Fohanno, C. V. Popa, and A. L. Bechec. 2009. "An experimental study of a confined and submerged impinging jet heat transfer using Al$_2$O$_3$-water nanofluid." *International Journal of Thermal Sciences* no. 48 (2):401–411. doi:10.1016/j.ijthermalsci.2008.10.007.

Nguyen, C. T., G. Roy, C. Gauthier, and N. Galanis. 2007. "Heat transfer enhancement using Al$_2$O$_3$-water nanofluid for an electronic liquid cooling system." *Applied Thermal Engineering* no. 27 (8–9):1501–1506. doi:10.1016/j.applthermaleng.2006.09.028.

Ollivier, E., J. Bellettre, M. Tazerout, and G. C. Roy. 2006. "Detection of knock occurrence in a gas SI engine from a heat transfer analysis." *Energy Conversion and Management* no. 47 (7–8):879–893. doi:10.1016/j.enconman.2005.06.019.

Otanicar, T. P., and J. S. Golden. 2009. "Comparative environmental and economic analysis of conventional and nanofluid solar hot water technologies." *Environmental Science and Technology* no. 43 (15):6082–6087.

Pak, B. C., and Y. I. Cho. 1998. "Hydrodynamic and heat transfer study of dispersed fluids with submicron metallic oxide particles." *Experimental Heat Transfer* no. 11 (2):151–170.

Palm, S. J., G. Roy, and C. T. Nguyen. 2006. "Heat transfer enhancement with the use of nanofluids in radial flow cooling systems considering temperature-dependent properties." *Applied Thermal Engineering* no. 26 (17–18):2209–2218.

Pantzali, M. N., A. A. Mouza, and S. V. Paras. 2009. "Investigating the efficacy of nanofluids as coolants in plate heat exchangers (PHE)." *Chemical Engineering Science* no. 64 (14):3290–3300.

Patankar, S. V. 1980. *Numerical Heat Transfer and Fluid Flow*. Washington, DC: Hemisphere.

Peng, H., G. L. Ding, W. T. Jiang, H. T. Hu, and Y. F. Gao. 2009. "Heat transfer characteristics of refrigerant-based nanofluid flow boiling inside a horizontal smooth tube." *International Journal of Refrigeration—Revue Internationale Du Froid* no. 32 (6):1259–1270.

Pletcher, R. H., J. C. Tannehill, and D. Anderson. 2013. *Computational Fluid Mechanics and Heat Transfer*, 3rd ed. Boca Raton, FL: CRC Press.

Poesio, P., G. Ooms, A. ten Cate, and J. C. R. Hunt. 2006. "Interaction and collisions between particles in a linear shear flow near a wall at low Reynolds numbers." *Journal of Fluid Mechanics* no. 555:113–130.

Prasher, R., D. Song, J. L. Wang, and P. Phelan. 2006. "Measurements of nanofluid viscosity and its implications for thermal applications." *Applied Physics Letters* no. 89 (13), Article ID 133108, 3pp.

Putra, N. Y., and F. N. Iskandar. 2011. "Application of nanofluids to a heat pipe liquid-block and the thermoelectric cooling of electronic equipment." *Experimental Thermal and Fluid Science* no. 35 (7):1274–1281.

Putra, N., W. Roetzel, and S. K. Das. 2003. "Natural convection of nanofluids." *Heat and Mass Transfer* no. 39:775–784.

Roberts, N. A., and D. G. Walker. 2010. "Convective performance of nanofluids in commercial electronics cooling systems." *Applied Thermal Engineering* no. 30 (16):2499–2504.

Roy, G., I. Gherasim, F. Nadeau, G. Poitras, and C. T. Nguyen. 2012. "Heat transfer performance and hydrodynamic behavior of turbulent nanofluid radial flows." *International Journal of Thermal Sciences* no. 58:120–129.

Roy, G. C., C. T. Nguyen, and M. Comeau. 2006. "Numerical investigation of electronic component cooling enhancement using nanofluids in a radial flow cooling system." *Journal of Enhanced Heat Transfer* no. 13 (2):1–15.

Roy, G., C. T. Nguyen, and P. R. Lajoie. 2004. "Numerical investigation of laminar flow and heat transfer in a radial flow cooling system with the use of nanofluids." *Superlattices and Microstructures* no. 35 (3–6):497–511.

Saffman, P. G. 1965. "The lift on a small sphere in a slow shear flow." *Journal of Fluid Mechanics* no. 22 (2): 385–400.

Sajadi, A. R., and M. H. Kazemi. 2011. "Investigation of turbulent convective heat transfer and pressure drop of TiO$_2$-water nanofluid in circular tube." *International Communications in Heat and Mass Transfer* no. 38:1474–1478.

Sarkar, J. 2011. "A critical review on convective heat transfer correlations of nanofluids." *Renewable and Sustainable Energy Reviews* no. 15 (6):3271–3277.

Sarkar, J., and R. Tarodiya. 2013. "Performance analysis of louvered fin tube automotive radiator using nanofluids as coolants." *International Journal of Nanomanufacturing* no. 9 (1):51–65.

Schlichting, H. 1979. *Boundary-Layer Theory*, 7th ed. New York: McGraw-Hill.

Selimefendigil, F., and H. F. Oztop. 2014. "Pulsating nanofluids jet impingement cooling of a heated horizontal surface." *International Journal of Heat and Mass Transfer* no. 69:54–65.

Sohn, C. W., and M. M. Chen. 1981. "Microconvective thermal conductivity in disperse two-phase mixtures as observed in a low velocity Couette flow experiment." *Journal of Heat Transfer* no. 103:45–51.

Syam Sundar, L., and K. V. Sharma. 2010. "Turbulent heat transfer and friction factor of Al$_2$O$_3$ nanofluid in circular tube with twisted tape inserts." *International Journal of Heat and Mass Transfer* no. 7–8:1409–1416.

Tanaka, T., and J. K. Eaton. 2008. "Classification of turbulence modification by dispersed spheres using a novel dimensionless number." *Physical Review Letters* no. 101 (11), Article ID 114502, 4pp.

Tavassoli, H., S. H. L. Kriebitzsch, M. A. van der Hoef, E. A. J. F. Peters, and J. A. M. Kuipers. 2013. "Direct numerical simulation of particulate flow with heat transfer." *International Journal of Multiphase Flow* no. 57:29–37.

Taylor, R. A., P. E. Phelan, T. P. Otanicar, R. Adrian, and R. Prasher. 2011. "Nanofluid optical property characterization: Towards efficient direct absorption solar collectors." *Nanoscale Research Letters* no. 6 (1):225.

Timofeeva, E. V. 2011. "Nanofluids for heat transfer—Potential and engineering strategies." In *Two Phase Flow, Phase Change and Numerical Modeling*, edited by A. Ahsan. Rijeka, Croatia: InTech, 435–450.

Tseng, S.-C., C. W. Lin, and K. Huang. 2005. "Heat transfer enhancement of nanofluids in rotary blade coupling of four-wheel-drive vehicles." *Acta Mechanica* no. 179:11–23.

Tyagi, H., P. Phelan, and R. Prasher. 2007. Predicted efficiency of a nanofluid-based direct absorption solar receiver. *Proceedings of the 2007 Energy Sustainability Conference*, June 27–30, 2007, Long Beach, CA. New York: American Society of Mechanical Engineers.

Vajjha, R. S., D. K. Das, and P. K. Namburu. 2010. "Numerical study of fluid dynamic and heat transfer performance of Al$_2$O$_3$ and CuO nanofluids in the flat tubes of a radiator." *International Journal of Heat and Fluid Flow* no. 31 (4):613–621.

Wang, X.-Q., and A. S. Mujumdar. 2008a. "A review on nanofluids—Part I: Theoretical and numerical investigations." *Brazilian Journal of Chemical Engineering* no. 25 (4):613–630.

Wang, X.-Q., and A. S. Mujumdar. 2008b. "A review on nanofluids—Part II: Experiments and applications." *Brazilian Journal of Chemical Engineering* no. 25 (4):631–648.

Warsi, Z. U. A. 1999. *Fluid Dynamics Theoretical and Computational Approaches*, 2nd ed. Boca Raton, FL: CRC Press.

Wasp, J. E., Kenny, P. J., and Ghandy, R. L. 1977. *Solid-Liquid Slurry Pipeline Transportation*. Clusthal, Germany: Trans Tech Publishing, 224pp.

Wen, D. S., and Y. L. Ding. 2004. "Experimental investigation into convective heat transfer of nanofluids at the entrance region under laminar flow conditions." *International Journal of Heat and Mass Transfer* no. 47 (24):5181–5188.

Wen, D. S., and Y. L. Ding. 2005. "Effect of particle migration on heat transfer in suspensions of nanoparticles flowing through minichannels." *Microfluidics and Nanofluidics* no. 1 (2):183–189.

Wen, D., G. Lin, S. Vafaei, and K. Zhang. 2009. "Review of nanofluids for heat transfer applications." *Particuology* no. 7 (2):141–150.

Xie, H., H. Lee, W. Youn, and M. Choi. 2003. "Nanofluids containing multiwalled carbon nanotubes and their enhanced thermal conductivities." *Journal of Applied Physics* no. 94 (8):4967–4971.

Xuan, Y. M., and Q. Li. 2000. "Heat transfer enhancement of nanofluids." *International Journal of Heat and Fluid Flow* no. 21 (1):58–64. doi:10.1016/S0142-727x(99)00067-3.

Xuan, Y. M., and Q. Li. 2003. "Investigation on convective heat transfer and flow features of nanofluids." *Journal of Heat Transfer—Transactions of the ASME* no. 125 (1):151–155.

Xuan, Y. M., and W. Roetzel. 2000. "Conceptions for heat transfer correlation of nanofluids." *International Journal of Heat and Mass Transfer* no. 43 (19):3701–3707.

Yang, Y.-T., and F.-H. Lai. 2010. "Numerical study of heat transfer enhancement with the use of nanofluids in radial flow cooling system." *International Journal of Heat and Mass Transfer* no. 53 (25–26):5895–5904.

Yang, Y.-T., and F.-H. Lai. 2011. "Numerical investigation of cooling performance with the use of Al_2O_3/water nanofluids in a radial flow system." *International Journal of Thermal Sciences* no. 50 (1):61–72.

Yang, Y., Z. G. Zhang, E. A. Grulke, W. B. Anderson, and G. Wu. 2005. "Heat transfer properties of nanoparticle-in-fluid dispersions (nanofluids) in laminar flow." *International Journal of Heat and Mass Transfer* no. 48 (6):1107–1116.

Yousefi, T., F. Veysi, E. Shojaeizadeh, and S. Zinadini. 2012. "An experimental investigation on the effect of Al_2O_3-H_2O nanofluid on the efficiency of flat-plate solar collectors." *Renewable Energy* no. 39 (1):293–298.

Yu, W., D. M. France, E. V. Timofeeva, D. Singh, and J. L. Routbort. 2012. "Comparative review of turbulent heat transfer of nanofluids." *International Journal of Heat and Mass Transfer* no. 55 (21–22):5380–5396.

Yu, W., and H. Xie. 2012. "A review on nanofluids: Preparation, stability mechanisms, and applications." *Journal of Nanomaterials* no. 2012, Article ID 435873, 17pp.

Zhou, S.-Q., and R. Ni. 2008. "Measurement of the specific heat capacity of water-based Al_2O_3 nanofluid." *Applied Physics Letters* no. 92:1–3.

CHAPTER 6

Experimental Study of Convective Heat Transfer in Nanofluids

Ehsan B. Haghighi, Adi T. Utomo,
Andrzej W. Pacek, and Björn E. Palm

CONTENTS

6.1 INTRODUCTION

The term nanofluid describing diluted suspensions of metal or metal oxide nanoparticles in water was first introduced by Choi and Eastman in 1995 during the International Mechanical Engineering Congress and Exhibition (ASME) congress [1]. They claimed, based on rather limited number of experimental data, that diluted suspensions of metal/metal oxide nanoparticles in water have unusually high thermal conductivity, much higher than expected based on the commonly used effective medium theory [2]. This chapter and the concept of "exceptional" nanofluids went rather unnoticed, and between 1995 and 2001, there were only a few papers published on nanofluids. In 2001, a US patent was granted [3], supported by two graphs with no error bars, claiming that thermal conductivity of fluids can be substantially increased by the addition of small amounts of metal/metal oxide nanoparticles. Since then the research on nanofluids has steeply accelerated with more than 2500 papers published between 2001 and 2014. It needs to be stressed here that despite this exponential growth in the number of publications, only part of heat transfer research community accepted the claims about exceptional properties of nanofluids. One of the authors attended a conference on heat transfer in nanofluids [4], where nearly 50% of the participants were highly skeptical about exceptional thermal properties of the nanofluids. The acceleration in research and very strong interest in nanofluids were not surprising. If the claims made in the US patent were correct and stable suspensions of nanoparticles with high thermal conductivity and relatively low viscosity could be produced at reasonable cost, this would be a serious breakthrough in a wide range of processes, in which the heat transfer is frequently a limiting step such as engine cooling, cooling of electronic devices, and nuclear systems cooling, to name a few [5].

The investigations on nanofluids reported in the literature can be divided into two types: (1) investigations of thermal conductivity and (2) investigations of heat transfer coefficients. In both cases, the studies have been mainly experimental. However, there were several attempts to develop a model to explain the alleged enhancement of thermal conductivity invoking Brownian motion of nanoparticles [6], agglomeration of nanoparticles increasing heat flux [7], high thermal conductivity of the liquids on the surface of nanoparticles [8], ballistic phonon transport in nanoparticles [9], or near-field radiation [10]. Each of these models was tested against selected experimental data, but none of them have been found to predict all experimental data well.

It appears that the argument on the unusual enhancement of thermal conductivity was put to rest by Buongiorno et al. [11] reporting data obtained in more than 30 institutions worldwide, showing that thermal conductivity of the nanofluids is well correlated by the models based on the effective medium theory. This means that the thermal conductivity cannot be drastically enhanced by the presence of nanoparticles in low concentrations.

In this chapter, we focus on the heat transfer coefficients of nanofluids in laminar and turbulent flows. In the past 10 years, it has been frequently reported, as briefly discussed subsequently, that the heat transfer coefficients of nanofluids are substantially higher than in the base liquids, but no plausible theoretical explanation has been offered. Typically, the experimental heat transfer coefficients are compared with the heat transfer coefficients of the base fluid (i.e., the fluid in which the nanoparticles are dispersed in) and with the heat transfer coefficients calculated from well-known correlations from the literature (very often incorrectly called "theoretical"). The majority of measurements reported in the literature were carried out by one research group/researcher in one experimental rig, and therefore there has always been a possibility of systematic experimental errors that can be very difficult to detect. All the results discussed in this work were obtained independently by two research groups [The Royal Institute of Technology (KTH), Stockholm, Sweden, and University of Birmingham (UB), Birmingham, West Midlands), and they were also scrutinized by 10 institutions involved in the European Project NanoHex [12].

6.2 LITERATURE REVIEW

The concept of heat transfer coefficient was introduced by Newton in the eighteenth century as a coefficient relating heat flux to the temperature difference between a surface and the surrounding fluid. There are a few, simple flow geometries where the heat transfer coefficients can be calculated from the solution of the fundamental equations (energy, mass, and momentum balances), but in most cases the heat transfer coefficients are determined from measurements. Such measurements resulted in semiempirical correlations for a specific flow type (internal or external), flow geometry, and flow regime (laminar or turbulent). The accuracy of those correlations (even the most common) should not be expected to be better than 15%.

Recently, many researchers have measured the heat transfer coefficients of different types of nanofluids in different flow regimes, and the majority reported enhancements of the heat transfer coefficients (compared with those of the base liquids) ranging from 15% to 300%. Some representative papers reporting the heat transfer coefficients in laminar [13,14] and turbulent flows are summarized in Table 6.1 and are briefly discussed subsequently.

Unfortunately, comparisons of the base fluids are often done at equal Reynolds numbers. As the increase in viscosity due to the addition of nanoparticles is usually much higher than the increase in thermal conductivity,

TABLE 6.1   Some Sources Reporting on Heat Transfer Coefficients in Nanofluids

Authors	Nanofluid	Dimension, Re	Enhancement of h_{nf} and Comments	Basis of Comparison
Wen and Ding [15]	Al_2O_3/water 1.6 vol.%	$D_i = 4.5$ mm $L/D_i = 216$ Re = 500–2100	47% near the inlet region, 14% near the discharge region	Equal Re
Hwang et al. [16]	Al_2O_3/water 0.3 vol.%	$D_i = 1.8$ mm $L/D_i = 1390$ Re = 400–700	8% in the developed region	Equal Re
Rea et al. [17]	Al_2O_3/water 6 vol.% ZrO_2/water 1.3 vol.%	$D_i = 4.5$ mm $L/D_i = 224$ Re = 140–1900	27% for alumina and 3% for zirconia	Equal velocity
Anoop et al. [18]	Alumina 4 wt.%	$D_i = 4.75$ mm $L/D_i = 253$ Re = 700–2000	25% for 45 nm particle size, 11% for 150 nm particle size	Equal Re
Liu and Yu [19]	Al_2O_3/water 5 vol.%	$D_i = 1.09$ mm $L/D_i = 280$ Re = 600–4500	19% near the entrance region, 9% near the discharge region	Equal Re
Vafaei and Wen [20]	Al_2O_3/water 1–7 vol.%	$D_i = 0.51$ mm $L/D_i = 600$	100% at high flow rate, no enhancement at low flow rate	Equal velocity
He et al. [21]	TiO_2/water 0.2–1.1 vol.%	$D_i = 3.97$ mm $L/D_i = 462$ Re = 900–5900	12% in laminar flow, 40% in turbulent flow	Equal Re
Ding et al. [22]	CNT/water 0.5 wt.%	$D_i = 4.5$ mm $L/D_i = 216$ Re = 800–1200	350% in developing flow	Equal Re
Garg et al. [23]	CNT/water 1 wt.%	$D_i = 1.55$ mm $L/D_i = 590$ Re = 600–1200	32% in developing flow	Equal Re
Liao and Liu [24]	CNT/water 0.5–2 wt.%	$D_i = 1.02$ mm $L/D_i = 217$ Re = 500– 10,000	18%–25% for 2 wt.% 49%–56% for 1 wt.%	Equal velocity
Pak and Cho [25]	γ-Al_2O_3, TiO_2/water 1–3 vol.%	$D_i = 10.66$ mm $L/D_i = 440$ Re = 104–105	12% decrease for 3 vol.%, γ-Al_2O_3/water	Equal velocity

(Continued)

TABLE 6.1 (*Continued*)   Some Sources Reporting on Heat Transfer Coefficients in
Nanofluids

Authors	Nanofluid	Dimension, Re	Enhancement of h_{nf} and Comments	Basis of Comparison
Kulkarni et al. [26]	TiO$_2$/ (EG–water) 2–10 vol.%	D_i = 3.14 mm L/D_i = 380 Re = 3000– 12,000	16% enhancement for 10 vol.% at Re = 10,000	Equal Re
Yu et al. [27]	SiC/water 3.7 vol.%	D_i = 2.27 mm L/D_i = 255 Re = 3300– 13,000	7% decrease	Equal velocity
Duangthongsuk and Wongwises [28]	TiO$_2$/water 0.2–2.0 vol.%	D_i = 9.53 mm L/D_i = 157 Re = 3000– 18,000	20%–32% enhancement at 1.0 vol.%	Equal Re
Fotukian and Nasr Esfahany [29]	CuO/water less than 0.24 vol.%	D_i = 5 mm L/D_i = 200 Re = 6000–31,000	Maximum 25% enhancement	Equal Re
Fotukian and Nasr Esfahany [30]	Al$_2$O$_3$/water <0.2 vol.%	D_i = 5 mm L/D_i = 360 Re = 5000– 30,000	48% enhancement at Re = 10,000 and 0.054 vol.%	Equal Re
Sajadi and Kazemi [31]	TiO$_2$/water ~0.25 vol.%	D_i = 5 mm L/D_i = 185 Re = 700–2050	22% enhancement at 0.25% Re = 5000	Equal Re
Suresh et al. [32]	Al$_2$O$_3$/water 0.3–0.5 vol.%	D_i = 5 mm L/D_i = 400 Re = 600– 16,000	10%–48% enhancement	Equal Re
Kayhani et al. [33]	TiO$_2$/water 0.1–2.0 vol.%	D_i = 3.7 mm L/D_i = 400 Re = 6000– 16,000	8% enhancement at 2.0 vol.% Re = 11,800	Equal Re
Haghighi et al. [34]	Al$_2$O$_3$, TiO$_2$, ZrO$_2$/water 9 wt.%	D_i = 3.7 mm L/D_i = 400 Re = 2000– 10,000	63%, 17%, and 52% decrease for Al$_2$O$_3$, TiO$_2$, and ZrO$_2$, respectively	Equal pumping power
Haghighi et al. [35]	Al$_2$O$_3$, TiO$_2$, CeO$_2$/water 9 wt.%	D_i = 0.5 mm L/D_i = 600 Re = 400–1000	Maximum 3.9% increase	Equal pumping power

CNT, carbon nanotube; EG, ethylene glycol.

the Prandtl numbers of the nanofluids are higher than those of the base fluids. Classical correlations for turbulent flow and developing laminar flow would therefore predict higher Nusselt numbers and higher heat transfer coefficients for the nanofluids than for the base fluids. This shows that comparisons at equal Reynolds numbers are irrelevant. Moreover, to maintain a given Reynolds number with a more viscous fluid, a higher velocity and a higher pumping power are required. As the pumping power is representing the "cost" of achieving the heat transfer, it would be most relevant to compare experimental results at equal pumping power, but this is rarely done in the literature. It should be noted that increased heat transfer coefficients can always be achieved in turbulent flow just by increasing the flow rate at the cost of increased pumping power.

As some of the reported enhancements of the heat transfer coefficients cannot be explained by standard heat transfer mechanisms (conduction, advection/convection, and radiation), several new mechanisms were proposed in the literature to explain this anomalous behavior. Wen and Ding [15] and Hwang et al. [16] suggested that in laminar flow the enhancement of the heat transfer coefficients of alumina–water nanofluids was due to the particle movement caused by the shear, viscosity gradient, thermophoresis, and Brownian motion. Wen and Ding [15] argued that particle migration created nonuniform thermal conductivity profiles across the channel, whereas Hwang et al. [16] postulated that particle migration toward the center of a pipe led to nonuniform viscosity profiles, which flatten the velocity profile and therefore increases the heat transfer coefficient. However, Buongiorno [36] claimed that in turbulent flow nanoparticles migrate toward the tube wall (into the boundary layer) due to the Brownian motion and thermophoresis, which might increase the heat transfer coefficients. For carbon nanotube (CNT) nanofluids, the formation of a three-dimensional network of nanotubes was also suggested as a reason for heat transfer coefficient enhancement [22,23]. These models agree with experimental data selected by the authors but often contradict experimental data of other investigators.

6.3 EXPERIMENTAL INVESTIGATION

6.3.1 Materials

During the recently completed EU-sponsored NanoHex project [12] aimed at the investigation and the potential exploitation of the enhanced thermal properties of nanofluids, many of the nanofluids reported in literature were tested: (1) water-based nanofluids with Al_2O_3, TiO_2, ZrO_2, SiC, CeO_2, SiO_2, China clay, carbon black, CNT, Ag, diamond, and Fe_2O_3 and (2) water/ethylene glycol solution-based nanofluids with Al_2O_3, TiO_2, CeO_2, Ag, and SiC. Preliminary tests did not show any significant enhancement neither of thermal conductivities nor of heat transfer coefficients. However, in certain flow configurations, heat transfer coefficients of water-based CNTs, alumina, and titania nanofluids were close to 15% higher than those of the base fluids at equal pumping power. Therefore, well-coordinated investigations of those nanofluids were carried out by KTH and UB.

The alumina and titania nanofluids were prepared by diluting concentrated suspensions supplied by ItN Nanovation AG, Saarbrücken, Germany, and Evonik AG, Essen, Germany. The pH, crystal phase, primary particle size, solid content, and concentration of stabilizers (surfactant and/or polymer) in those suspensions are summarized in Table 6.2 and the images of nanoparticles are shown in Figure 6.1.

The primary nanoparticles were irregular and they tend to aggregate into 120–220 nm agglomerates of approximately log-normal size distributions.

The concentrated suspensions were diluted keeping pH constant and homogenized using ultrasound. Diluted alumina and titania nanofluids

TABLE 6.2   pH, Primary Particle and Aggregate Sizes, Solid Content, and Additive Concentration of Investigated Suspensions of Nanoparticles

Nanofluids	Abbreviation	pH	Primary Particle Size (nm)[a]	Most Common Aggregate Size[b] (nm)	Solid Content (%)	g Additive/ g Solid
			Alumina			
ItN Nanovation	ITN-AL	9.1	100–200	200	43.2	0.018
Evonik (Aerodisp 440)	EVO-AL	4.1	10–20	150	40.6	0.014
			Titania			
ItN Nanovation	ITN-TI	7.8	20–50	140 (UB) 220 (KTH)	29.3	0.208
Evonik (Aerodisp W740X)	EVO-TI	6.7	20–50	130	40.6	0.030

[a] Based on TEM images.
[b] Based on dynamic light scattering (HPPS, Malvern, Worcestershire) measurement.

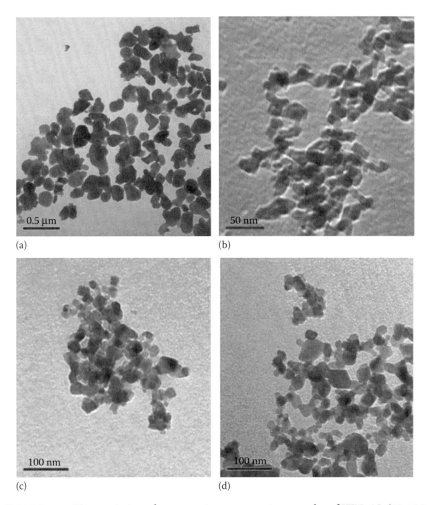

FIGURE 6.1 Transmission electron microscope micrographs of ITN-AL (50,000 magnification) (a), EVO-AL (500,000 magnification) (b), ITN-TI (300,000 magnification) (c), and EVO-TI (300,000 magnification) (d).

were very stable and no sedimentation was observed in several days. The manufacturers did not give the exact composition of additives used to stabilize the nanofluids.

The multiwalled CNT nanofluids were prepared by diluting water-based CNT paste (NanoAmor, Los Alamos, NM) in a 0.5 wt.% polyvinylpyrrolidone (molecular weight = 10,000) solution to 0.3 wt.% and homogenizing using ultrasound. Small aggregates were detected in CNT nanofluids even after prolonged sonication (Figure 6.2).

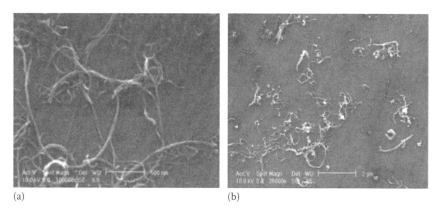

(a)   (b)

FIGURE 6.2   Scanning electron microscope images of CNTs at different magnifications: (a) 100,000; (b) 25,000.

6.4 EXPERIMENTAL SETUP

The heat transfer coefficients of the nanofluids were measured in horizontal stainless steel tubes with constant heat flux on the walls. Schematic diagrams of the experimental rigs at KTH and UB are shown in Figure 6.3a and b. The entrance sections prior to the test sections ensured that in each rig flow was hydrodynamically fully developed and the static mixers installed at the end of the heating sections ensured uniform temperature at the outlet.

In both rigs, the local external wall temperatures at 16 points (KTH) and 9 points (UB) along the test sections, the temperatures at the inlet and outlet, the flow rates, and the power supply were measured, and the data were logged by a data acquisition system and sent to a PC. Technical details of both rigs are summarized in Table 6.3.

The local heat transfer coefficients (h_x) were calculated from the values of the local wall and fluid temperatures and the local heat flux:

$$h_x = \frac{q''}{T_{w,i}(x) - T_b(x)} \tag{6.1}$$

From these values, the local Nusselt numbers were calculated.

The average Nusselt numbers were calculated as the area averages of the local Nusselt numbers.

$$Nu_{av} = \frac{\sum_{j}^{n} 0.5(Nu_j + Nu_{j+1})(x_{j+1} - x_j)}{x_n - x_1} \tag{6.2}$$

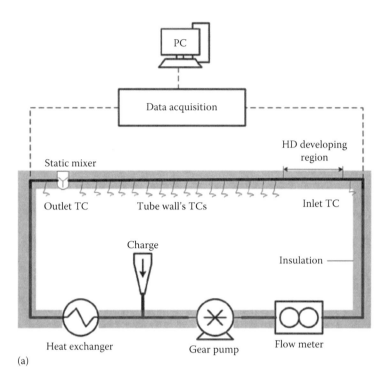

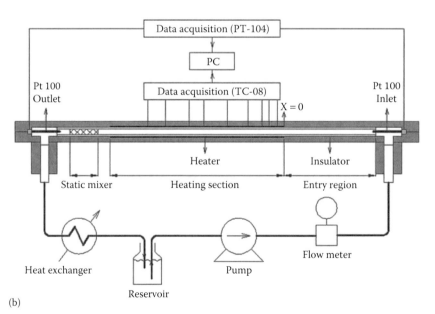

FIGURE 6.3 Experimental rigs in KTH (a) and Birmingham University (b). TC, thermocouple; HD, hydrodynamically.

TABLE 6.3   Details of the Test Sections in KTH and UB

Parameters	KTH	UB
Pipe (material, D_i, and wall thickness)	SS, 3.70 mm, and 0.15 mm	SS, 4.57 mm, and 0.89 mm
Entrance, heating, and mixing sections	250, 1468, and 80 mm	650, 1220, and 100 mm
Temperature recording: wall, inlet, and outlet	16 T-type (0.25 mm) thermocouples for the wall, 3 T-type (0.5 mm) for the inlet and the outlet	9 T-type (0.08 mm) thermocouples for the wall, 2 Pt 100 RTD (3 mm) for the inlet and the outlet
Thermocouple positions	18, 41, 66, 110, 210, 320, 423, 528, 632, 735, 840, 945, 1050, 1160, 1260, and 1360 mm from the start of heating section	45, 105, 158, 255, 400, 562, 664, 830, and 956 mm from the start of heating section
Accuracy of temperature measurement	Better than 0.1°C	0.03°C at 0°C for the Pt 100 RTD and 0.1°C for the thermocouples
Type of heater and power	Direct electric current through tube wall (DC), 3000 W	Electric tape heater, 300 W
Type of insulation and heat loss from the system	Armaflex foam (with $k \leq 0.036$ W/mK) and fiber glass insulation (with $k = 0.035$ W/mK), less than 5%	Phenolic foam insulator (with $k = 0.02$ W/mK), less than 5%
Pump	Gear pump (MCP-Z, Ismatec, Glattbrugg, Switzerland) with pump head (170-000, Micropump, Vancouver, WA)	Peristaltic pump (Watson-Marlow 520, Wilmington, DE)
Flow meter	Coriolis mass flow meter (CMFS015 with 2700 transmitter, Micro Motion, Ede, The Netherlands)	Coriolis mass flow meter (Optimass 3000-S3, KROHNE, Wellingborough, UK)
Cooling jacket	1.7 m double pipe, a plate heat exchanger, and a small chiller (180 W cooling capacity)	2 m double-pipe heat exchanger, chiller (400 W cooling capacity)

The heat supplied to the system (q) was calculated from the mass flow, the specific heat, and the temperature change of the fluid, $\dot{m}c_p(T_{out} - T_{in})$. The inner wall temperature, $T_{w,i}(x)$, at UB was calculated from the measured outer wall temperature, $T_{w,o}(x)$, and the solution of the steady-state, one-dimensional energy balance with constant heat flux boundary condition at the outer wall:

$$T_{w,i}(x) = T_{w,o}(x) - \frac{q}{2\pi L k_w} \ln \frac{D_i}{D_o} \tag{6.3}$$

At KTH, the inner wall temperature was also calculated from the solution of the energy equation but in this case with a volumetric energy source (heat generation) and perfect insulation at the outside as boundary conditions:

$$T_{w,i}(x) = T_{w,o}(x) - \frac{S}{16k_w}\left[D_o^2 \ln\left(\frac{D_i}{D_o}\right) - \left(D_o^2 - D_i^2\right)\right] \tag{6.4}$$

The local average bulk temperature, $T_b(x)$, was calculated from

$$T_b(x) = T_{in} + \frac{qx}{\dot{m}c_p L} \tag{6.5}$$

The measured values of local Nu_j and average Nu_{av} numbers in laminar flow were compared with the Shah correlation [37]:

$$Nu_j = \begin{cases} 1.302 x_j^{*-(1/3)} - 1 & x_j^* \leq 0.00005 \\ 1.302 x_j^{*-(1/3)} - 0.5 & 0.00005 \leq x_j^* \leq 0.0015 \\ 4.364 + 8.68(1000 x_j^*)^{-0.506} \exp(-41 x_j^*) & x_j^* > 0.0015 \end{cases} \tag{6.6}$$

$$Nu_{av} = \begin{cases} 1.953 L^{*-1/3} & L^* \leq 0.03 \\ 4.364 + 0.0722 L^{*-1} & L^* > 0.03 \end{cases} \tag{6.7}$$

where:

$$x^* = \frac{x/D_i}{RePr}$$

$$L^* = \frac{(x_n - x_1)/D_i}{RePr}$$

In the turbulent flow, the experimental results were compared to the values calculated from the Gnielinski correlation [38]:

$$Nu = \frac{(f/8)(Re-1000)Pr}{1+12.7(f/8)^{0.5}(Pr^{0.66}-1)}\left[1+\left(\frac{d}{L}\right)^{2/3}\right]\left(\frac{Pr}{Pr_w}\right)^{0.11} \tag{6.8}$$

To assess the accuracy of the measurements in both experimental rigs, the Nusselt numbers in laminar and turbulent flows of distilled water were measured in both rigs and the results are compared with the values calculated from Equations 6.6 and 6.8 in Figure 6.4. These results show the good agreement between experimental data obtained at UB and KTH. The Nusselt numbers in laminar and turbulent flows and at inlet temperatures of 25°C and 40°C agree in most cases with the calculated values within ±10%, which is the typical error band for this type of measurements.

6.5 THERMOPHYSICAL PROPERTIES OF NANOFLUIDS

The thermal conductivities of the nanofluids were measured at 20°C with KD2 Pro (Decagon Devices, Inc., Pullman, Washington; transient hot-wire method) and TPS 2500 (HotDisk AB, Göteborg, Sweden; transient plane source method) instruments at UB and KTH, respectively. Both instruments were tested with several pure liquids (water, ethylene glycol, and glycerol) and the deviations from reference values were within ±2%.

There was also a very good agreement between the measured thermal conductivities and values calculated from the Hamilton and Crosser model [39].

$$\frac{k_{nf}}{k_f} = \frac{k_p+(n-1)k_f+(n-1)\phi(k_p-k_f)}{k_p+(n-1)k_f-\phi(k_p-k_f)} \tag{6.9}$$

The thermal conductivities of ItN Nanovation (ITN-AL) and Evonik (EVO-TI) were also measured at 40°C and the experimental data followed the theoretical predictions based on the excluded volume model. This suggests that the effect of the nanoparticles' Brownian motion on the thermal conductivity of nanofluids is negligible. None of the investigated nanofluids show anomalously high thermal conductivity, and this is in agreement with results reported by Buongiorno et al. [11].

The viscosity of nanofluids was measured at shear rates between 10 and 120 per second using rotational rheometers with a plate and cone geometry (AR 1000, TA Instruments, New Castle, DE) and a coaxial cylinder

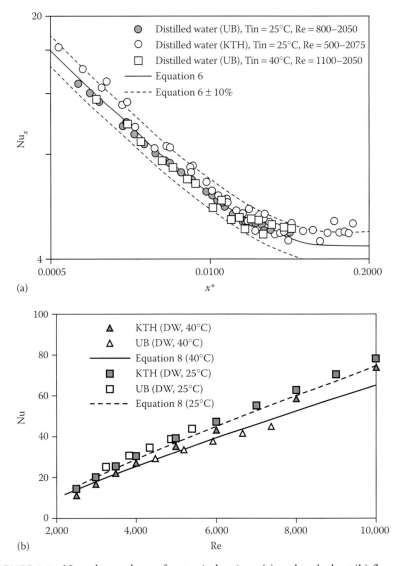

FIGURE 6.4 Nusselt numbers of water in laminar (a) and turbulent (b) flows in pipe. DW, Distilled water.

geometry (Brookfield DV-II + Pro, Middleboro, MA) at UB and KTH, respectively. All metal oxide nanofluids were Newtonian, whereas the CNT nanofluids were very weakly shear thinning ($n = 0.95$). Therefore, they were also treated as Newtonian. The viscosity of all nanofluids was higher than that predicted by the Einstein–Batchelor model, which can be explained by formation of aggregates increasing the effective volume fraction of the suspended solid [40,41]. The effective volume fraction of aggregates (ϕ_a) can be estimated by using the fractal theory [40]:

$$\phi_a = \phi \left(\frac{r_a}{r_p} \right)^{3-df} \tag{6.10}$$

The above equation indicates that the smaller the ratio of the aggregate sizes to the primary particle size, the lower the effective volume fraction of the aggregates. High viscosity of CNT nanofluids was most likely due to entanglement between CNT fibers [23,24]. The rheology of nanoparticle suspensions is rather complex, and it is affected by repulsive electrostatic forces, attractive van der Waals forces, steric stabilization, and bridging. Therefore, a detailed discussion of the effect of morphology on rheology is well outside the scope of this text.

The specific heat and density of nanofluids were calculated by using mass fraction (x) and volume fraction (ϕ) weighted averages of specific heats and densities of the particles and water:

$$c_{p,nf} = x_p c_{p,p} + (1 - x_p) c_{p,bf} \tag{6.11}$$

$$\rho_{nf} = \phi_p \rho_p + (1 - \phi_p) \rho_{bf} \tag{6.12}$$

The accuracy of Equations 6.9 through 6.12 for nanofluids was tested [14,42,43] and the deviations from experimental values were within 5%. The physical properties of all investigated nanofluids are summarized in Table 6.4.

At 40°C all physical properties of the investigated nanofluids were practically the same as those at 20°C. The fact that the changes of physical properties of nanofluids are marginal compared to water itself supports the hypothesis that the heat transfer coefficient in the water-based nanofluids in the same flow regime should not be very different from the heat transfer coefficient in water.

6.6 HEAT TRANSFER COEFFICIENTS IN LAMINAR FLOW

The heat transfer coefficients were measured at solid concentrations of 9 wt.% (≈ 2.4 vol.%) for metal oxide nanofluids and 0.3 wt.% (~ 0.14 vol.%) for CNT nanofluids. These values are within the typical range of solid concentrations reported in the literature. The local Nusselt numbers of all nanofluids in laminar flow are summarized in Figure 6.5 and compared with the predictions from Equation 6.6. Figure 6.5a–c shows the local Nusselt numbers for each nanofluid calculated from measured

TABLE 6.4 Thermophysical Properties of Investigated Nanofluids Relative to Water

Property	ITN-AL 9 wt.% 20°C	EVO-AL 9 wt.% 20°C	ITN-TI 9 wt.% 20°C	EVO-TI 9 wt.% 20°C	CNTs 0.3 wt.% 20°C
Thermal Conductivity					
UB	1.074	1.044	1.034	1.050	1.019
KTH	1.088	1.058	1.035	1.058	–
Average deviation	$1.081 \pm 0.7\%$	$1.051 \pm 0.7\%$	$1.035 \pm 0.1\%$	$1.054 \pm 0.4\%$	–
Equation 6.9	1.072	1.086	1.061	1.062	1.027[a]
Viscosity					
Measured by					
UB	1.100	1.410	1.410	1.190	1.49
KTH	1.190	1.440	1.970	1.280	–
Average deviation	$1.145 \pm 3.9\%$	$1.425 \pm 1.0\%$	–	$1.235 \pm 3.6\%$	–
Einstein–Batchelor model	1.065	1.078	1.065	1.067	1.004
Specific Heat Capacity					
Equation 6.11	0.926	0.926	0.925	0.925	0.997
Density					
UB	1.071	1.067	1.073	1.071	1.003
KTH	1.077	1.071	1.077	1.075	–
Average ± deviation	$1.074 \pm 0.3\%$	$1.069 \pm 0.2\%$	$1.075 \pm 0.2\%$	$1.073 \pm 0.2\%$	–
Equation 6.12	1.072	1.067	1.072	1.071	1.002

[a] The aspect ratio is assumed to be 100.

local temperatures and the local heat fluxes. In Figure 6.5d, the average Nusselt numbers for all investigated nanofluids and water are summarized and compared with the predictions from Equation 6.6. In all cases, the great majority of the experimental data agree with Equation 6.6 within ±10%, strongly suggesting that the heat transfer coefficients of the nanofluids can be predicted from the classical correlations developed for single-phase liquids, although nanofluids are solid–liquid suspensions. These results are also in very good agreement with the results reported by Rea et al. [17] for alumina and zirconia nanofluids in laminar flows.

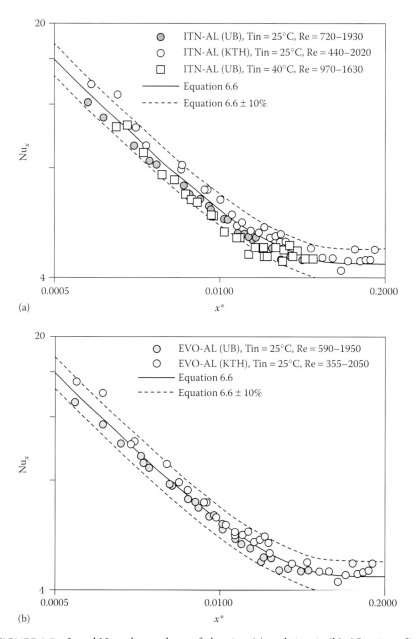

FIGURE 6.5   Local Nusselt numbers of alumina (a) and titania (b). (*Continued*)

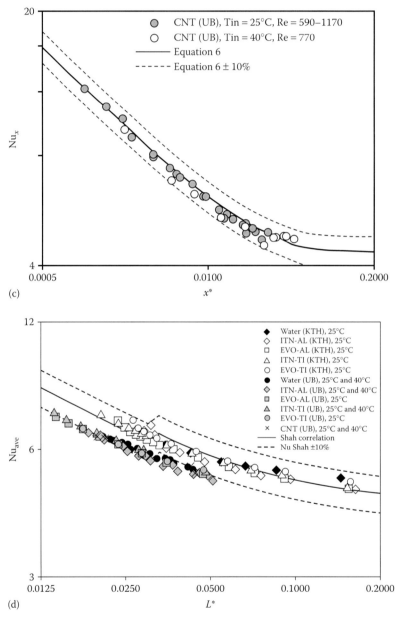

FIGURE 6.5 *(Continued)* Local Nusselt numbers of CNT (c) nanofluids in laminar flow and average Nusselt numbers (d) for all investigated nanofluids and water (for comparison).

At a higher inlet temperature of 40°C, the local Nusselt numbers of alumina, titania, and CNT nanofluids also follow the correlation in Equation 6.6, suggesting that the Brownian motion of nanoparticles does not affect the heat transfer coefficients of nanofluids even at this slightly higher temperature. These results again agree well with Buongiorno [36]. Hwang et al. [16] also concluded that the effect of nanoparticles' Brownian diffusion and thermophoresis on the heat transfer in nanofluids is negligible because the diffusion coefficients for nanoparticles' Brownian motion and thermophoresis (on the order of 10^{-10}–10^{-12} m²/s) are too small compared to the thermal diffusivity of the base fluid (~10^{-7} m²/s for water). The thermophoretic velocity of nanoparticles, which is on the order of 10^{-7} m/s [37], is also too small compared to the average fluid velocity in this study, which is about 0.1–0.5 m/s to affect the convective heat transfer in radial direction. Even in microchannels as small as 50 μm × 190 μm, the movement of nanoparticles is negligible, and Escher et al. [44] reported that silica nanofluids behave as homogeneous mixtures.

The experimental data from both KTH and UB agree with the prediction of the average Nusselt numbers from Equation 6.6, in most cases within ±10% as shown in Figure 6.5.

6.7 HEAT TRANSFER COEFFICIENTS IN TURBULENT FLOW

For the analysis of the effect of nanoparticles on heat transfer in turbulent flow, we may first consider the fact that the Kolmogorov scale (k_k) in typical turbulent pipe flow (Re from 10,000 to 20,000) is between 2 μm and 10 μm, which means that the nanoparticles are more than 100 times smaller than the smallest scale of turbulent fluctuations. In simple terms, nanoparticles "do not know" that they are in turbulent flow as they are exposed to the laminar flow inside the smallest turbulent eddies. However, considering the fact that several researchers [21,26,28,30–33] reported enhancement of the heat transfer coefficients in turbulent flow, it was decided to measure the heat transfer coefficients in such flows as well.

The average Nusselt numbers are summarized in Figure 6.6a and b and compared with the predictions from Equation 6.8. Figure 6.6a and b shows similar results for ITN-AL and EVO-TI nanofluids. The physical properties of Nu and Re numbers were calculated from Equations 6.11 and 6.12, whereas viscosity was measured. Most of the experimental data for both

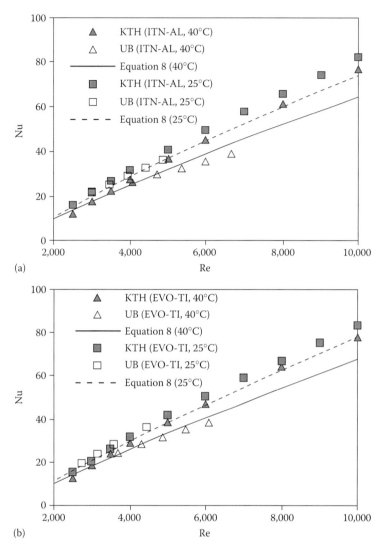

FIGURE 6.6 Average Nusselt numbers of alumina (a) and titania (b) nanofluids in the transitional and turbulent flows.

nanofluids agree with the predicted values within ±10% and no points deviate more than 20%.

According to the results for both fluids, the heat transfer coefficient correlation suggested by Gnielinski [38] for single-phase turbulent flow accurately predicts heat transfer in nanofluids, provided correct thermophysical properties of the nanofluids are used. Similar results were reported by Williams et al. [45], Liu and Yu [19], Chandrasekar et al. [32], and Kulkarni et al. [26]. Clearly none of the nanofluids investigated in this work showed anomalous heat transfer coefficient enhancement relative to that of the base fluid.

6.8 CONCLUSIONS

The heat transfer coefficients in laminar and turbulent flows of Al_2O_3, TiO_2, and CNT nanofluids have been measured independently by UB and KTH. The measurements were carried out in very similar, specially designed experimental rigs, and the accuracy of the measurements was assessed by the measurements of heat transfer coefficients of water. Most data points for water were within 10% of the predictions of commonly used correlations.

The heat transfer coefficients of the nanofluids measured at both universities were within similar error bands, and they were also in good agreement with the values calculated from literature correlations for laminar and turbulent flows, if the correct properties of the nanofluids were used. Based on these results, it can be concluded that, as far as macroscopic thermal and hydrodynamic behaviors are concerned, the nanofluids investigated in this work behave as homogeneous mixtures and their thermal performance can be predicted from classical correlations from the literature. No unusual enhancement of the thermal conductivity or heat transfer coefficients could thus be found.

NOMENCLATURE

av	average
c_p	specific heat (kJ kg^{-1} K^{-1})
L	length of the tube (m)
D	diameter (m)
f	Darcy friction factor
h	heat transfer coefficient (Wm^{-2}K^{-1})
i	inside
k	thermal conductivity (Wm^{-1}K^{-1})
$k_k = (v^3/\varepsilon)^{0.25}$	Kolmogorov length scale (m)
$\dot{m}$	mass flow rate (kg/s)
$Nu_j = (h_j D)/k$	local Nusselt number
Nu_{av}	average Nusselt number
o	outside
$q = \dot{m}c_p(T_{out} - T_{in})$	heat supplied to the system (W)
q''	heat flux (Wm^{-2})
$S = 4q/\left[\pi L\left(D_o^2 - D_i^2\right)\right]$	volumetric heat source (Wm^{-3})

T	temperature (°C)
w	wall
x	thermocouple position, distance from the inlet (m)
x_p	solid mass fraction
ε	energy dissipation rate (m²s⁻³)
ν	kinematic viscosity (m²s⁻¹)
ρ	density (kg m⁻³)
$\phi = (x_p/\rho_p)/[x_p/\rho_p + (1-x_p)/\rho_{bf}]$	solid volume fraction

SUBSCRIPTS

bf	base fluid
in	inlet
nf	nanofluid
out	outlet
p	particle
w,i	wall, inside
w,o	wall, outside

REFERENCES

1. S. U. S. Choi, and J. A. Eastman. Enhancing thermal conductivity of fluids with nanoparticles. In D. A. Siginer and H. P. Wang (eds.), *Developments and Applications of Non-Newtonian Flows*, FED-vol. 231/MD-vol. 66, ASME International Mechanical Engineering, New York, 1995, pp. 99–105.
2. M. Chandrasekar, and S. Suresh. A review on the mechanisms of heat transport in nanofluids. *Heat Transfer Engineering* 30 (14), 2009, 1136–1150.
3. S. U. S. Choi, and J. A. Eastman. Enhanced heat transfer using nanofluids, US Patent No. U.S. 6,221,275 B1, 2001.
4. Nanofluids: Fundamentals and applications, *ECI Conference*, September 16–20, 2007, Copper Mountain, CO.
5. S. K. Das, S. U. S. Choi, W. Yu, and T. Pradeep. *Nanofluids: Science and Technology*, Wiley, USA, 2008, pp. 1–13.
6. S. P. Jang, and S. U. S. Choi. Role of Brownian motion in the enhanced thermal conductivity of nanofluids. *Applied Physics Letters* 84 (21), 2004, 4316.
7. W. Evans, R. Prasher, J. Fish, P. Meakin, P. Phelan, and P. Keblinski. Effect of aggregation and interfacial thermal resistance on thermal conductivity of nanocomposites and colloidal nanofluids. *International Journal of Heat and Mass Transfer* 51 (5/6), 2008, 1431–1438.
8. W. Yu, and S. U. S. Choi. The role of interfacial layers in the enhanced thermal conductivity of nanofluids: A renovated Maxwell model. *Journal of Nanoparticle Research* 5 (1/2), 2003, 167–171.

9. P. Keblinski, S. R. Phillpot, S. U. S. Choi, and J. A. Eastman. Mechanisms of heat flow in suspensions of nano-sized particles (nano fluids). *International Journal of Heat and Mass Transfer* 45, 2002, 855–863.

10. G. Domingues, S. Volz, K. Joulain, and J.-J. Greffet. Heat transfer between two nanoparticles through near field interaction. *Physical Review Letters* 94 (8), 2005, 2–5.

11. J. Buongiorno, D. C. Venerus, N. Prabhat, T. McKrell, J. Townsend, R. Christianson, and Y. V. Tolmachev. A benchmark study on the thermal conductivity of nanofluids. *Journal of Applied Physics* 106 (9), 2009, 4312.

12. NanoHex-enhanced nano-fluid heat exchange, EU project, FP7, grant agreement 22882.

13. A. T. Utomo, P. Robbins, H. Poth, and A. W. Pacek. Experimental and theoretical study of thermal conductivity, viscosity and heat transfer coefficient of titania and alumina nanofluids. *International Journal of Heat and Mass Transfer* 55, 2012, 7772–7781.

14. A. T. Utomo, E. B. Haghighi, A. I. T. Zavareh, M. Ghanbarpourgeravi, H. Poth, R. Khodabandeh, B. Palm, and A. W. Pacek. The effect of nanoparticles on laminar heat transfer in a horizontal tube. *International Journal of Heat and Mass Transfer* 69 (2014), 2013, 77–91.

15. D. Wen, and Y. Ding. Experimental investigation into convective heat transfer of nanofluids at the entrance region under laminar flow condition. *International Journal of Heat and Mass Transfer* 47 (24), 2004, 5181–5188.

16. K. S. Hwang, S. P. Jang, and S. U. S. Choi. Flow and convective heat transfer characteristics of water-based Al_2O_3 nanofluids in fully developed laminar flow regime. *International Journal of Heat and Mass Transfer* 52 (1–2), 2009, 193–199.

17. U. Rea, T. McKrell, L. W. Hu, and J. Buongiorno. Laminar convective heat transfer and viscous pressure loss of alumina-water and zirconia-water nanofluids. *International Journal of Heat and Mass Transfer* 52 (7–8), 2009, 2042–2048.

18. K. B. Anoop, T. Sundararajan, and S. K. Das. Effect of particle size on the convective heat transfer in nanofluid in the developing region. *International Journal of Heat and Mass Transfer* 52 (9–10), 2009, 2189–2195.

19. D. Liu, and L. Yu. Experimental investigation of single-phase convective heat transfer of nanofluids in a mini-channel. *Proceedings of the 14th International Heat Transfer Conference*, IHTC 14-23018, ASME, Washington, DC, August 8–13, 2010.

20. S. Vafaei, and D. Wen. Convective heat transfer of alumina nanofluids in a microchannel. *Proceedings of 14th International Heat Transfer Conference*, IHTC14-22206, ASME, Washington, DC, August 8–13, 2010, vol. 6, pp. 585–589.

21. Y. He, Y. Jin, H. Chen, Y. Ding, D. Cang, and H. Lu. Heat transfer and flow behavior of aqueous suspensions of TiO_2 nanoparticles (nanofluids) flowing upward through a vertical pipe. *International Journal of Heat and Mass Transfer* 50 (11–12), 2007, 2272–2281.

22. Y. L. Ding, H. Alias, D. S. Wen, and R. A. Williams. Heat transfer of aqueous suspensions of carbon nanotubes (CNT nanofluids). *International Journal of Heat and Mass Transfer* 49 (1–2), 2006, 240–250.
23. P. Garg, J. L. Alvarado, C. Marsh, T. A. Carlson, D. A. Kessler, and K. Annamalai. An experimental study on the effect of ultrasonication on viscosity and heat transfer performance of multi-wall carbon nanotube-based aqueous nanofluids. *International Journal of Heat and Mass Transfer* 52 (21–22), 2009, 5090–5101.
24. L. Liao, and Z. H. Liu. Forced convective flow drag and heat transfer characteristics of carbon nanotube suspensions in a horizontal small tube. *Heat Mass Transfer* 45 (8), 2009, 1129–1136.
25. B. C. Pak, and Y. I. Cho. Hydrodynamic and heat transfer study of dispersed fluids with submicron metallic oxide particles. *Experimental Heat Transfer* 11 (2), 1998, 151–170.
26. D. P. Kulkarni, P. K. Namburu, H. E. Bargar, and D. K. Das. Convective heat transfer and fluid dynamic characteristics of SiO_2-ethylene glycol/water nanofluid. *Heat Transfer Engineering* 29, 2008, 1027–1035.
27. W. Yu, D. M. France, D. S. Smith, D. Singh, E. V. Timofeeva, and J. L. Routbort. Heat transfer to a silicon carbide/water nanofluid. *International Journal of Heat and Mass Transfer* 52 (15–16), 2009, 3606–3612.
28. W. Duangthongsuk, and S. Wongwises. An experimental study on the heat transfer performance and pressure drop of TiO_2-water nanofluids flowing under a turbulent flow regime. *International Journal of Heat Mass Transfer* 53 (1–3), 2010, 334–344.
29. S. Fotukian, and M. Nasr Esfahany. Experimental study of turbulent convective heat transfer and pressure drop of dilute CuO/water nanofluid inside a circular tube. *International Communication in Heat and Mass Transfer* 37 (2), 2010, 214–219.
30. S. Fotukian, and M. Nasr Esfahany. Experimental investigation of turbulent convective heat transfer of dilute γ-Al_2O_3/water nanofluid inside a circular tube. *International Journal of Heat and Fluid Flow* 31 (4), 2010, 606–612.
31. A. Sajadi, and M. Kazemi. Investigation of turbulent convective heat transfer and pressure drop of TiO_2/water nanofluid in circular tube. *International Communication in Heat and Mass Transfer* 38 (10), 2011, 1474–1478.
32. S. Suresh, P. Selvakumar, M. Chandrasekar, and V. Srinivasa Raman. Experimental studies on heat transfer and friction factor characteristics of Al_2O_3/water nanofluid under turbulent flow with spiraled rod inserts. *Chemical Engineering and Processing* 53, 2012, 24–30.
33. M. H. Kayhani, H. Soltanzadeh, M. M. Heyhat, M. Nazari, and F. Kowsary. Experimental study of convective heat transfer and pressure drop of TiO_2/water nanofluids. *International Communication in Heat and Mass Transfer* 39, 2012, 456–462.
34. E. B. Haghighi, M. Saleemi, N. Nikkam, R. Khodabandeh, M. S. Toprak, M. Muhammed, and B. Palm. Accurate basis of comparison for convective heat transfer in nanofluids. *International Communication in Heat and Mass Transfer* 52, 2014, 1–7.

35. E. B. Haghighi, M. Saleemi, N. Nikkam, Z. Anwar, I. Lumbreras, M. Behi, S. A. Mirmohammadi et al. Cooling performance of nanofluids in a small diameter tube. *Experimental Thermal and Fluid Science* 49, 2013, 114–122.
36. J. Buongiorno. Convective transport in nanofluid. *Journal of Heat Transfer* 128 (3), 2006, 240–250.
37. R. K. Shah, and A. London. *Laminar Flow Forced Convection in Ducts, Supplement 1 to Advances in Heat Transfer*, Academic Press, New York, 1978, pp. 128–129.
38. V. Gnielinski. New correlations for heat and mass transfer in turbulent pipe and channel flow. *Forsch Ingenieurwes* 41 (1), 1975, 8–16.
39. R. L. Hamilton, and O. K. Crosser. Thermal conductivity of heterogeneous two component systems. *Industrial and Engineering Chemistry* 1 (3), 1962, 187–191.
40. P. Ding, and A. W. Pacek. Effect of pH on de-agglomeration and rheology/morphology of aqueous suspensions of goethite nano-powder. *Journal of Colloid and Interface Science* 325, 2008, 165–172.
41. R. Prasher, D. Song, J. Wang, and P. Phelan. Measurement of nanofluid viscosity and its implications for thermal applications. *Applied Physics Letters* 89 (13), 2006, 133108.
42. M. N. Pantzali, A. A. Mouza, and S. V. Paras. Investigating the efficacy of nanofluids as coolants in plate heat exchangers (PHE). *Chemical Engineering Science* 64 (14), 2009, 3290–3300.
43. H. O'Hanley, J. Buongiorno, T. McKrell, and L. W. Hu. Measurement and model validation of nanofluid specific heat capacity with differential scanning calorimetry. *Advances in Mechanical Engineering* 2012, 2012, 181079.
44. W. Escher, T. Brunschwiler, N. Shalkevich, A. Shalkevich, T. Burgi, B. Michel, and D. Poulikakos. On the cooling of electronics with nanofluids. *Journal of Heat Transfer* 133 (05), 2011, 051401.
45. W. Williams, J. Buongiorno, and L. W. Hu. Experimental investigation of turbulent convective heat transfer and pressure loss of alumina/water and zirconia/water nanoparticle colloids (nanofluids) in horizontal tubes. *Journal of Heat Transfer* 130 (4), 2008, 042412.

CHAPTER 7

Performance of Heat Exchangers Using Nanofluids

Bengt Sundén and Zan Wu

CONTENTS

7.1 INTRODUCTION

Heat exchangers are equipments being used for transfer of heat between two or more fluids at different temperatures (Sundén 2012). They are widely used in various fields, including power plants, automotives, space heating, refrigeration and air-conditioning systems, chemical plants, petrochemical processes, electronic cooling, and environment engineering. Many heat transfer enhancement techniques have been developed for heat exchangers to improve their thermal efficiency by means of surface area enlargement and boundary layer modification. Conventional heat transfer fluids such as water, ethylene glycol, and engine oil have relatively low thermal conductivity values, which thus limit the heat transfer rates. Due to recent progress in nanotechnology, thermal conductivity values can be increased by adding nanometer-sized structures (e.g., particles, fibers, tubes) in conventional heat transfer fluids to form the so-called nanofluids.

Property characterization of nanofluids has been reviewed extensively by Fan and Wang (2011), Kleinstreuer and Feng (2011), and Taylor et al. (2013). The increase in thermal conductivity depends on particle volume fraction, morphology (e.g., primary size, shape, agglomeration in the form of clusters and aggregates), dispersion, colloidal stability, and fluid temperature. Large divergence exists in reported thermal conductivity enhancement in the literature and no agreement has been achieved on enhancement mechanisms. As stated by Wu, Feng et al. (2014) based on the classic effective medium theory, there is still a lot of room for thermal conductivity enhancement of nanofluids. Further thermal conductivity enhancement can be obtained by manipulating the nanostructure morphology in nanofluids to form elongated and/or percolated aggregates. Rheology behavior of nanofluids is different from that of base fluids. The increase in viscosity of the nanofluid requires larger pumping cost and thus neutralizes the benefit from the thermal conductivity enhancement. Besides, nanofluids of relatively large particle volume fractions or carbon nanotube nanofluids present non-Newtonian behavior, which may cause problems for heat exchangers designed for Newtonian fluids. Therefore, understanding of nanofluid properties is critical for heat transfer applications. More investigations should be conducted to achieve large thermal conductivity enhancement with relatively low viscosity increase, for example, by modifying the nanostructure morphology in nanofluids.

The main objective of this chapter is to first give a state-of-the-art overview of nanofluid performance in tubes and heat exchangers, and then present an experimental investigation of aqueous alumina and multiwalled carbon nanotube (MWCNT) nanofluids in a double-pipe helical heat exchanger for possible heat transfer enhancement. Besides, different performance evaluation criteria are checked to evaluate the nanofluid performance. Challenges and future research needs are also presented.

7.1.1 Performance of Nanofluids in Tubes and Heat Exchangers

Heat transfer characteristics of nanofluids in straight tubes have been extensively studied, as shown in reviews of Dalkilic et al. (2012) and Hussein et al. (2014). However, no agreement on anomalous heat transfer enhancement has been achieved to date. Sergis and Hardalupas (2011) stated statistically that most of the previous studies indicated low heat transfer enhancement; 11% of the sample showed deterioration of the heat transfer coefficient and 3% indicated no enhancement at all, as shown in Figure 7.1. Prabhat et al. (2011), Williams et al. (2008), and Yu et al. (2010)

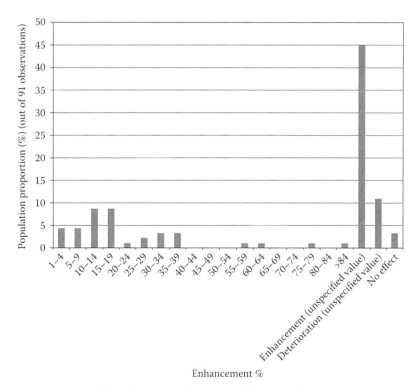

FIGURE 7.1 Probability function of convective heat transfer enhancement. (Data from Sergis, A., and Y. Hardalupas, *Nano. Res. Lett.*, 6, 391, 2011.)

concluded that there seems to be anomalous heat transfer enhancement in the entrance region in nanofluid laminar flow, although previous single-phase correlations can accurately reproduce the turbulent convective heat transfer behavior of nanofluids in tubes by adopting the measured temperature- and loading-dependent thermal conductivities and viscosities of the nanofluids in the analysis. Yang et al. (2013) obtained exact solutions for fully developed laminar flow in straight channels and tubes and concluded that (1) the anomalous heat transfer enhancement was captured, especially for the case of titania–water nanofluids in a tube when the nanoparticle volume fractions are larger than 2% and (2) the maximum Nusselt number based on the bulk mean nanofluid thermal conductivity is achieved at $N_{BT} \approx 0.5$, although it becomes lower than that of the pure fluid at $N_{BT} < 0.3$. The parameter N_{BT} indicates the ratio of Brownian and thermophoretic diffusivities.

So far, there have been few studies on heat transfer characteristics of nanofluids in complex geometries (e.g., microchannels, helically coiled tubes, enhanced tubes) and heat exchangers. Huminic and Huminic (2012) reviewed about 20 published papers on application of nanofluids in various types of heat exchangers. Here, we only summarize recent investigations in this area. Escher et al. (2011) experimentally investigated laminar flow thermal performance of silica nanofluids with volume fractions up to 31% in microchannels and demonstrated that previous standard correlations can be used to estimate the convective heat transfer coefficient by using measured thermal conductivities and viscosities. Similarly, Haghighi et al. (2013) reported heat transfer performance in laminar flow for five nanofluids in a microchannel of 0.5 mm inner diameter and found that the local Shah correlation (for heat transfer coefficient) and the Darcy equation (for friction factor) are still valid for the tested nanofluids as long as the correct thermophysical properties are employed. Seyf and Mohammadian (2011) numerically simulated laminar nanofluid flow in a counterflow microchannel heat exchanger and suggested that nanofluid can enhance the heat exchanger effectiveness by Brownian motion of nanoparticles. In this study, a single-phase approach was used for nanofluid modeling and arbitrary thermal conductivity and viscosity values were adopted in the simulation. Tiwari et al. (2013) experimentally studied the hydraulic and thermal performance of various nanofluids in a plate heat exchanger. It showed that the CeO_2/water nanofluid at optimum volume fraction yields the best performance index ratio. The performance index is defined as the ratio of heat transferred between the fluids to pumping power. An experimental investigation was carried out by Taws et al. (2012) to determine the performance of CuO/water nanofluids in a chevron-type industrial plate heat exchanger. Results showed no significant Nusselt number enhancement for the nanofluid with 2.0% particle volume fraction based on the same Reynolds number, whereas for the nanofluid with 4.65% volume fraction, a decrease in the Nusselt number was observed.

Nanofluids can also be applied in evaporators. They have potential to enhance heat transfer coefficient and critical heat flux in pool boiling and flow boiling. Many studies (Kim 2009; Wen et al. 2011; White et al. 2011) have established that the boiling heat transfer coefficient and the critical heat flux enhancement by nanofluids result from a thin porous nanoparticle deposition layer on the heater surface, which serves to improve the wettability and capillarity of the boiling surface. The main disadvantages of using nanofluids for boiling heat transfer improvement are the nanoparticle pollution and poor nanoparticle deposition strength. Besides, it is hard to control the thickness of the nanoparticle deposition layer. Thick layer presents an additional thermal resistance and thus deteriorates heat transfer. Therefore, direct usage of nanofluids in evaporators may not be beneficial. However, boiling investigations using nanofluids offered inspiring insights into the engineering of a boiling surface by nanoscale features. The progress in nanoengineering opens new room for nanoscale surface modification without a large change in the surface topography at the microscale (see Sundén and Wu [2014]; Wu and Sundén [2014]).

7.1.2 Discussion on Convective Heat Transfer Enhancement of Nanofluids

Based on the above literature review, there is controversy on whether a nanofluid can present anomalous heat transfer enhancement over the base fluid. The anomalous heat transfer enhancement means that the existing classic correlations underpredict the heat transfer behavior of nanofluids. It has been wrongly used in the literature to indicate a Nusselt number enhancement at an equal Reynolds number. Most convective heat transfer studies present Nusselt number increase based on the equal Reynolds number. However, as stated by Yu et al. (2010) and Wu et al. (2013), the equal Reynolds number basis can be misleading because the net result for the equal Reynolds number comparison is a combination of the nanofluid property effect and the flow velocity effect. Due to the higher viscosity of the nanofluid, the flow velocity in the nanofluid is generally higher than that of the base fluid at the same Reynolds number, which provides an advantage for the nanofluid compared to the base fluid. If the base fluid is to be pumped at the same flow velocity as the nanofluids, it may approach or exceed the thermal performance of the nanofluid. The result based on the equal Reynolds number will be more misleading at higher estimated or measured relative viscosity. Therefore, comparison of heat transfer enhancement should be based on equal flow velocity or equal pumping power for nanofluids. Because many investigations indicated misleading heat transfer enhancement based on the equal Reynolds number, the actual enhancement proportion and enhancement value are overestimated in Figure 7.1.

Nanofluid properties are other factors that may add complexity to the enhancement analysis. For example, scatter in heat transfer enhancement data in the literature might be due to the divergence in nanofluid properties, even for the same nanofluid with the same volume fraction. In addition, as different nanofluid property equations were used in numerical analysis, the simulated heat transfer enhancement might be very different from the real value and the stated anomalous enhancement could be an analysis artifact. Therefore, more evidence is needed to achieve a conclusion.

7.2 EXPERIMENTAL INVESTIGATION OF AQUEOUS ALUMINA AND MWCNT NANOFLUIDS IN A DOUBLE-PIPE HELICAL HEAT EXCHANGER

7.2.1 Experimental Apparatus and Method

A schematic illustration of the experimental setup is shown in Figure 7.2a. It consists of two loops for the cold and hot fluids, respectively. The hot water or nanofluid runs in the hot closed loop, whereas the cold water is forced in the cold open loop. Water or nanofluid is heated in a 50 L reservoir by an imbedded electric heater of 6 kW fixed at the bottom of the reservoir. The tested double-pipe helical heat exchanger was constructed by copper tubes and standard copper connections. The inner helically coiled tube, shown in Figure 7.2b, has an inner diameter (d_i) of 13.28 mm. The outer surface of the inner tube was enhanced by circular fin arrays (not shown in Figure 7.2b) with a fin height of 3.2 mm. The ratio of the outer surface area (A_o) to the inner surface area (A_i) of the inner tube is 4.83. The outer helically coiled tube has an inner diameter of 26 mm. The approximate hydraulic diameter of the annulus side (d_a) is 8 mm (fin arrays not considered). The number of turns (n) of the

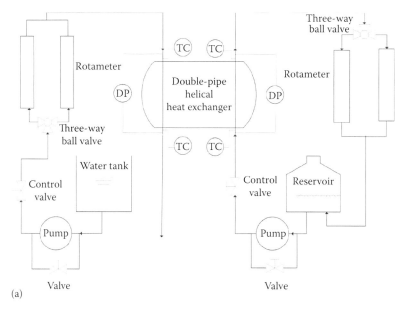

(a)

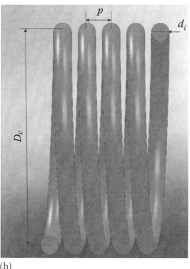

(b)

FIGURE 7.2 Schematic illustrations of experimental rig (a) and helically coiled tube (b). DP, differential pressure transducer; TC, thermocouple.

helical coils is 4.5, and each coil has a coil diameter of curvature (D_c, measured from the center of the inner tube) of 254 mm. The pitch of the helical coil (p) is 34.5 mm. The total length of the tested helical heat exchanger is 3.591 m. See Wu et al. (2013) and Wu, Wang et al. (2015) for more details on measurements, uncertainties, and experimental conditions.

7.2.2 Nanofluid Preparation and Properties

Untreated concentrated γ-Al$_2$O$_3$/water nanofluid with spherical alumina nanoparticles of 40 nm mean diameter was purchased from a commercial company (Nanophase Technologies Corporation, Romeoville, Illinois). No surfactants were added in the nanofluid. Different amounts of concentrated nanofluid were diluted in water to obtain nanofluids with low mass fractions. The diluted nanofluid mixture was mechanically stirred for 0.5 hour followed by ultrasonic vibration for 3 hours. The pH value of the prepared alumina nanofluid is about 3 ~ 3.5, which is far away from the isoelectric point of alumina nanofluid to maintain colloidal stability. Each prepared nanofluid was immediately tested in the double-pipe helical heat exchanger within 2 days after the nanofluid preparation.

An aqueous MWCNT suspension of 1.0% mass fraction was purchased from a commercial company (Nanocyl, Sambreville, Belgium). According to the vendor's specification, the suspension consists of thin MWCNTs dispersed in deionized water (97% mass fraction) and surfactant sodium dodecyl benzene sulfonate (2.0% mass fraction), and it is said to be stable for several months. The MWCNT, produced via the catalytic carbon vapor deposition process, has an average length of 1.5 µm and an average diameter of 9.5 nm, with an average aspect ratio of 158. The surface area of the MWCNT is 250–300 m^2 g^{-1}. The carbon purity of the MWCNTs is 90%, whereas the remaining 10% is metal oxide. Similar to alumina nanofluid, different amounts of concentrated nanofluid were diluted in water to obtain MWCNT/water nanofluids with different fractions. The diluted nanofluid mixture was mechanically stirred for 0.5 hour followed by ultrasonic vibration for 3 hours. The pH values of the prepared MWCNT/water nanofluids are in the range of 7.0 ~ 8.0.

Five alumina nanofluids and three MWCNT nanofluids are listed in Table 7.1. All the eight nanofluids were tested in the double-pipe helical heat exchanger. Water experiments were conducted between nanofluid tests in the same double-pipe helical heat exchanger to verify the nanofluid stability. The water experimental data points before and after the nanofluid tests present very similar thermal behavior, indicating very

TABLE 7.1　Mass and Volume Fractions of the Eight Tested Nanofluids

	Alumina/Water Nanofluid					MWCNT/Water Nanofluid		
Mass fraction (%)	0.78	2.18	3.89	5.68	7.04	0.02	0.05	0.1
Volume fraction (%)	0.20	0.56	1.02	1.50	1.88	0.0111	0.0278	0.0555

small and negligible deposition of nanoparticles during the nanofluid tests (Wu et al. 2013). Afterward, the experimental rig was flushed by water for 1 day to take off possible nanoparticle deposition in the whole rig before next nanofluid measurements.

The density of the nanofluid was calculated by

$$\rho_{nf} = \rho_{bf}(1-\varphi) + \rho_p \varphi \tag{7.1}$$

The specific heat of the nanofluid was calculated by

$$\rho_{nf} c_{p,nf} = \rho_{bf} c_{p,bf}(1-\varphi) + \rho_p c_{p,p} \varphi \tag{7.2}$$

where $c_{p,nf}$, $c_{p,bf}$, and $c_{p,p}$ are specific heats of the nanofluid, the base fluid, and the particle, respectively. Effective dynamic viscosity and thermal conductivity of nanofluids were measured by a rotational rheometer HAAKE RS6000 (Thermo Fisher Scientific Inc., Waltham, Massachusetts) and a thermal constants analyzer (TPS 2500S from Hot Disk AB, Uppsala, Sweden), respectively (see Wu, Feng et al. [2014]).

7.2.3 Data Analysis

The apparent Darcy friction factor was calculated by the following equation:

$$f_{app} = 2 \frac{d_i}{L} \frac{\Delta P}{\rho u^2} \tag{7.3}$$

The heat flux q was averaged between the heat transferred by the inner hot fluid q_h and the heat absorbed by the annulus cold water q_c:

$$q = \left(\frac{q_h + q_c}{2}\right) = \left[\frac{c_{p,h} m_h (T_{hi} - T_{ho}) + c_{p,c} m_c (T_{co} - T_{ci})}{2}\right] \tag{7.4}$$

The deviation in energy balance between the hot loop and the cold loop is less than 1.0%. The logarithmic mean temperature difference (LMTD) was determined by the following equation (Sundén 2012):

$$LMTD = \frac{(T_{hi} - T_{co}) - (T_{ho} - T_{ci})}{\ln\left[\frac{(T_{hi} - T_{co})}{(T_{ho} - T_{ci})}\right]} \tag{7.5}$$

Assuming no fouling resistance and ignoring the wall thermal resistance due to the thin wall, large tube length, and high thermal conductivity of copper, the inner tube heat transfer coefficient h was determined by

$$h = \frac{1}{A_i\left[(\mathrm{LMTD}/q)-(1/h_a A_o)\right]} \tag{7.6}$$

The annulus thermal resistance in the above equation was also neglected because of the following reasons: (1) the annulus heat transfer coefficient h_a is relatively large due to the intensive turbulence induced by the fins on the outer surface of the inner tube; (2) $A_o/A_i = 4.83$; (3) the volumetric flow rate on the annulus side was kept relatively large during the experiments; (4) a small change in h was noticed for a 20% change of the annulus flow rate during the experiments. Thus, Equation 7.6 can be simplified as

$$h = \frac{q}{A_i \cdot \mathrm{LMTD}} \tag{7.7}$$

The inner tube heat transfer coefficient can also be obtained from modified Wilson plot techniques, for example, see Rose (2004) and Van Rooyen et al. (2012). However, based on the above reasons in the tested heat exchanger, it was not needed to apply a modified Wilson plot technique. In addition, because the annulus thermal resistance $h_a A_o$ is negligible compared to the dominant inner tube thermal resistance hA_i at relatively high annulus flow rates, the modified Wilson plot technique is no longer valid for the calculation of annulus heat transfer coefficient. The modified Wilson plot technique might be used to estimate the annulus heat transfer coefficient in the helical heat exchanger at very low annulus flow rates (when thermal resistances of the both sides are on the same order of magnitude) but with a relatively high uncertainty (due to the narrow range of flow rates in laminar flow). In this study, only the inner tube heat transfer coefficient was investigated and evaluated. The estimated uncertainties for apparent friction factor and heat transfer coefficient are 3.3% and 3.2%, respectively. See Wu et al. (2013) for uncertainty values of all primary measurements and dependent quantities.

7.3 RESULTS AND DISCUSSION

7.3.1 Pressure Drop

The relationship between the apparent Darcy friction factor f_{app} calculated from Equation 7.3 and the Reynolds number Re_b for water and eight nanofluids is illustrated in Figure 7.3. The apparent friction factor decreases with Re_b when $\mathrm{Re}_b < 6000$, whereas it increases slowly when $\mathrm{Re}_b > 6000$. In this study, a

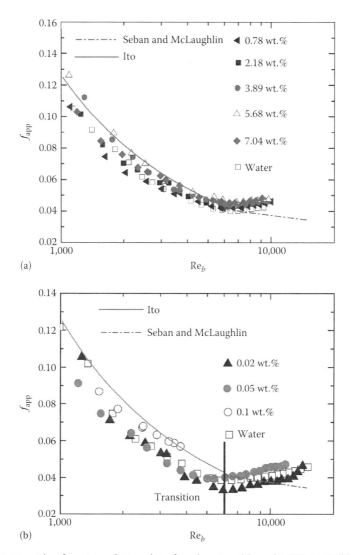

FIGURE 7.3 The f_{app}–Re_b relationship for alumina (a) and MWCNT (b) nanofluids. The solid line indicates the Ito equation. The dash-dot line indicates the Seban and McLaughlin equation. (Data from Ito, H., *Appl. Math. Mech.*, 11, 653–663, 1969; Seban, R.A., and McLaughlin E.F., *Int. J. Heat Mass Transf.*, 6, 387–395, 1963; Wu, Z. et al., *Appl. Therm. Eng.*, 60, 1, 266–274, 2013; Wu, Z. et al., *Appl. Therm. Eng.*, in press.)

critical Reynolds number of approximately 6000 was assumed, which agrees with the transition value of 6494 calculated by the transition criterion in the work of Ito (1969). The transition from laminar flow to turbulent flow for all the tested fluids occurs almost at the same Reynolds number. Therefore, the transitional velocity of the nanofluids will be larger than that of the base fluid due to the larger viscosity of the former compared to the latter. The nanoparticles may stabilize the flow in helically coiled tubes. However, more data are needed to verify this phenomenon. No obvious difference exists among the five alumina nanofluids, especially in laminar flow. For MWCNT/water nanofluid in laminar flow, the apparent friction factor for the 0.1 wt.% nanofluid is the largest among the three tested MWCNT nanofluids mainly due to its higher increase in viscosity. As shown in Figure 7.3, the Ito (1969) equation and the Seban and McLaughlin (1963) equation can predict the experimental data relatively well for laminar and turbulent flows, respectively. For turbulent flow, the Seban and McLaughlin equation predicts a decreasing trend with Reynolds number, whereas our data points show an increasing trend. A possible reason might be that the data points are still located in the transition region. Besides, the Seban and McLaughlin equation also underestimates the friction factor for water in the previous literature for turbulent flow. Therefore, a new predictive equation for the friction factor in the transition region in helically coiled tubes needs to be developed.

7.3.2 Heat Transfer in Laminar Flow

Figure 7.4 demonstrates the relationship between $\text{Nu}_b(\text{Pr}_b)^{-0.4}$ and the inner tube Dean number De_b [$= \text{Re}_b(d_i/D_c)^{0.5}$] for laminar flow. The subscript b indicates the average bulk temperature. All properties used in the dimensionless numbers were calculated at the average bulk temperature. The average bulk temperature was estimated from the inner tube inlet and outlet temperatures. For nanofluids, the nanofluid properties were used instead of those of the base fluids. Temperature effects were accounted for in the Prandtl number Pr_b. As shown in Figure 7.4, the Nusselt number increases with the Dean number. Water, the five alumina nanofluids, and the three MWCNT nanofluids present very similar heat transfer characteristics. This indicates that the net effect of nanoparticles (alumina or MWCNT) on the heat transfer performance in helically coiled tubes is probably insignificant. The thermal conductivity increase by nanoparticles is beneficial for heat transfer, whereas the secondary flow intensity induced by centrifugal forces may be reduced by nanoparticles due to the larger viscosity and density of the nanofluid compared to those of the base fluid.

In laminar flow, the Wu et al. (2013) correlation, originally developed based on water and alumina/water nanofluid data in helically coiled tubes, can also predict the heat transfer performance of MWCNT/water nanofluid very well. The Wu et al. correlation is given as

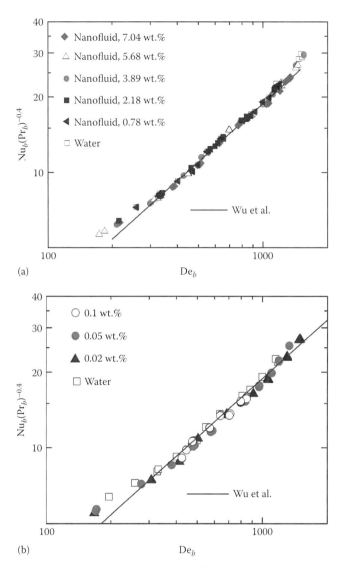

FIGURE 7.4 $\mathrm{Nu}_b(\mathrm{Pr}_b)^{-0.4}$ vs. De_b for laminar flow against the Wu et al. correlation: (a) alumina nanofluid; (b) MWCNT nanofluid. (Data from Wu, Z. et al., *Appl. Therm. Eng.*, 60, 1, 266–274, 2013; Wu, Z. et al., *Appl. Therm. Eng.*, in press.)

$$\mathrm{Nu}_b = 0.089 \times \mathrm{De}_b^{0.775}\, \mathrm{Pr}_b^{0.4} \tag{7.8}$$

It is found that the Wu et al. correlation gives an excellent estimation for both alumina and MWCNT nanofluids on the ground that all data points are located within a $\pm 5\%$ error band, except those with $\mathrm{De}_b < 200$. The applicable range of the Wu et al. correlation is as follows: $200 < De_b < 1300$, $4.0 < Pr_b < 7.0$, $\varphi < 2.0\%$.

7.3.3 Heat Transfer in Turbulent Flow

Similar to the laminar flow, Figure 7.5 presents the relationship between $Nu_b(Pr_b)^{-0.4}$ and the inner tube Reynolds number Re_b for the turbulent

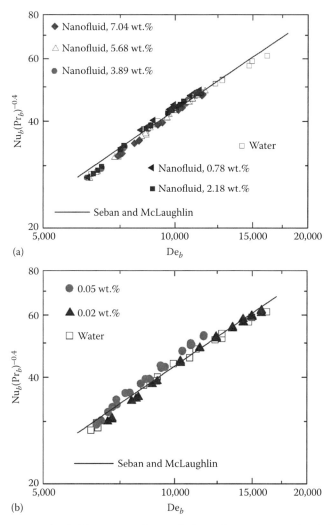

FIGURE 7.5 $Nu_b(Pr_b)^{-0.4}$ vs. De_b for turbulent flow against the Seban and McLaughlin correlation: (a) alumina nanofluid; (b) MWCNT nanofluid. The solid line indicates the Seban and McLaughlin correlation. (Data from Seban, R. A., and McLaughlin, E. F., *Int. J. Heat Mass Transf.*, 6, 387–395, 1963; Wu, Z. et al., *Appl. Therm. Eng.*, 60, 1, 266–274, 2013; Wu, Z. et al., *Appl. Therm. Eng.*, in press.)

flow. It is clear that $Nu_b(Pr_b)^{-0.4}$ increases with Re_b and the eight nanofluids and water show very similar trends. The Seban and McLaughlin (1963) correlation can predict the thermal behavior of water and nanofluids very accurately, with a mean absolute error and a standard deviation of 2.63% and 3.25%, respectively. The existing correlation can accurately reproduce the turbulent convective heat transfer behavior of nanofluids in helically coiled tubes by adopting the properties of the nanofluids in the analysis.

Based on the above experimental analysis, no anomalous heat transfer enhancement exists in our case for both laminar and turbulent flows. Buongiorno (2006) stated that Brownian diffusion and thermophoresis may become important as slip mechanisms. The two slip mechanisms were checked in the work of Wu et al. (2013) based on the order of magnitude analysis, and it was concluded that these mechanisms were negligible for alumina nanoparticles of primary particle size of 40 nm. Due to attractive van der Waals forces, particles may agglomerate to form aggregates. Wu, Feng et al. (2014) estimated an aggregate size of 86 nm for the tested alumina nanofluids. Dynamic light scattering measurements show a peak value located within 80–95 nm, which confirmed the estimated aggregate size. Elliptical aggregates probably form in the MWCNT nanofluids and the average aggregate length is about 200 nm, which was also confirmed by dynamic light scattering measurements (Wu, Feng et al. 2014). Movement of elliptical MWCNT aggregates is more restricted due to larger flow resistance compared to that of alumina aggregates. Therefore, the tested alumina and MWCNT nanofluids can be treated as homogeneous fluids. Additional effects of nanoparticles, for example, Brownian motion and thermophoresis, on the convective heat transfer characteristics of the nanofluids are negligible compared to the dominant thermophysical properties of the nanofluids.

7.4 PERFORMANCE COMPARISON CRITERIA FOR NANOFLUIDS

As stated in Section 7.1.2, heat transfer enhancement comparison for nanofluids over their base fluids based on the equal Reynolds number can be misleading and tends to overestimate the actual heat transfer enhancement largely. Figure 7.6 demonstrates two performance comparison criteria for the tested alumina nanofluids, that is, based on the equal Reynolds number and the equal flow velocity. The viscosity in the Reynolds number of Figure 7.6a was calculated by the Williams et al. (2008) correlation. The Williams et al. correlation gives a larger viscosity increase for nanofluids than the measured value, which induces a more obviously misleading

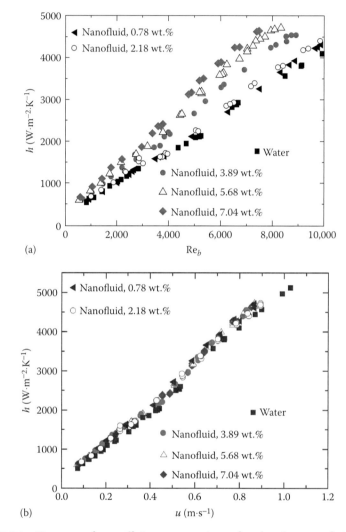

FIGURE 7.6 Heat transfer coefficient comparisons for alumina nanofluids: (a) h vs. Re_b for the equal Reynolds number basis with properties calculated from the Williams et al. correlation; (b) h vs. u for the equal flow velocity basis. (Data from Wu, Z. et al., *Appl. Therm. Eng.*, 60, 1, 266–274, 2013.)

result of the heat transfer enhancement. Over 40% heat transfer enhancement can be obtained for the 7.04 wt.% nanofluid compared to water for the equal Reynolds number basis, as shown in Figure 7.6a. However, from Figure 7.6b at the same flow velocity, the heat transfer enhancement of nanofluids over the base fluid is much less than that based on the equal Reynolds number. Thus, the anomalous heat transfer enhancement shown

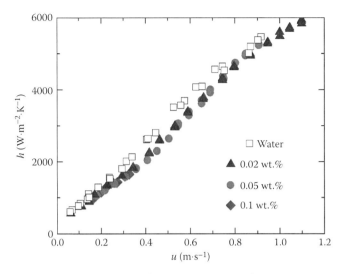

FIGURE 7.7 h vs. u for the equal flow velocity basis for MWCNT nanofluids. (Data from Wu, Z. et al., *Appl. Therm. Eng.*, in press.).

in Figure 7.6a is just an analysis artifact. The method based on the equal Reynolds number comparison should not be used. For MWCNT nanofluids, heat transfer deterioration was observed based on the equal flow velocity, as shown in Figure 7.7. The heat transfer degradation in the tested MWCNT nanofluids is mainly due to the low enhancement of the thermal conductivity and the high increase of the viscosity.

Based on the equal flow velocity basis, the following equations for heat transfer enhancement evaluation can be generated from Equation 7.8 and the Seban and McLaughlin correlation:

For laminar flow,

$$\frac{h_{nf}}{h_{bf}} = \left(\frac{\rho_{nf}}{\rho_{bf}}\right)^{0.775} \left(\frac{k_{nf}}{k_{bf}}\right)^{0.6} \left(\frac{\mu_{bf}}{\mu_{nf}}\right)^{0.375} \left(\frac{c_{p,nf}}{c_{p,bf}}\right)^{0.4} \tag{7.9}$$

For turbulent flow,

$$\frac{h_{nf}}{h_{bf}} = \left(\frac{\rho_{nf}}{\rho_{bf}}\right)^{0.85} \left(\frac{k_{nf}}{k_{bf}}\right)^{0.6} \left(\frac{\mu_{bf}}{\mu_{nf}}\right)^{0.45} \left(\frac{c_{p,nf}}{c_{p,bf}}\right)^{0.4} \tag{7.10}$$

The heat transfer coefficient ratio h_{nf}/h_{bf} is a function of the properties of the nanofluid and water. Thus, heat transfer enhancement of the nanofluid relative to the base fluid depends on the temperature, base fluid type, nanoparticle properties, nanoparticle concentrations, and other factors that affect the nanofluid properties, such as surfactants and nanoparticle agglomeration. According to Equations 7.9 and 7.10 based on the equal flow velocity, the largest enhancement of the heat transfer coefficient will be 3.25% and 3.30% for the 7.04 wt.% alumina/water nanofluid among the five tested alumina nanofluids at an average bulk temperature of 20°C for laminar and turbulent flows, respectively. These low values of enhancement are not very clear in Figure 7.6b, due to the measurement uncertainties and the enhancement decrease induced by the reduction of the secondary flow intensity by the nanoparticles. Secondary flow intensity mitigation may neutralize the benefit from the thermal conductivity increase. For MWCNT nanofluids, similar to the heat transfer deterioration shown in Figure 7.7, the heat transfer coefficient ratio h_{nf}/h_{bf} is about 0.937 and 0.924 for laminar and turbulent flows, respectively, according to Equations 7.9 and 7.10 based on the equal flow velocity. If based on the equal pumping power, the heat transfer coefficient ratio h_{nf}/h_{bf} for the above nanofluids will be even lower than that based on the equal flow velocity. Overall, the tested alumina nanofluids and MWCNT nanofluids are not promising for heat transfer purposes when other costs such as the pumping power cost and the cost of nanoparticles need to be considered.

Besides the above-mentioned comparison criteria, Lee (2009) obtained a boundary line for laminar flow using thermal conductivity and viscosity to determine if a nanofluid is more or less effective than the base fluid. The method was adopted to analyze the performance of nanofluids in a microchannel heat sink. A microchannel heat sink can be considered as a passive heat exchanger that cools a device by dissipating heat into the surrounding medium. It was selected for analysis due to its simplicity and its potential in electronics cooling. The microchannel heat sink consists of N circular channels, each with diameter d, as shown in Figure 7.8. The total channel width W is a constant and can be approximately calculated as $W = N \times d$.

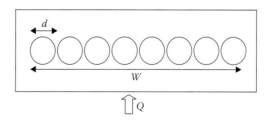

FIGURE 7.8   A schematic of a microchannel heat sink.

Assume a constant heat flux boundary condition for all channels and the flow is hydrodynamically and thermally fully developed. The difference in temperature between the surface wall temperature and the local bulk fluid temperature is

$$\Delta T = T_w - T_m = \frac{Q}{N\pi dLh} = \frac{Q}{(W/d)\pi L}\frac{1}{k\mathrm{Nu}} \qquad (7.11)$$

The pressure drop for a channel of channel length L is as follows:

$$\Delta P = f\frac{L}{d}\frac{\rho u^2}{2} \qquad (7.12)$$

For laminar flow, $f = 64/\mathrm{Re}$.

$$\Delta P = \frac{128\mu(V/N)L}{\pi d^4} = \frac{128\mu VL}{W\pi d^3} \qquad (7.13)$$

The above equation can be rewritten with respect to d as

$$d = \left(\frac{128\mu VL}{W\pi\Delta P}\right)^{1/3} \qquad (7.14)$$

Therefore, at a fixed total volume flow rate, a fixed pumping power, and a fixed channel length, the following relation can be obtained for laminar flow with a constant Nusselt number:

$$\Delta T(x) = \frac{Q}{W\pi LkNu}\left(\frac{128\mu VL}{W\pi\Delta P}\right)^{1/3} \propto \frac{\mu^{1/3}}{k} \qquad (7.15)$$

If nanofluids are more effective than their base fluids, the wall temperature using nanofluids should be less than the wall temperature using base fluids.

$$\Delta T_{bf}(x) > \Delta T_{nf}(x) \qquad (7.16)$$

From Equation 7.16, we can obtain

$$\frac{k_{nf}}{k_{bf}} > \left(\frac{\mu_{nf}}{\mu_{bf}}\right)^{1/3} \qquad (7.17)$$

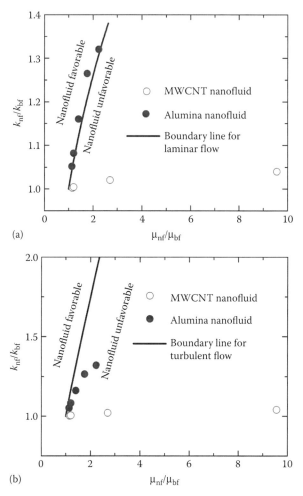

FIGURE 7.9 The boundary line for performance of nanofluids at fixed total volume flow rate and fixed pumping power: (a) laminar flow; (b) turbulent flow. The alumina and MWCNT nanofluids measured in Wu et al. correlation are shown for evaluation. (Data from Wu, Z. et al., *Fronti. Heat Mass Transf.*, 5, 18, 2014.)

The above reasoning shows that nanofluids are effective as long as the thermal conductivity enhancement is larger than the one-third power of the viscosity increase under the present constraints for laminar flow. The boundary line for performance of nanofluids in laminar flow is shown in Figure 7.9a. The thermal conductivity and viscosity ratios of alumina nanofluids and MWCNT nanofluids measured in the work of Wu, Feng et al. (2014) are also plotted in Figure 7.9. Alumina nanofluids can provide a better performance than base fluids for laminar flow, whereas the tested MWCNT nanofluids are very unfavorable.

The boundary line developed by Lee (2009) is only applicable for laminar flow. A relation for turbulent flow will be developed in this chapter. For turbulent flow, one can use the Blasius' relation to calculate the Darcy friction factor and the Dittus–Boelter equation to calculate the Nusselt number (Sundén 2012). Thus, for turbulent flow,

$$\Delta P = \frac{1.79\mu^{0.25}\rho^{0.75}(V/N)^{1.75}L}{\pi^{1.75}d^{4.75}} = \frac{1.79\mu^{0.25}\rho^{0.75}V^{1.75}L}{\pi^{1.75}d^{3}W^{1.75}} \tag{7.18}$$

Therefore,

$$d = \left(\frac{1.79\mu^{0.25}\rho^{0.75}V^{1.75}L}{\pi^{1.75}W^{1.75}\Delta P}\right)^{1/3} \tag{7.19}$$

The difference between the wall temperature and the local bulk fluid temperature is

$$\Delta T = \frac{Q}{(W/d)\pi L}\frac{1}{k\mathrm{Nu}}$$

$$= \frac{Q}{0.07\pi^{0.2}V^{0.8}W^{0.2}L}\frac{\mu^{0.4}}{k^{0.6}c_{p}^{0.4}\rho^{0.8}}\left(\frac{1.79\mu^{0.25}\rho^{0.75}V^{1.75}x}{\pi^{1.75}W^{1.75}\Delta P}\right)^{1/3} \tag{7.20}$$

$$\propto \frac{\mu^{0.483}}{k^{0.6}c_{p}^{0.4}\rho^{0.55}}$$

Based on the same reasoning, one obtains

$$\frac{k_{\mathrm{nf}}}{k_{\mathrm{bf}}}\left(\frac{c_{p,\mathrm{nf}}}{c_{p,\mathrm{bf}}}\right)^{0.667}\left(\frac{\rho_{\mathrm{nf}}}{\rho_{\mathrm{bf}}}\right)^{0.917} > \left(\frac{\mu_{\mathrm{nf}}}{\mu_{\mathrm{bf}}}\right)^{0.805} \tag{7.21}$$

If the contributions from specific heat and density are neglected, the above equation can be reduced to the following expression:

$$\frac{k_{\mathrm{nf}}}{k_{\mathrm{bf}}} > \left(\frac{\mu_{\mathrm{nf}}}{\mu_{\mathrm{bf}}}\right)^{0.805} \tag{7.22}$$

From the above equation, at fixed total volume flow rate and fixed pumping power, nanofluids may be effective only when the thermal conductivity enhancement is larger than the 0.805 power of the viscosity increase for turbulent flow. The boundary line for performance of nanofluids in turbulent flow is shown in Figure 7.9b. Both the tested alumina nanofluids and MWCNT nanofluids are located under the boundary line and in the nanofluid unfavorable region, that is, under the present constraints, the tested nanofluids are not effective. Comparing Equation 7.17 for laminar flow and Equation 7.22 for turbulent flow, the enhancement boundary line for turbulent flow is much stricter than that for laminar flow. In other words, for a specific nanofluid, heat transfer enhancement is present much easier for laminar flow than for turbulent flow. Stable nanofluids might be effective for microchannels as the main flow regime is laminar flow. Therefore, more nanofluid studies should be focused on laminar flow rather than turbulent flow.

The above comparison criteria are derived from the first law of thermodynamics. Entropy generation is another approach to evaluate system efficiency based on the second law of thermodynamics. According to Bejan (1995), there are two types of coupled losses in a heat exchanger. One is associated with the heat transfer across a finite fluid-to-fluid temperature difference. Another one is plagued by the frictional pressure drop in its channels. These two losses are signs of heat exchanger irreversibility, which can be analyzed in terms of entropy generation. Due to coupling of temperature difference and pressure drop on the two sides of heat exchangers, expressions for entropy generation are relatively complicated, which can be found in the works of Bejan (1995) and Shah and Sekulić (2003). For simplicity, the microchannel heat sink in Figure 7.8 is discussed. Therefore, only one side needs to be considered. Assuming uniform heat flux and mass flux distributions, the entropy generation for each channel can be expressed as follows:

$$S_{gen} = \left(S_{gen}\right)_T + \left(S_{gen}\right)_F = \frac{c_p Q^2}{\pi^3 T_{bulk}^2 NukW^2 L^2} + \frac{8 f \rho^2 V^3}{\pi W^3 T_{bulk}} \tag{7.23}$$

where $(S_{gen})_T$ and $(S_{gen})_F$ indicate entropy generation by temperature difference and entropy generation by frictional pressure drop, respectively. For laminar flow, Nu = 4.36 and $f = 64/Re$, then the above equation can be rewritten as

$$S_{gen} = \left(S_{gen}\right)_T + \left(S_{gen}\right)_F = \frac{c_p Q^2}{135.2 T_{bulk}^2 kW^2 L^2} + \frac{128\rho V^2 d\mu}{W^3 T_{bulk}} \tag{7.24}$$

At low flow velocities, the entropy generation is mainly from the contribution of the temperature difference, that is, the term $(S_{gen})_T$ in the above equation is the main source of entropy generation. When adding nanoparticles into the base fluid, $(S_{gen})_T$ tends to decrease due to thermal conductivity enhancement. Therefore, nanofluid has potential for heat transfer enhancement at low flow velocities in laminar flow. However, at relatively large flow velocities and Reynolds numbers, the total entropy generation will increase due to viscosity increase. The results from entropy generation analysis seem to be consistent with the above discussion.

Performance durability of nanofluids needs to be considered when better heat exchanger performance is achieved by using nanofluids. Properties of nanofluids may change with elapsed time. For example, Figure 7.10 shows that the average thermal conductivity decreases with elapsed time. A 7.0% reduction in thermal conductivity averaged for each day can be seen in Figure 7.10 after 55 days. Possible reasons for the thermal conductivity reduction are the formation of relatively large nanoparticle clusters and deposition. Therefore, the heat transfer performance deteriorates accordingly and may not be better than that for the base fluid.

Thermal conductivity of nanofluids is measured at static conditions, whereas viscosity of nanofluids is measured at shear flow conditions. Normally, thermal conductivity values measured at static conditions and viscosity values at high shear rates are used to evaluate the performance of nanofluids. However, thermal conductivity at flowing conditions might be different from that at static conditions. For example, shear flow in heat exchangers can change the nanofluid morphology. For rodlike particles or

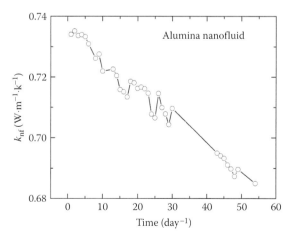

FIGURE 7.10   The change of thermal conductivity with elapsed time at 20°C at static conditions. (Data from Wu, Z. et al., *Fronti. Heat Mass Transf.*, 5, 18, 2014.)

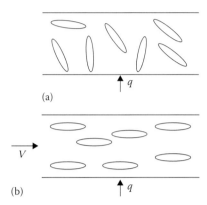

FIGURE 7.11 Possible aggregate morphology at static (a) and flowing (b) conditions.

aggregates, the thermal conductivity enhancement at flowing conditions may be lower than that at static conditions. As shown in Figure 7.11, the randomly oriented nanoparticle aggregates (might be percolated) tend to be parallel with the flow direction due to the liquid inertia. The long axis of the aggregate is prone to be perpendicular to the temperature gradient under flowing conditions, and therefore degrades the thermal conductivity enhancement, especially at a relatively high flow velocity. Therefore, the heat transfer enhancement for nanofluids is prone to be overestimated based on the static thermal conductivity values.

Besides, thermal resistances of a heat exchanger should be considered properly when people want to use nanofluids effectively in heat exchangers. Similar to electric resistances, when several thermal resistances are connected in series, the overall thermal resistance is given by their sum. If a thermal resistance is much greater than the others, it can be considered dominant and enhancement techniques should be focused on this thermal resistance. As nanofluid is liquid, it is not suitable for the gas side. Nanofluid can be applied to improve heat exchanger performance when the heat exchanger has a dominant convective thermal resistance in the liquid side. Nanoparticles can be added into the liquid side to decrease the dominant convective thermal resistance. For a well-designed heat exchanger, convective thermal resistances of the both sides are almost equal. Thus, nanofluid might not be fully effective for a well-designed liquid-to-gas heat exchanger, whereas it might be beneficial for a well-designed liquid-to-liquid heat exchanger if nanofluids are used in both the liquid sides to decrease the overall thermal resistance.

7.5 CONCLUSIONS

An experimental investigation was presented to reveal the pressure drop and heat transfer characteristics of alumina and MWCNT nanofluids of different nanoparticle volume fractions in a double-pipe helical heat exchanger. A brief literature review was also provided.

For both laminar and turbulent flows, no anomalous heat transfer enhancement was found. The heat transfer enhancement of the alumina nanofluids compared to water is less than about 3.0% according to the equal flow velocity basis, whereas the heat transfer deterioration was observed for MWCNT nanofluids. The main reason for low heat transfer enhancement or even heat transfer deterioration is the low thermal conductivity enhancement and the relatively large viscosity increase for the tested nanofluids, especially for MWCNT nanofluids. The existing correlations can approximately reproduce the laminar and the turbulent pressure drop and convective heat transfer behaviors of nanofluids in helically coiled tubes by adopting the measured nanofluid properties in the analysis. The net effect of nanoparticles (alumina or MWCNT) on the heat transfer performance in helically coiled tubes is probably insignificant. No obvious multiphase phenomenon was found and the tested alumina and MWCNT nanofluids might be treated as homogeneous fluids.

Heat transfer enhancement comparison for nanofluids versus their base fluids based on the equal Reynolds number can be misleading and tends to overestimate the actual heat transfer enhancement largely. Therefore, the equal Reynolds number basis, although used extensively in the previous literature, should not be applied for performance comparison. Equal flow velocity or pumping power basis can be used for heat transfer comparison. The equal pumping power basis is stricter than the equal flow velocity basis.

Besides, the performance of a microchannel heat sink using nanofluids was evaluated. The boundary lines for performance comparison were presented for both laminar and turbulent flows. For laminar flow, nanofluids are effective as long as the thermal conductivity enhancement is larger than the one-third power of the viscosity increase at fixed volume flow rate and fixed pumping power (Lee 2009). For turbulent flow, nanofluids may be effective only when the thermal conductivity enhancement is larger than the 0.805 power of the viscosity increase under similar constraints. The enhancement boundary line for turbulent flow is much stricter than that for laminar flow. Entropy generation analysis based on the second law of thermodynamics gives similar suggestions and concludes that nanofluids can be promising for heat transfer enhancement at low flow velocities in laminar flow. Therefore, more nanofluid studies should be focused on laminar flow rather than turbulent flow. To achieve better heat exchanger performance by using nanofluids, nanoparticle morphologies that can enhance the thermal conductivity greatly but increase the viscosity slowly will be very promising. Besides, thermal resistances of a heat exchanger should be analyzed before nanoparticle addition.

ACKNOWLEDGMENT

Financial support from the Swedish Energy Agency and the Swedish Research Council is gratefully acknowledged.

NOMENCLATURE

Variables

A_i	inner surface area of the inner tube (m²)
A_o	outer surface area of the inner tube (m²)
c_p	specific heat at constant pressure (J kg⁻¹ K⁻¹)
d	diameter (m)
De	Dean number, $Re_b(d_i/D_c)^{0.5}$
f	friction factor
h	heat transfer coefficient (W m⁻² K⁻¹)
k	thermal conductivity (W m⁻¹ K⁻¹)
L	length (m)
LMTD	logarithmic mean temperature difference (K)
m	mass flow rate (kg s⁻¹)
N	number of channels
N_{BT}	the ratio of Brownian and thermophoretic diffusivities
Nu	Nusselt number, hd/k
ΔP	pressure drop (Pa)
Pr	Prandtl number, $c_p\mu/k$
Q	power (W)
q	heat flux (W m⁻²)
Re	Reynolds number, $\rho u d/\mu$
S_{gen}	entropy generation per unit length (W K⁻¹ m⁻¹)
T	temperature (K)
ΔT	temperature difference (K⁻¹)
u	velocity (m s⁻¹)
V	total volume flow rate (m³ s⁻¹)

W total channel width (m)

w weight concentration

Greek symbols

μ dynamic viscosity (Pa s)

ρ density (kg m^{-3})

φ volume fraction

Subscripts

a annulus

bf base fluid

c cold side

ci cold side inlet

co cold side outlet

h hot side

hi hot side inlet

ho hot side outlet

i inside

nf nanofluid

p nanoparticles

REFERENCES

Bejan, A. 1995. *Entropy Generation Minimization: The Method of Thermodynamic Optimization of Finite-Size Systems and Finite-Time Processes.* Florida: CRC Press.

Buongiorno, J. 2006. Convective transport in nanofluids. *ASME J. Heat Transf.* 128(3): 240–250.

Dalkilic, S., Kayaci, N., Celen, A., Tabatabaei, M., Yildiz, O., Daungthongsuk, W., and S. Wongwises. 2012. Forced convective heat transfer of nanofluids—A review of the recent literature. *Curr. Nanosci.* 8(6): 949–969.

Escher, W., Brunschwiler, T., Shalkevich, N., Shalkevich, A., Burgi, T., Michel, B., and D. Poulikakos. 2011. On the cooling of electronics with nanofluids. *ASME J. Heat Transf.* 133(5): 051401.

Fan, J., and L. Wang. 2011. Review of heat conduction in nanofluids. *ASME J. Heat Transf.* 133(4): 040801.

Haghighi, E. B., Saleemi, M., Nikkam, N. et al. 2013. Cooling performance of nanofluids in a small diameter tube. *Exp. Therm. Fluid Sci.* 49: 114–122.

Huminic, G., and A. Huminic. 2012. Application of nanofluids in heat exchangers: A review. *Renew. Sust. Energy Rev.* 16(8): 5625–5638.

Hussein, A. M., Sharma, K. V., Bakar, R. A., and K. Kadirgama. 2014. A review of forced convection heat transfer enhancement and hydrodynamic characteristics of a nanofluid. *Renew. Sust. Energy Rev.* 29: 734–743.

Ito, H. 1969. Laminar flow in curved pipes. *Appl. Math. Mech.* 11: 653–663.

Kim, S. J. 2009. Subcooled boiling heat transfer and critical heat flux in water based nanofluids at low pressure. PhD Thesis, Massachusetts: MIT.

Kleinstreuer, C., and Y. Feng. 2011. Experimental and theoretical studies of nanofluid thermal conductivity enhancement: A review. *Nano. Res. Lett.* 6(1): 1–13.

Lee, J. H. 2009. Convective performance of nanofluids for electronics cooling. PhD Thesis, CA: Stanford University.

Prabhat, N., Buongiorno, J., and L. W. Hu. 2011. Convective heat transfer enhancement in nanofluids: Real anomaly or analysis artifact? *Proceedings of the ASME/JSME 2011 8th Thermal Engineering Joint Conference*, March 13–17, Honolulu, HI. New York: ASME.

Rose, J. W. 2004. Heat-transfer coefficients, Wilson plots and accuracy of thermal measurements. *Exp. Therm. Fluid Sci.* 28(2): 77–86.

Seban, R. A., and E. F. McLaughlin. 1963. Heat transfer in tube coils with laminar and turbulent flow. *Int. J. Heat Mass Transf.* 6: 387–395.

Sergis, A., and Y. Hardalupas. 2011. Anomalous heat transfer modes of nanofluids: A review based on statistical analysis. *Nano. Res. Lett.* 6: 391.

Seyf, H. R., and S. K. Mohammadian. 2011. Thermal and hydraulic performance of counterflow microchannel heat exchangers with and without nanofluids. *ASME J. Heat Transf.* 133(8): 081801.

Shah, R. K., and D. P. Sekulić. 2003. *Fundamentals of Heat Exchanger Design.* NJ: Wiley.

Sundén, B. 2012. *Introduction to Heat Transfer.* Southampton: WIT Press.

Sundén, B., and Z. Wu. 2014. Advanced heat exchangers for clean and sustainable technology. In *Handbook of Clean Energy Systems*, ed. J. Yan. NJ: Wiley.

Taws, M., Nguyen, C. T., Galanis, N., and I. Gherasim. 2012. Experimental investigation of nanofluid heat transfer in a plate heat exchanger. *Proceedings of the ASME 2012 Heat Transfer Summer Conference*, July 8–12, Puerto Rico.

Taylor, R., Coulombe, S., Otanicar, T. et al. 2013. Small particles, big impacts: A review of the diverse applications of nanofluids. *J. Appl. Phys.* 113(1): 011301.

Tiwari, A. K., Ghosh, P., and J. Sarkar. 2013. Heat transfer and pressure drop characteristics of CeO_2/water nanofluid in plate heat exchanger. *Appl. Therm. Eng.* 57(1): 24–32.

Van Rooyen, E., Christians, M., and J. R. Thome. 2012. Modified Wilson plots for enhanced heat transfer experiments: Current status and future perspectives. *Heat Transf. Eng.* 33(4–5), 342–355.

Wen, D., Corr, M., Hu, X., and G. Lin. 2011. Boiling heat transfer of nanofluids: The effect of heating surface modification. *Int. J. Therm. Sci.* 50(4): 480–485.

White S. B., Shih A. J., and K. P. Pipe. 2011. Boiling surface enhancement by electrophoretic deposition of particles from a nanofluids. *Int. J. Heat Mass Transf.* 54: 4370–4375.

Williams, W., Hu, L. W., and J. Buongiorno. 2008. Experimental investigation of turbulent convective heat transfer and pressure loss of alumina/water and zirconia/water nanoparticle colloids (nanofluids) in horizontal tubes. *ASME J. Heat Transf.* 130(4): 042412.

Wu, Z., Feng, Z., Sundén, B., and L. Wadsö. 2014. A comparative study on thermal conductivity and rheology properties of alumina and multi-walled carbon nanotube nanofluids. *Frontiers Heat Mass Transfer.* 5(1): 18.

Wu, Z., B. Sundén. 2014. On further enhancement of single-phase and flow boiling heat transfer in micro/minichannels. *Renew. Sust. Energy Rev.* 40, 11–27.

Wu, Z., Wang, L., and B. Sundén. 2013. Pressure drop and convective heat transfer of water and nanofluids in a double-pipe helical heat exchanger. *Appl. Therm. Eng.* 60(1): 266–274.

Wu, Z., Wang, L., Sundén, B., and L. Wadsö. 2015. Aqueous carbon nanotube nanofluids and their thermal performance in a helical heat exchanger. *Appl. Therm. Eng.*

Yang, C., Li, W., Sano, Y., Mochizuki, M., and A. Nakayama. 2013. On the anomalous convective heat transfer enhancement in nanofluids: A theoretical answer to the nanofluids controversy. *ASME J. Heat Transf.* 135(5): 054504.

Yu, W., France, D. M., Timofeeva, E. V., Singh, D., and J. L. Routbort. 2010. Thermophysical property-related comparison criteria for nanofluid heat transfer enhancement in turbulent flow. *Appl. Phys. Lett.* 96(21): 213109.

CHAPTER 8

Thermal Nanofluid Flow in Microchannels with Applications

Clement Kleinstreuer and Zelin Xu

CONTENTS

8.1 INTRODUCTION AND OVERVIEW

8.1.1 Nanofluids as Coolants

The demand for high-efficiency performance and compact design of devices in modern mechanical, chemical, and biomedical engineering puts a premium on effective *microsystem cooling*. Still, limited heat transfer rates can be one of the major obstacles for microsystems, which may generate significant heat fluxes. One way to overcome this difficulty is to use novel coolants with better thermal performance than convectional fluids such as oil, water, or ethylene glycol. For example, adding solid nanoparticles (NPs) to the liquids at low volume fractions creates a new type of fluid–particle mixture that may substantially enhances heat transfer rates. Thus, the resulting *nanofluid*, that is, a dilute NP suspension in liquids, is deemed promising as a solution to the cooling problem of microsystems.

Nanofluids with different types of NPs, including metal oxide particles and carbon nanotubes, in different base fluids such as water, ethylene glycol, and oil, have been prepared to investigate their thermal properties as well as heat transfer performance under various conditions. Although results are somewhat scattered, most experimental studies found augmented thermal conductivities and viscosities over the base fluids. Specifically, prevailing experimental evidence indicates enhanced heat transfer performance and a slight elevated pressure drop of nanofluid flow compared to the base fluid. To evaluate the effectiveness of thermal nanofluid flow in microchannels, the enhancement of the heat transfer coefficient (or the Nusselt number) is more suitable than the thermal conductivity, because other base fluid properties such as viscosity, density, and specific heat are also altered by the dispersed NPs. The convective heat transfer coefficient of nanofluid flow depends on a number of dimensionless groups, including the Reynolds, the Prandtl, and the Peclet numbers, as well as parameters such as NP size, volume fraction, shape, and mixture temperature.

Hence, comprehensive considerations are needed when determining the performance of nanofluids as coolants in microchannels.

Nanofluid convective heat transfer characteristics have been recently reviewed by Kleinstreuer et al. (2013), Wu and Zhao (2013), Salman et al. (2013), Prabhat et al. (2012), Yu et al. (2012), Godson et al. (2010), and Kakaç and Pramuanjaroenkij (2009). Specific case studies of nanofluid flow and heat transfer considered magnetic nanofluids (Nkurikiyimfura et al., 2013), porous media flow (Mahdi et al., 2013), heat pipe flow (Sureshkumar et al., 2013), and micro heat exchanger flow (Gunnasegaran et al., 2012; Mohammed et al., 2011). Concerning numerical nanofluid flow and heat transfer characteristics, Bahiraei (2014) and Kamyar et al. (2012) compared the different methods adopted by various studies. Sarkar (2011) and Sundar and Singh (2013) summarized the convective heat transfer correlations of nanofluids developed for both laminar and turbulent flows. Mahian et al. (2013) provided an overview of entropy generation in nanofluid flow. The heat transfer applications with nanofluids as coolants are also discussed in several books, for example, Kleinstreuer (2014), Kumar (2010), Das et al. (2008), and Li (2008).

This chapter provides an updated review on nanofluid flow with applications and discusses future directions. It is organized as follows: first, the thermal nanofluid flow characteristics are summarized based on experimental and numerical studies. General trends as well as discrepancies are discussed. Then, the computational development and cooling application of nanofluid flow in a shell and tube heat exchanger is demonstrated to show the effect of different thermal conductivity and effective viscosity models on convective heat transfer and pressure drop. In addition, a real-world application of nanofluid cooling of concentration photovoltaic (CPV) cells is presented to gain improved electrical energy conversion efficiency. Reduction of entropy generation, resulting in an improved system design, is also provided.

8.1.2 Experimental Evidence

Because NPs are ultrafine, low-concentration nanofluids are expected to cause little penalty in pressure drop. Based on experimentally demonstrated augmentation in thermal conductivity, nanofluids are considered promising in enhancing forced convection heat transfer in microchannels and related heat-generating microsystems. A system's heat transfer performance is expressed with the convective heat transfer coefficient or the Nusselt number:

$$h = \frac{q_w}{T_w - T_f} \tag{8.1}$$

$$\mathrm{Nu}_x = \frac{h \cdot x}{k} \tag{8.2}$$

where:

q_w is the local wall heat flux

Nu_x is the local Nusselt number at location x

T_w and T_f are the average temperatures of the wall and the fluid, respectively

k is the thermal conductivity

8.1.2.1 Laminar Flow

Experimental results showed that the Nusselt number increases with the Reynolds and Prandtl numbers as well as NP volume fraction. The enhancement is particularly significant in the entrance region, where local particle concentration was suggested to be the reason. Moreover, the ratio of convective heat transfer coefficients between the nanofluid and the base fluid was found to increase with the Peclet number (i.e., Re·Pr) as well as NP volume fraction. The heat transfer enhancement reported in most of the studies lies in the range of 5%–45% for spherical NPs; however, for high aspect ratio NPs such as carbon nanotubes, the enhancement in both thermal conductivity and convective heat transfer coefficient are much greater. Some of the most recent results are summarized subsequently.

Heris et al. (2013) conducted an experimental study of convective heat transfer of Al_2O_3–water nanofluid in a square duct under constant heat flux in laminar flow. The convective heat transfer coefficient ratio of the nanofluid to water was found to increase with the Peclet number up to 27.6% at 2.5% volume fraction. The h_{nf}/h_{bf} ratio increased further with particle loading, especially at high flow rates. Heyhat et al. (2013) experimentally investigated the heat transfer coefficient and the friction factor of Al_2O_3–water nanofluid flowing in a horizontal tube under laminar flow conditions. They found that within volume fractions of 0.1%–2%, the coefficient ratio h_{nf}/h_{bf} increased with the Reynolds number and particle concentration. At the same time, the pressure drop elevated dramatically by increasing the particle volume fraction. Yu et al. (2012) experimentally investigated the convective heat transfer of Al_2O_3–polyalphaolefin nanofluids containing spherical and rodlike NPs. The results indicate that the aspect ratio and the particle alignment and orientation are important factors affecting the heat transfer and pressure drop of such nanofluid flow. Madhesh et al. (2014) used water-based copper–titania hybrid nanofluids, with particle loading ranging from 0.1% to 2% to investigate the heat transfer characteristics in a shell-and-tube heat exchanger. The results show that the convective heat transfer coefficient and the Nusselt number increase with higher Reynolds and Peclet numbers, augmented further with NP concentrations up to 1%. Edalati et al. (2012) experimentally investigated the heat transfer of a triangular duct, employing CuO–water nanofluid in laminar flow and under constant heat flux conditions. They found that the ratio of heat transfer coefficients between the nanofluid and the base fluid increased with the NP volume fraction and the Peclet number.

It should be noted that the enhancement of the heat transfer coefficient is often found to be larger than that of the thermal conductivity for the same NP volume fraction. Particle migration effects, for example, shear stress, viscosity gradient, thermophoresis, and/or Brownian motion, have been proposed to account for the enhanced heat transfer performance of nanofluids over the base fluids in the laminar flow regime. It might lead to nonuniform thermal conductivity and viscosity profiles and reduce boundary layer thicknesses. Again, the results obtained from different studies disagree and often lack physical explanations.

8.1.2.2 Turbulent Flow

For turbulent convective heat transfer, elevated, unchanged, and reduced h_{nf} values have all been reported in comparison with h_{bf}. No consensus has been reached as of whether the dispersion of NPs in the base fluid enhances the convective heat transfer, although the number of studies that detected increased convective heat transfer coefficients with the Reynolds number and particle volume fraction is in the majority. Specifically, most studies that found elevated h_{nf} values attributed that to increased NP loadings. Another widely recognized character is the NP size effect on the convective heat transfer coefficient. Larger particles seem to give better h_{nf} results for the same particle volume fractions.

To measure the performance of thermal nanofluid flow, it is argued that the Nusselt number is not a proper comparison basis because it excludes the effect of increased pumping cost due to the dispersion of NPs in the base fluid. Therefore, the thermal performance factor was adopted by some researchers, which can be expressed as follows:

$$\eta = \frac{\left(\mathrm{Nu}_{nf}/\mathrm{Nu}_{bf}\right)}{\left(f_{nf}/f_{bf}\right)^{1/3}} \tag{8.3}$$

Heyhat et al. (2012) measured Al_2O_3–water nanofluid flow in a circular tube under constant wall temperature under turbulent flow conditions. The heat transfer coefficient was found to increase with NP volume fraction in the range between 0.1% and 2%. The Reynolds number hardly influenced the heat transfer enhancement. Sahin et al. (2013) investigated Al_2O_3–water nanofluid flow in a circular tube under constant heat flux boundary condition. They found that for particle loadings up to 4% the highest heat transfer enhancement was achieved at 0.5% concentration. Azmi et al. (2013) found that for the SiO_2–water nanofluid investigated with up to 4% particle volume fraction, the heat transfer coefficient peaked at 3% loading for turbulent flow in a tube with constant wall heat flux. Arani and Amani (2013) studied the effect of particle size on convective heat transfer and pressure drop for fully developed turbulent flow of TiO_2–water nanofluid, having a volume fraction of 0.01–0.02. They found that among all particle sizes investigated, that is, from 10 to 40 nm mean diameter, 20 nm particles give the best convective heat transfer performance based on the thermal performance factor. Meriläinen et al. (2013) used water-based Al_2O_3, SiO_2, and MgO nanofluids with an NP volume fraction up to 4% to study the turbulent convective heat transfer in an annulus tube under constant wall temperature condition. They showed that the average convective heat transfer coefficients of their nanofluids enhanced compared to the based fluid for the same Reynolds number. However, when taking pressure loss into consideration, only the SiO_2-based nanofluid at a small concentration showed noticeable improvement in heat transfer compared to the base fluid. Esfe et al. (2014) studied the heat transfer behavior of a MgO–water turbulent flow in a circular pipe, where the volume fraction of NPs was less than 1%. The thermal performance factor was found to increase with particle concentration.

To explain the enhancement of heat transfer using nanofluids, some researchers suggested that thermophoresis as well as Brownian diffusion of NPs lead to increased particle–fluid slip. They suggested that there is an intensification of turbulence due to particle motion. Also, the particle concentration is smaller in the boundary layer than in the bulk due to radial flow, which reduces the viscosity and suppresses the thickness of the boundary layer, thus promoting the heat transfer. Still, some argue that at high Reynolds numbers, the heat transfer is dominated by convection, implying that NPs provide only a small contribution to the overall heat transfer.

8.1.3 Numerical Studies

In the numerical simulation of fluid flow and heat transfer behavior of nanofluids, the methods used in the literature can be categorized into three groups: single-phase approach, two-phase approach, and lattice Boltzmann method (LBM).

8.1.3.1 Single-Phase Approach

In this approach, it is assumed that the suspended NPs are in thermal equilibrium with the liquid phase and that the relative velocity between the two phases is negligible. The reason is that NPs are so small that they follow the streamlines of the fluid exactly, making the mixture behave like a homogeneous mixture. Thus, a set of governing equations for pure fluids can be used with the effective thermophysical properties of nanofluids replacing the fluid properties. The single-phase approach is easy to implant and requires less computational time; the predicted results concerning convective heat transfer characteristics of nanofluids agree well with experiments. However, the results depend strongly on the selected thermophysical property models, especially those for thermal conductivity and viscosity. Different thermophysical models can lead to quite different convective heat transfer characteristics. Moreover, particle migration, which cannot be modeled by the single-phase model, can significantly affect the heat transfer results by creating nonuniform concentration fields, most prominently in entrance regions and boundary layers, especially when the Peclet number is large.

To improve the single-phase model, some modifications have been applied to include the slip between the particles and the base fluid by adding a virtual term in the thermal conductivity expression. Also, solving the NP mass transfer equation together with the momentum and energy equations to account for nonuniform concentration distributions can also increase the accuracy. Nevertheless, the single-phase model provides acceptable results with low computational time requirements.

8.1.3.2 Two-Phase Approach

Though the ultrafine particles may easily be fluidized, several factors such as Brownian motion, thermophoresis, and aggregation can significantly change the particle motion and lead to velocity differences between the two phases. Hence, it is argued that for a more accurate numerical result, two-phase models, which solve one set of governing equations for each of the phases, are more suitable. Most of the numerical studies used the so-called Eulerian–Eulerian approach to investigate the fluid flow and heat transfer characteristics of nanofluids, where both phases are considered as "fluids" with volume fractions summing up to unity (Bahiraei, 2014). Alternatively, the Eulerian–Lagrangian framework can be employed, which tracks the particle trajectories in the fluid phase.

Two-phase models have been used only recently, but have shown better accuracy than the single-phase model compared to experimental evidence.

Also, the comparisons between different two-phase models favor the mixture model over the Eulerian two-phase model. Although the two-phase models are capable of capturing the nonuniform concentration fields, it should be noted that most studies assumed that the turbulence of the fluid phase is not directly affected by the presence of the NP phase.

8.1.3.3 Lattice Boltzmann Method

In addition to the classical numerical methods, an alternative approach to analyze thermal nanofluid flow is the LBM, which fills the gap between microscopic and macroscopic phenomena (Kamyar et al., 2012). In the LBM, the microscopic interactions between particles are first modeled using a collision model; the Navier–Stokes equations are then solved to provide macroscopic information. Next, the microscopic and macroscopic quantities of components, for example, density and velocity, are connected using a relationship. Despite the high computational cost, the LBM has shown satisfactory results in thermal nanofluid flow applications. Due to the limited studies available, more research may be needed using this method to see the applicability as well as the accuracy.

8.1.4 Fluid Flow and Heat Transfer Correlations

Nanofluids have shown superior heat transfer capabilities than base fluids in both laminar and turbulent regimes. The heat transfer coefficient appears to increase more than the thermal conductivity. A natural question following this phenomenon would be whether the Nusselt number and friction factor correlations developed for pure fluids still hold for nanofluids. Most of the experimental studies showed that the friction factor correlations developed for pure fluids can still be used for nanofluids in both laminar and turbulent flow regimes, given that the corresponding nanofluid properties are used. However, whether the conventional correlations for the Nusselt number can accurately predict the nanofluid heat transfer characteristics using the measured nanofluid properties is under debate (Sarkar, 2011; Sundar and Singh, 2013). New correlations have been proposed to fit the experimental data for nanofluid Nusselt numbers. Unfortunately, in both laminar and turbulent flow regimes, strong disagreements have been observed between proposed correlations, indicating that the correlations have limited applicability. For example, Yu et al. (2012) reviewed experimental data sets for turbulent heat transfer of nanofluids in horizontal tubes. They concluded that if properly compared, the actual heat transfer coefficient enhancement of nanofluids over their base fluids can be predicted quite accurately using the Dittus–Boelter equation.

8.1.5 Entropy Generation in Thermal Nanofluid Flow

The use of thermal nanofluid flow for enhanced heat transfer has been the major focus for most studies. However, for efficient removal of high heat fluxes and proper optimization of thermal systems, not only the heat transfer has to be maximized, but the increase in entropy generation has to be minimized as well. Entropy generation is the measure of process irreversibilities caused by both heat transfer and friction effects. It can be employed as a criterion to assess the performance of thermal devices. For example, Li and Kleinstreuer (2010) numerically investigated entropy generation of CuO–water nanofluid flow in trapezoidal microchannels. In the laminar flow regime, they found that adding NPs to the base fluid reduces entropy generation, with an optimal volume fraction corresponding to minimized entropy generation. Feng and Kleinstreuer (2010) investigated the entropy generation in a radial cooling system and found that nanofluid decreases the total generated entropy of the system without significantly increasing the required pumping power. Mahian et al. (2013) have recently reviewed theoretical and computational studies on entropy generation due to flow and heat transfer of nanofluids. From a perspective of the second law of thermodynamics, adding NPs to the base fluid can be very beneficial, because it decreases the entropy generation rate, especially in the low-to-medium Reynolds number regime. When the Reynolds number is very high, the frictional entropy generation contributes more to the total entropy generation than the thermal entropy generation, pushing up the entropy generation rate due to the augmented nanofluid viscosity. Clearly, the thermophysical properties of nanofluids play an important role in determining the entropy generation rate.

8.1.6 Summary

The performance of thermal nanofluid flow has been reviewed based on 2010–2014 publications, considering experimental as well as numerical results. Nanofluids with enhanced thermal conductivities are believed to generate better heat transfer performances. However, many problems are still unsolved. Disagreements prevail for experimental studies, thus requiring standardized protocols for preparation of NPs, nanofluids, and measurement methods. In addition, identifying the mechanisms for the enhanced thermal conductivity as well as the viscosity of nanofluids is imperative. For computational studies, accurate thermophysical property models seem to be the key to obtain reliable results. Still, more work is needed to compare different computational approaches. Similar to the correlations developed for pure fluids, new sets of friction factors and Nusselt numbers are needed for different NP–liquid pairings, flow regimes, and channel cross sections. Finally, in analyzing the thermal nanofluid flow, it is important to consider the entropy generation of the system in order to obtain optimal operational and geometrical system conditions.

8.2 THEORY

In this chapter, two examples will be discussed to demonstrate the performance of thermal nanofluid flow. In the first example, heat transfer and fluid flow characteristics of laminar Al_2O_3–water nanofluid flow in a double tube heat exchanger are discussed. In the second example, turbulent nanofluid flow for the application of cooling of CPV cells for improved efficiency is presented. For the selected test cases, it is assumed that the continuum hypothesis is valid, considering a channel hydraulic diameter $D_h > 100\ \mu m$. In Section 8.1.3, the single-phase method provides acceptable results for thermal nanofluid flow while requiring much less computational time than the two-phase method. Thus, this method has been adopted for the two test cases. For steady, incompressible nanofluid flow, the continuity, momentum, and energy equations have to be solved, considering temperature- and volume fraction-dependent mixture properties. According to Section 8.1.5, the entropy generation analysis is a powerful tool in evaluating the thermal nanofluid flow. Hence, in addition to the conservation laws and models for the mixture properties, the second law of thermodynamics has to be formulated for system optimization via reduction in the entropy generation rate.

Continuity equation:

$$\frac{\partial}{\partial x_i}(\rho_{nf} u_i) = 0 \tag{8.4}$$

Momentum equations:

For laminar flow:

$$\frac{\partial}{\partial x_j}(\rho_{nf} u_i u_j) = -\frac{\partial p}{\partial x_i} + \frac{\partial}{\partial x_j}\left[\mu_{nf}\left(\frac{\partial u_i}{\partial x_j} + \frac{\partial u_j}{\partial x_i}\right)\right] \tag{8.5}$$

For turbulent flow:

$$\frac{\partial}{\partial x_j}(\rho_{nf} u_i u_j) = -\frac{\partial p}{\partial x_i} + \frac{\partial}{\partial x_j}\left[\mu_{nf}\left(\frac{\partial u_i}{\partial x_j} + \frac{\partial u_j}{\partial x_i}\right) - \rho_{nf}\overline{u_i u_j}\right] \tag{8.6}$$

Energy equations:

For laminar flow:

$$\frac{\partial}{\partial x_i}\left(\rho_{nf} u_i h_{tot}\right) = \frac{\partial}{\partial x_i}\left(k_{nf}\frac{\partial T}{\partial x_i}\right) + \mu_{nf}\Phi \tag{8.7}$$

For turbulent flow:

$$\frac{\partial}{\partial x_i}\left(\rho_{nf} u_i h_{tot}\right) = \frac{\partial}{\partial x_i}\left(k_{nf}\frac{\partial T}{\partial x_i} - \rho_{nf}\overline{u_i h}\right) + \mu_{nf}\Phi - \frac{\partial u_i}{\partial x_j}\left(\rho_{nf}\overline{u_i u_j}\right) \tag{8.8}$$

Here the energy dissipation term reads as follows:

$$\Phi = \left(\frac{\partial u_i}{\partial x_j} + \frac{\partial u_j}{\partial x_i}\right)\frac{\partial u_i}{\partial x_j} \tag{8.9}$$

where:

u and x are the velocity and direction components $i, j = 1, 2, 3$

ρ denote the density

μ denote viscosity

k denote thermal conductivity

nf denotes nanofluid

h_{tot} denotes the mean total enthalpy

$\rho_{nf}\overline{u_i h}$ and $\rho_{nf}\overline{u_i u_j}$ are additional terms due to turbulent fluxes

8.2.1 Mixture Properties

The basic coolant mixture properties are functions of NP volume fraction φ and mixture temperature T. NPs are assumed to be spherical, monodisperse, and forming a homogeneous dilute suspension, although the effects of NP aggregation on k_{nf} was considered. In the case of low NP volume fractions, the nanofluid density and heat capacity can be expressed as follows (Li and Kleinstreuer, 2010):

$$\rho_{nf} = \varphi\rho_p + (1-\varphi)\rho_{bf} \tag{8.10a}$$

$$(\rho c_p)_{nf} = \varphi(\rho c_p)_p + (1-\varphi)(\rho c_p)_{bf} \tag{8.10b}$$

where the subscripts bf and p indicate the base fluid and the particle, respectively.

The temperature-dependent properties of the base fluid, that is, water in this study, can be expressed as follows (Xu and Kleinstreuer, 2014a):

$$\rho_{water} = 1000 \cdot \left[1 - \frac{(\tilde{T}+15.7914)}{5,08,929.2 \cdot (\tilde{T}-205.0204)} \cdot (\tilde{T}-277.1363)^2 \right] \left(\frac{kg}{m^3} \right) \quad (8.11a)$$

$$c_{p,water} = 9616.873445 - 48.7364833 \cdot \tilde{T} + 0.1444662 \cdot \tilde{T}^2$$
$$-0.000141414 \cdot \tilde{T}^3 \left(\frac{J}{m^3 \cdot K} \right) \quad (8.11b)$$

$$\mu_{water} = A_1 \cdot 10^{A_2/(T-A_3)} \left(\frac{kg}{m \cdot s} \right) \quad (8.11c)$$

$$k_{water} = -1.1245 + 0.009734 \cdot \tilde{T} - 0.00001315 \cdot \tilde{T}^2 \left(\frac{W}{m \cdot K} \right) \quad (8.11d)$$

where:
$\tilde{T} = T/[1(K)]$ is the nondimensional temperature
A_1, A_2, and A_3 in Equation 8.11c are constants with values of 2.414×10^{-5}, 247.8, and 140, respectively (Fox et al., 2004)

8.2.1.1 Effective Thermal Conductivity

For the mixture thermal conductivity, a newly proposed model is employed (Xu, 2014). This model is a modification of the Feng–Kleinstreuer (F–K) model (Kleinstreuer and Feng, 2012) by including the NP aggregation effect and thermal contact resistance. It divides the thermal conductivity k_{nf} of the nanofluid into two parts: a static part k_{static} similar to Maxwell's model (Maxwell, 1881) and a micromixing part k_{mm} due to Brownian motion of NP aggregates. Specifically,

$$k_{nf} = k_{static} + k_{mm} \quad (8.12)$$

The static part is a modification of the Maxwell model:

$$k_{static} = k_{bf} \cdot \left\{ 1 + \frac{3 \left[(k_a/k_{bf}) - 1 \right] \varphi}{\left[(k_a/k_{bf}) + 2 \right] - \left[(k_a/k_{bf}) - 1 \right] \varphi} \right\} \quad (8.13)$$

The micromixing part is given by

$$k_{mm} = 19,631 \cdot C_c \varphi \frac{\kappa_B \tau_a}{m_a} (\rho c_p)_{nf} (\overline{T} \cdot \ln \overline{T} - \overline{T}) \quad (8.14)$$

where κ_B is the Boltzmann constant.

The NP thermal conductivity in the original Maxwell model is replaced by the effective thermal conductivity of particle aggregate unit k_a:

$$k_a = k_{de} \cdot \frac{3 + \varphi_b \left[2\beta_{11}\left(1 - L_{11}\right) + \beta_{33}\left(1 - L_{33}\right) \right]}{3 - \varphi_b \left(2\beta_{11} L_{11} + \beta_{33} L_{33} \right)} \tag{8.15}$$

where:

$$k_{de} = (3\varphi_{de} - 1)k_{peff} + (2 - 3\varphi_{de})k_{bf}$$
$$+ \sqrt{[(3\varphi_{de} - 1)k_{peff}]^2 + [(2 - 3\varphi_{de})k_{bf}]^2 + 2[2 + 9\varphi_{de}(1 - \varphi_{de})]k_{peff} k_{bf}} \tag{8.16}$$

The effective thermal conductivity of the NPs including the thermal contact resistance, k_{peff}, is given as follows:

$$k_{peff} = \frac{k_p}{1 + \left(R_K k_p / d_p \right)} \tag{8.17}$$

In Equations 8.15 through 8.17, k_{de} is the effective thermal conductivity of the mixture of base fluid and dead end particles, φ_{de} is the volume fraction of dead end particles in an aggregate unit, φ_b is the volume fraction of backbone particles in an aggregate unit, L_{ii} and β_{ii} are parameters appearing in the derivation of the effective thermal conductivity of arbitrary isotropic particulate composites with interfacial thermal resistance. The readers are referred to Xu and Kleinstreuer (2014b) Section 2.5 for a complete explanation of the meanings of these parameters. Moreover, R_K is the Kapitza resistance, ρ is the density, c_p is the specific heat capacity, and $\bar{T}$ is the time-averaged temperature. The subscripts p and a denote NP and aggregate unit, respectively. C_c is a correction factor that differs among different NP–base fluid pairings, but the value is around unity. For Al_2O_3–water nanofluids, $C_c = 1.1$. The characteristic time interval τ_a is expressed as follows:

$$\tau_a = \frac{m_a}{3\pi\mu_{bf} R_g} \tag{8.18}$$

where R_g is the average radius of the aggregate units:

$$R_g = 2.5 d_p \sqrt{\frac{\varphi}{0.01}} \cdot \sqrt{\frac{T_0}{\bar{T}}} \tag{8.19}$$

Here, d_p is the NP diameter and $T_0 = 273$ K is the reference temperature.

In light of experimental evidence, this model is suitable for several types of metal oxide NPs in water with volume fractions up to 10% and mixture temperatures below 350 K.

8.2.1.2 Effective Dynamic Viscosity

Conventional viscosity models for mixtures often fail to predict the viscosities of nanofluids very well. For example, Brinkman (1952) proposed a mixture viscosity model as a function of particle concentration:

$$\mu_{nf} = \frac{\mu_{bf}}{(1-\varphi)^{2.5}} \tag{8.20}$$

Corcione (2011) obtained an empirical correlation for the effective viscosity of nanofluids with NP diameters ranging from 25 to 200 nm, volume concentrations of 0.01%–7.1%, and temperatures of 293–333 K. It is expressed as follows:

$$\mu_{nf} = \frac{\mu_{bf}}{1 - 34.87 \left(d_p/d_{bf} \right)^{-0.3} \varphi^{1.03}} \tag{8.21}$$

where d_{bf} is the equivalent diameter of a base fluid molecule, given by

$$d_{bf} = \left(\frac{6M}{N\pi\rho_{bf0}} \right)^{1/3} \tag{8.22}$$

where:
   M is the molecular weight of the base fluid
   N is the Avogadro number
   ρ_{bf0} is the mass density of the base fluid calculated at temperature $T_0 = 293$ K

According to Section 8.1.3, the flow and heat transfer characteristics of nanofluids depend on the selected thermophysical property models. Hence, to better understand the performance of thermal nanofluid flow in microchannels, the theoretical model proposed by Brinkman (1952) and the empirical nanofluid viscosity obtained by Corcione (2011) were adopted. Figure 8.1 compares the two models as a function of volume fraction. As already noticed by Nguyen et al. (2007), the conventional Brinkman model underpredicts the nanofluid viscosity. Furthermore, the functional dependence $\mu_{nf}(\varphi)$ is highly nonlinear in the model by Corcione (2011). Such

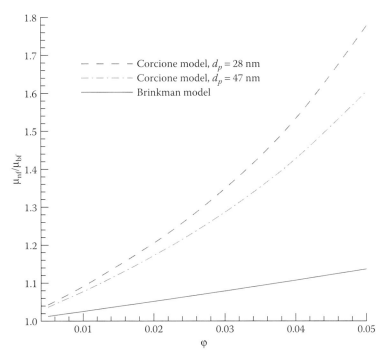

FIGURE 8.1   Dynamic viscosity models for nanofluids: Viscosity ratio versus volume fraction.

a viscosity enhancement of nanofluids may increase the pressure drop or the requirement of pumping power. The viscosity ratios predicted by both models are independent of mixture temperature.

8.2.2 Entropy Generation

Entropy generation minimization aims at obtaining optimal geometry and operational conditions through minimization of the amount of work wasted during operation (Ratts and Raut, 2004). For convection heat transfer, the local entropy generation rate (S_{gen} in W/K) can be expressed as follows (Kleinstreuer, 2010, 2014):

$$S_{gen} = S_{gen}(\text{thermal}) + S_{gen}(\text{friction}) \tag{8.23}$$

where:

$$S_{gen}(\text{thermal}) = \frac{k}{T^2}\left[\left(\frac{\partial T}{\partial x}\right)^2 + \left(\frac{\partial T}{\partial y}\right)^2 + \left(\frac{\partial T}{\partial z}\right)^2\right] \tag{8.24}$$

$$S_{\text{gen}}(\text{frictional}) = \frac{\mu \Phi}{T} = \frac{\mu}{T}\left(\overline{\Phi} + \Phi'\right)$$

$$= \frac{\mu}{T}\left[\left(\frac{\partial \overline{u_i}}{\partial x_j} + \frac{\partial \overline{u_j}}{\partial x_i}\right)\frac{\partial \overline{u_i}}{\partial x_j} + \overline{\frac{\partial u_i}{\partial x_j}\frac{\partial u_i}{\partial x_j}}\right]$$

(8.25)

Clearly, Equation 8.12 encapsulates the irreversibilities due to heat transfer and frictional effects.

Bejan (1996) pointed out that for turbulent flow the dissipation due to mean flow is roughly the same as that due to turbulent fluctuation. In fact, the frictional entropy generation rate accounts for a negligible part of S_{gen}, unless the Reynolds number becomes very large. Thus, for turbulent nanofluid flow, the correlation in the work of Bejan (1996) is curve-fitted with the frictional entropy generation rate expressed as a function of $\overline{S_{\text{gen}}(\text{frictional})}$ and Re.

8.3 NANOFLUID CONVECTIVE HEAT TRANSFER APPLICATIONS

8.3.1 Thermal Nanofluid Flow in a Double-Tube Heat Exchanger

The double-tube heat exchanger configuration is depicted in Figure 8.2. Hot water flows in the annulus; cold nanofluid flowing in the inner tube exchanges heat with water through a copper tube wall in between.

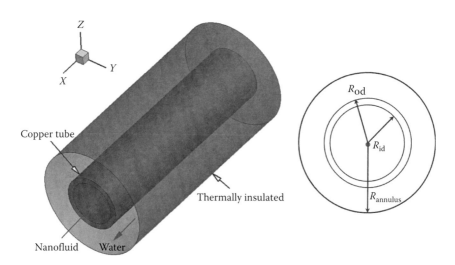

FIGURE 8.2   Model of the counterflow, double-tube heat exchanger.

8.3.1.1 Numerical Method

In addition to the heat convection, the steady-state heat conduction in the copper tube wall was calculated using Fourier's law:

$$k_{copper} \frac{\partial^2 T}{\partial x_i^2} = 0 \tag{8.26}$$

The thermal conductivity of the copper tube wall k_{copper} was assumed to be constant. The outer annulus wall was thermally insulated. Uniform velocities were applied at the inlets, whereas zero gauge pressures were applied at the outlets. The no-slip boundary condition was enforced at all solid walls. Due to symmetry, only a quarter of the geometry was used as the computational domain; hence, the symmetric boundary condition was applied where appropriate.

The governing equations were solved, employing the user-enhanced software package ANSYS-CFX 15 (ANSYS, Inc., Canonsburg, PA), which is based on the control volume method. The computations were performed on a local workstation (Dell Precision T7500, Dell Inc., Round Rock, TX). A typical case contains about 198,000 structured mesh elements with 224,000 nodes. The near-wall mesh was locally refined to better resolve the larger velocity and temperature gradients near the boundaries. The convergence criterion was 5×10^{-6} for the average residuals. Mesh independence was examined and verified by decreasing the mesh size to one-half, which produced a maximum result change of less than 0.6%.

8.3.1.2 Model Validation

To validate the numerical model, the velocity profiles of fully developed flow in the circular tube as well as in the annulus were compared with analytical solutions:

$$u(r) = 2V \left(\frac{r^2}{R_{id}^2} - 1 \right) \tag{8.27}$$

$$u(r) = - \frac{2V}{R_{annulus}^2 + R_{od}^2 - \left[\left(R_{annulus}^2 - R_{od}^2 \right) \middle/ \ln \left(R_{annulus}/R_{od} \right) \right]}$$
$$\left[r^2 - R_{annulus}^2 + \frac{R_{annulus}^2 - R_{od}^2}{\ln \left(R_{od}/R_{annulus} \right)} \ln \left(r/R_{annulus} \right) \right] \tag{8.28}$$

where:

V is the average flow velocity
r is the radial position

Figure 8.3 shows good agreements between them.

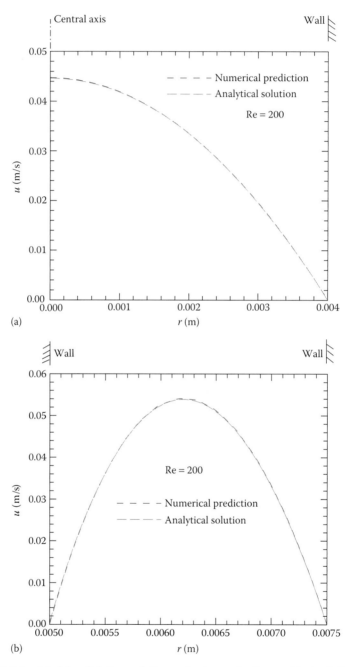

FIGURE 8.3 Model validation of dimensionless velocity profile in fully developed region: (a) flow in the circular tube; (b) flow in the annulus region.

Also, the Nusselt numbers of both the circular tube and the annulus in fully developed region were compared with the analytical values. For circular tubes, Nu = 4.36; for annulus, the Nu number calculated using the hydraulic diameter $D_{annular} - D_{od}$ can be compared with that of rectangular channels (Kays and Crawford, 1993). Here, due to the large aspect ratio of the annulus, the Nu number of the annulus in fully developed region is compared with that of the convective heat transfer between parallel plates. The latter has a value of Nu = 5.385 for the condition of one wall subjected to constant heat flux and one wall insulated. Figure 8.4 provides the Nusselt number comparisons at different Reynolds numbers. For both cases, the differences are within 1%. Thus, this model also predicts heat transfer accurately.

8.3.1.3 Results and Discussion

The friction factor and pressure drop as well as the convective heat transfer coefficient of alumina–water nanofluids were obtained for the counterflow, double-tube heat exchanger with hot water flowing in the annulus and cold nanofluid flowing in the inner tube. The convectional dynamic viscosity model by Brinkman (1952) and the empirical model by Corcione (2011) were compared. The improved F–K model for heat transfer enhancement was employed.

Figure 8.5 shows the comparison of the pressure gradients at different Reynolds numbers for water and nanofluids with different NP volume fractions. The conventional Brinkman viscosity model generates lower pressure gradients than the viscosity model developed based on experimental results. This difference becomes more prominent when the particle loading is high. Of practical interest here is the friction factor change with the Reynolds number using different viscosity models (see Figure 8.6). The Fanning friction factor used to evaluate the viscous effects of flow through microchannels can be expressed as follows:

$$f = \frac{\Delta p \cdot D_h}{2\rho U^2 L} \tag{8.29}$$

where:
Δp is the pressure drop between the inlet and the outlet
D_h and L are the hydraulic diameter and length of the channel
U is the average fluid velocity

It should be noted that the density ρ used in calculating the friction factor should be evaluated using the corresponding temperature and NP volume

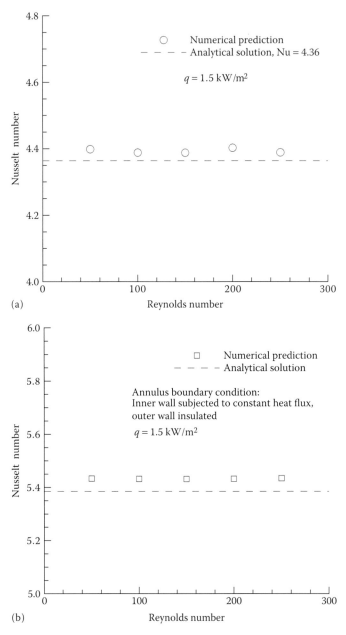

FIGURE 8.4   Model validation of Nusselt number in fully developed region: (a) Circular tube; (b) annulus.

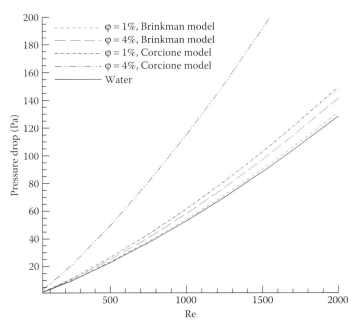

FIGURE 8.5   Pressure gradient versus Reynolds number for nanofluid flow using two viscosity models.

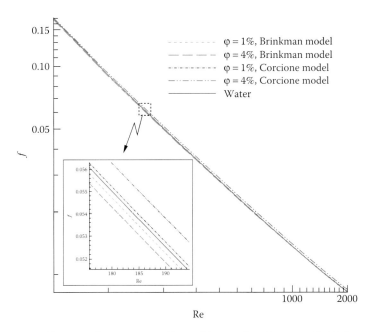

FIGURE 8.6   Friction factor versus Reynolds number for nanofluid flow using two viscosity models.

fraction. Hence, the effect of adding NPs to the base fluid on the mixture density was also considered. Figure 8.6 shows that the different viscosity models and NP volume fractions yield little change in the friction factor at the same Reynolds number. This indicates that nanofluids with NP volume fractions less than 4% may not cause large penalty in pumping power, even though the viscosity has been unfavorably changed. Clearly, the effect of volume fraction on the friction factor is larger using the Corcione model than using the Brinkman model.

In order to evaluate the thermal performance of nanofluid flow, the nanofluid convective heat transfer coefficients at various Reynolds numbers and pumping powers were compared for pure water and nanofluid flows with different volume fractions.

Nanofluid flow yields higher heat transfer coefficient than pure water flow at the same Reynolds number and under the same pumping power, as can be seen in Figure 8.7. Meanwhile, the heat transfer coefficient predicted using the Corcione model is higher than that using the Brinkman model. As discussed in Section 8.1.2, using the Reynolds number as a comparison basis is sometimes misleading. Because the average velocity of nanofluids will be larger than the water velocity at the same Re, which could mean increased pumping power. To eliminate this effect, constant pumping power is also used to show the performance comparison between water and nanofluids. Figure 8.7b demonstrates that under constant pumping powers, nanofluids still generate higher heat transfer coefficient than water, which suggests that the gain from increased thermal conductivity overweighs the loss from increased viscosity. Interestingly, when $P = $ constant, predictions by employing the two different viscosity models are very close at $\varphi = 0.04$, which is just the opposite of the case when Re = constant. This is because under the same pumping power the inlet velocity for the two cases differs less than when Re = constant, indicating a smaller Re using the Corcione model than using the Brinkman model. Overall, the average enhancement of thermal performance for Al_2O_3–water nanofluid with a volume fraction of 4% is about 20%.

8.3.2 Nanofluid Cooling of CPV Cells

For CPV cell cooling, the configuration described in Xu and Kleinstreuer (2014a) was adopted (see Figure 8.8). The improved F–K model for thermal conductivity (Xu, 2014) and the Brinkman model for viscosity (Brinkman, 1952) were used. An Al_2O_3–water nanofluid flows through the channel to maintain a high-efficiency cell operation and avoid excessive cell temperatures.

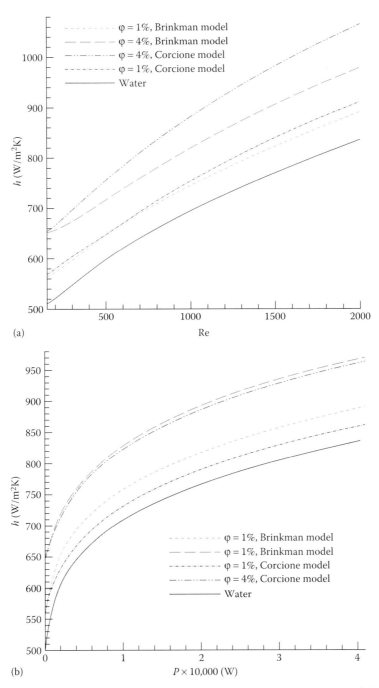

FIGURE 8.7 Nanofluid heat transfer coefficient using two viscosity models and improved F–K thermal conductivity model: (a) *h* versus Re; (b) *h* versus pumping power.

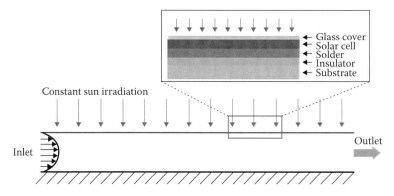

FIGURE 8.8   Simplified model of the cooling channel of the CPV module.

8.3.2.1 Numerical Method

Fully developed flow condition upon the entrance and zero gauge pressure at the outlet were applied. The no-slip boundary condition was enforced at all solid walls. To make the model computationally tractable, steady-state flow was assumed. Furthermore, neglecting variations in the span-wise (lateral) direction reduces the model to two dimensional. The bottom channel wall was thermally insulated.

The incoming concentrated solar radiation converts partly to electricity and partly to heat, which is dissipated through active cooling. Hence, the resulting heat flux is

$$q_{cell} = C \cdot q \cdot \alpha \cdot (1 - \eta) \tag{8.30}$$

where:

   C is the concentration ratio

   $q = 1000$ W/m^2 is the energy flux of 1 sun irradiation

   $\alpha = 0.9$ is the absorptivity

   η is the cell efficiency given by

$$\eta = \eta_{ref} \left[1 - \beta (T_{cell} - T_{ref}) \right] \tag{8.31}$$

The reference efficiency $\eta_{ref} = 0.2$ at a reference temperature of $T_{ref} = 298$ K; the temperature coefficient $\beta = 0.0045$ K^{-1}.

One-dimensional heat conduction from the cell surface to the forced convection interface was assumed, represented by an equivalent thermal circuit. Due to the space constraints, the readers are referred to Xu and Kleinstreuer (2014a) for a more detailed description of the model.

A typical case contains around 157,000 structured mesh elements with 240,000 nodes. The near-wall mesh was locally refined by a factor of 1.08. The y^+-coordinate was kept under unity. The convergence criterion was 1×10^{-6} for the average residual of mass, momentum, and heat transfer. Mesh independence was examined and verified by decreasing the mesh size to one-half, which produced a maximum result change of less than 1%.

8.3.2.2 Model Validation

To validate the numerical model, the velocity profile of the base fluid was compared to the experimental data of Laufer (1948). The near-wall velocity profile was also compared with the experimental data of Lindgren (1965). Also, the local Nusselt numbers at different axial locations were compared with the experimental results of Sparrow et al. (1957) for a circular tube. The comparisons showed that the current model provides good accuracy (see Xu [2014] for details).

8.3.2.3 Results and Discussion

Nanofluids generate better heat transfer performances than the base fluids alone, due to higher thermal conductivities resulting from Brownian motion effects, as discussed in Section 8.3.1. This means higher cell efficiencies (i.e., lower cell temperature) using nanofluid cooling. Figure 8.9 shows the better cooling performance of nanofluid over pure water. Clearly, the use of a nanofluid yields higher cell efficiency than pure water cooling, but this difference becomes smaller when the Reynolds number increases. This may be because of the viscosity model adopted.

Increasing the Reynolds number can effectively promote the cell efficiency. However, a very high Re is not necessarily demanded. Because in the lower Reynolds number range, heat transfer between the wall and the fluid is dominant by forced convection, which depends directly on the flow rate. At a higher Reynolds number range, the cell temperature is already close to the coolant temperature, so the gain from further increasing Re will be marginal. To determine an optimal Reynolds number, a gross power output, that is, P_{gross} = electrical power output – pumping power input, can be defined.

Figure 8.10 shows the suitable Reynolds numbers based on P_{gross} for different concentration ratios ranging from $C = 100$ to $C = 1000$. A suitable Reynolds number exists for any specific concentration ratio, above which further increases of Re harm the system efficiency. It should be noted that P_{gross} was normalized with values at different Reynolds numbers

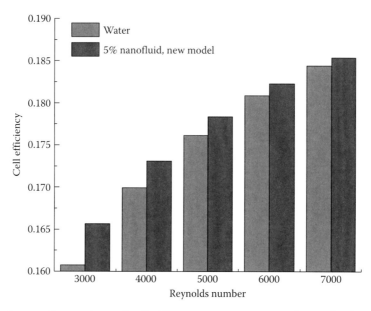

FIGURE 8.9 Comparison of cell efficiencies using water and nanofluid cooling under different inlet Reynolds numbers.

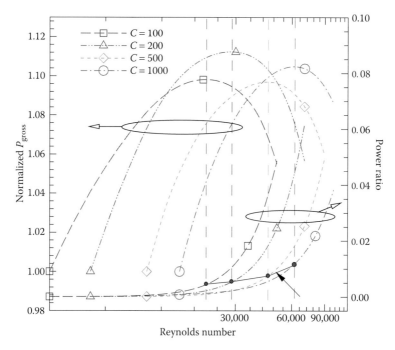

FIGURE 8.10 Variation of optimal Reynolds numbers in terms of maximum gross power output, as well as input/output power ratios under different concentration ratios.

for different concentration ratios. Therefore, the larger magnitudes of the normalized P_{gross} should not be interpreted as higher gross power output. The ratios of pumping power input to electrical power output are also shown in Figure 8.10. Interestingly, at suitable Reynolds numbers determined by P_{gross}, the power ratio increases with concentration ratio C, as pointed by the arrow in the figure. This is probably due to the rise of the temperature difference between the cell surface and the convective heat transfer interface under higher concentration ratios.

As mentioned in Section 8.1.5, in thermal system design, minimization of the entropy generation via operational and/or geometrical optimizations has been proved effective. Here, the effect of the NP volume fraction, the inlet Reynolds number, the channel height, as well as the nanofluid inlet temperatures on the entropy generation due to both friction and heat transfer is analyzed.

Figure 8.11 suggests that at low Reynolds numbers, S_{gen}(thermal) is several orders of magnitude greater than S_{gen}(frictional), which indicates that thermal entropy generation dominates the cooling process. The entropy generation rate decreases when more NPs are added to the base fluid, because the thermal conductivity of the nanofluid is enhanced with an increase in particle loading, generating lower temperature gradients.

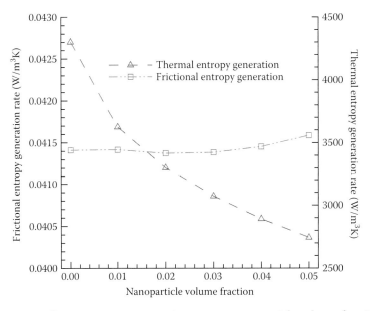

FIGURE 8.11   System entropy generation versus nanoparticle volume fraction.

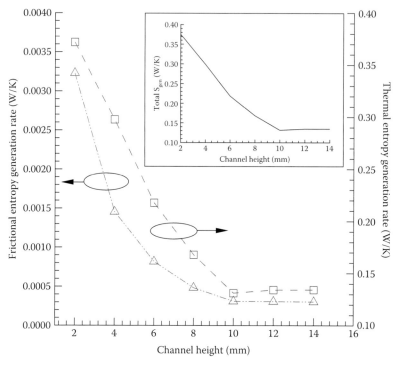

FIGURE 8.12 Thermal and frictional entropy generation rate versus channel height.

This again indicates that nanofluid cooling provides a better efficiency than pure water, as entropy generation is related to energy loss.

Figure 8.12 shows the thermal and frictional entropy generation rate in different channels. A Reynolds number for each channel is used in the attempt that the cell efficiency is kept at a constant. As a result, when increasing the channel height, the average velocity first drops before h reaches 10 mm, then slowly increases. Entropy generation is reduced when increasing the channel height for $h < 10$ mm, but after $h > 10$ mm, little change in S_{gen} is observed. This is because a higher Reynolds number caused lower temperature gradient in the bulk (see Table 8.1). Clearly, $h = 10$ is the desired channel configuration.

Figure 8.13 shows the plots of the entropy generation rate versus the inlet Reynolds number for nanofluids in an $h = 10$ mm channel with 4% particle loading. The frictional entropy generation rate increases with the Reynolds number due to an enhanced friction effect. On the contrary, S_{gen}(thermal) is smaller at higher Reynolds numbers. Hence, an optimal

TABLE 8.1   Applied Nanofluid Flow Conditions for Different Channel Heights

Channel height (mm)	2	4	6	8	10	12	14
Reynolds number	2,980	5,210	7,000	8,300	9,400	11,500	13,650
Average velocity (m/s)	1.6054	1.4034	1.2571	1.1178	1.0128	1.0326	1.0505

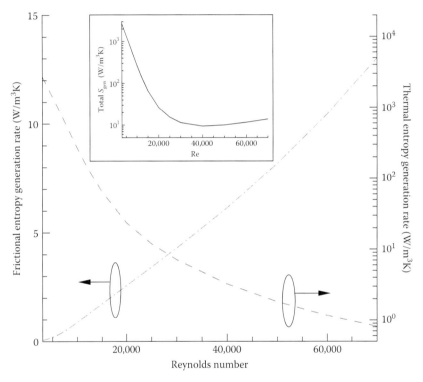

FIGURE 8.13   System entropy generation versus Reynolds number. $h = 10$ mm.

Reynolds number is identified corresponding to minimum total entropy generation. The thermal entropy generation rate increases with the nanofluid inlet temperature (see Figure 8.14). However, the frictional entropy generation rate decreases at the same time. In fact, as the temperature is elevated, the nanofluid viscosity drops, which requires lower flow rate because Re is constant, thus bringing down the S_{gen}(frictional). An optimal inlet temperature is found at which the total entropy generation rate is minimized.

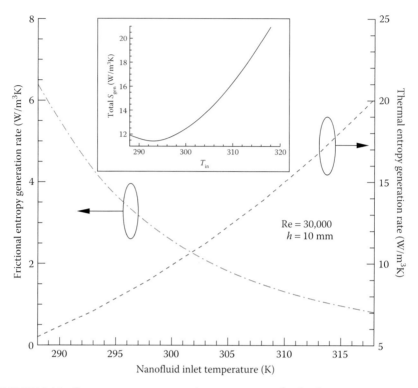

FIGURE 8.14 System entropy generation versus nanofluid inlet temperature.

8.4 CONCLUSION

The performance of thermal nanofluid flow in microchannels was summarized. Most recent experimental studies concerning nanofluid heat transfer and fluid flow characteristics were reviewed, whereas the current methods for numerical study of nanofluid convective heat transfer were discussed. In addition, the efforts in developing flow and heat transfer correlations as well as in analyzing the entropy generation of nanofluid flow were discussed. Nanofluids seem to provide better heat transfer performance than base fluids with little penalty in pressure drop. However, for a better understanding of nanofluid convective heat transfer, further experimental and numerical studies are needed to identify the heat transfer mechanisms and provide accurate predictions.

To demonstrate the characteristics of thermal nanofluid flow, two cases were numerically analyzed: nanofluid flow in a counterflow double-tube heat exchanger and nanofluid cooling of CPV cells. Entropy generation minimization was used to optimize the geometrical and operational conditions. Specifically, in the first case, two different viscosity models and a newly developed thermal conductivity model for laminar water flow and alumina–water nanofluid flow were compared after extensive computer model validations. The results show that nanofluids measurably enhance the thermal performance without causing obvious penalty in pressure drop. In the second case, turbulent nanofluid cooling was analyzed for improved efficiency of CPV cells with a focus on entropy generation minimization. Entropy generation reduces when using nanofluids due to their improved thermal transport mechanism.

ACKNOWLEDGMENTS

The authors are grateful for partial financial support from the National Science Foundation grant CBET-1232988 and for the use of ANSYS software as part of the North Carolina State University–ANSYS Professional Partnership.

REFERENCES

Abbasian Arani, A.A., and J. Amani. 2013. "Experimental investigation of diameter effect on heat transfer performance and pressure drop of TiO_2–water nanofluid." *Experimental Thermal and Fluid Science* 44:520–533.

Azmi, W.H., K.V. Sharma, P.K. Sarma, R. Mamat, S. Anuar, and V.D. Rao. 2013. "Experimental determination of turbulent forced convection heat transfer and friction factor with SiO_2 nanofluid." *Experimental Thermal and Fluid Science* 51:103–111.

Bahiraei, M. 2014. "A comprehensive review on different numerical approaches for simulation in nanofluids: Traditional and novel techniques." *Journal of Dispersion Science and Technology.* 35(7):984–996.

Bejan, A. 1996. *Entropy Generation Minimization: The Method of Thermodynamic Optimization of Finite-Size Systems and Finite-Time Processes.* Boca Raton, FL: CRC Press, Chap. 3.

Brinkman, H.C. 1952. The viscosity of concentrated suspensions and solutions. *The Journal of Chemical Physics* 20:571–571.

Corcione, M. 2011. "Empirical correlating equations for predicting the effective thermal conductivity and dynamic viscosity of nanofluids." *Energy Conversion and Management* 52(1):789–793.

Das, S.K., S.U.S. Choi, W. Yu, and T. Pradeep, ed. 2008. *Nanofluids: Science and Technology.* Hoboken, NJ: John Wiley & Sons.

Edalati, Z., S.Z. Heris, and S.H. Noie. 2012. "The study of laminar convective heat transfer of CuO/water nanofluid through an equilateral triangular duct at constant wall heat flux." *Heat Transfer Asian Research* 41:418–429.

Esfe, M.H., S. Saedodin, and M. Mahmoodi. 2014. "Experimental studies on the convective heat transfer performance and thermophysical properties of MgO–water nanofluid under turbulent flow." *Experimental Thermal and Fluid Science* 52:68–78.

Feng, Y., and C. Kleinstreuer. 2010. "Nanofluid convective heat transfer in a parallel-disk system." *International Journal of Heat and Mass Transfer* 53:4619–4628.

Fox, R.W., A.T. McDonald, and P.J. Pritchard. 2004. *Introduction to Fluid Mechanics*, 6th ed. New York: John Wiley & Sons.

Godson, L., B. Raja, D. M. Lal, and S. Wongwises. 2010. "Enhancement of heat transfer using nanofluids—An overview." *Renewable and Sustainable Energy Reviews* 14(2):629–641.

Gunnasegaran, P., N.H. Shuaib, H.A. Mohammed, M.F. Abdul Jalal, and E. Sandhita. 2012. "Heat transfer enhancement in microchannel heat sink using nanofluids." In *Fluid Dynamics, Computational Modeling and Applications*, ed. Hector Juarez, L. 287–326. Rijeka, Croatia: InTech.

Heris, S.Z., T.H. Nassan, S.H. Noie, H. Sardarabadi, and M. Sardarabadi. 2013. "Laminar convective heat transfer of Al_2O_3/water nanofluid through square cross-sectional duct." *International Journal of Heat and Fluid Flow* 44:375–382.

Heyhat, M.M., F. Kowsary, A.M. Rashidi, S. Alem Varzane Esfehani, and A. Amrollahi. 2012. "Experimental investigation of turbulent flow and convective heat transfer characteristics of alumina water nanofluids in fully developed flow regime." *International Communications in Heat and Mass Transfer* 39:1272–1278.

Heyhat, M.M., F. Kowsary, A.M. Rashidi, M.H. Momenpour, and A. Amrollahi. 2013. "Experimental investigation of laminar convective heat transfer and pressure drop of water-based Al_2O_3 nanofluids in fully developed flow regime." *Experimental Thermal and Fluid Science* 44:483–489.

Kakaç, S., and A. Pramuanjaroenkij. 2009. "Review of convective heat transfer enhancement with nanofluids." *International Journal of Heat and Mass Transfer* 52(13–14):3187–3196.

Kamyar, A., R. Saidur, and M. Hasanuzzaman. 2012. "Application of computational fluid dynamics (CFD) for nanofluids." *International Journal of Heat and Mass Transfer* 55(15–16):4104–4115.

Kays, W.M., and M.E. Crawford. 1993. *Convective Heat and Mass Transfer*. New York: McGraw-Hill.

Kleinstreuer, C. 2010. *Modern Fluid Dynamics: Basic Theory and Selected Applications in Macro- and Micro-Fluidics*. New York: Springer.

Kleinstreuer, C. 2014. *Microfluidics and Nanofluidics: Theory and Selected Applications*. Hoboken, NJ: John Wiley & Sons.

Kleinstreuer, C., and Y. Feng. 2012. "Thermal nanofluid property model with application to nanofluid flow in a parallel-disk system—Part I: A new thermal conductivity model for nanofluid Flow." *ASME Journal of Heat Transfer* 134(5):051002-1-11.

Kleinstreuer, C., J. Li, and Y. Feng. 2013. "Computational analysis of enhanced cooling performance and pressure drop for nanofluid flow in microchannels." In *Nanoparticle Heat Transfer and Fluid Flow*, eds. E.M. Minkowycz, W.J., Sparrow, and E.M.J.P. Abraham. Boca Raton, FL: CRC Press.

Kumar, C.S. ed. 2010. *Microfluidic Devices in Nanotechnology: Applications*. Hoboken, NJ: John Wiley & Sons.

Laufer, J. 1948. "Investigation of turbulent flow in a two-dimensional channel," PhD dissertation, California Institute of Technology, Pasadena, CA.

Li, D., ed. 2008. *Encyclopedia of Microfluidics and Nanofluidics.* Amsterdam, the Netherlands: Springer.

Li, J., and C. Kleinstreuer. 2010. "Entropy generation analysis for nanofluid flow in microchannels." *ASME Journal of Heat Transfer* 132:122401-1-8.

Lindgren, E.R. 1965. *Experimental Study on Turbulent Pipe Flows of Distilled Water.* Report 1 AD621071, Civil Engineering Department, Oklahoma State University, Stillwater, MN.

Madhesh, D., R. Parameshwaran, and S. Kalaiselvam. 2014. "Experimental investigation on convective heat transfer and rheological characteristics of Cu–TiO_2 hybrid nanofluids." *Experimental Thermal and Fluid Science* 52:104–115.

Mahdi, R.A., H.A. Mohammed, and K.M. Munisamy. 2013. "Improvement of convection heat transfer by using porous media and nanofluid: Review." *International Journal of Science and Research* 2(8):34–47.

Mahian, O., A. Kianifar, C. Kleinstreuer, M. Al-Nimr, I. Pop, A.Z. Sahin, and S. Wongwises. 2013. "A review of entropy generation in nanofluid flow." *International Journal of Heat and Mass Transfer* 65:514–532.

Maxwell, J.C. (1881). *A Treatise on Electricity and Magnetism.* Oxford: Clarendon Press.

Meriläinen, A., A. Seppälä, K. Saari, J. Seitsonen, J. Ruokolainen, S. Puisto, N. Rostedt et al. 2013. "Influence of particle size and shape on turbulent heat transfer characteristics and pressure losses in water-based nanofluids." *International Journal of Heat and Mass Transfer* 61:439–448.

Mohammed, H., G. Bhaskaran, N.H. Shuaib, and R. Saidur. 2011. "Heat transfer and fluid flow characteristics in microchannels heat exchanger using nanofluids: A review." *Renewable and Sustainable Energy Reviews* 15(3):1502–1512.

Nguyen, C.T., G. Roy, C. Gauthier, and N. Galanis. 2007. "Heat transfer enhancement using Al_2O_3-water nanofluid for an electronic liquid cooling system." *Applied Thermal Engineering* 27(8):1501–1506.

Nkurikiyimfura, I., Y. Wang, and Z. Pan. 2013. "Heat transfer enhancement by magnetic nanofluids—A review." *Renewable and Sustainable Energy Reviews* 21:548–561.

Prabhat, N., J. Buongiorno, and L.-W. Hu. 2012. "Convective heat transfer enhancement in nanofluids: Real anomaly or analysis artifact?" *Journal of Nanofluids* 1, 55–62.

Ratts, E.B., and A.G. Raut. 2004. "Entropy generation minimization of fully developed internal flow with constant heat flux." *ASME Journal of Heat Transfer* 126(4):656–659.

Sahin, B., G.G. Gültekin, E. Manay, and S. Karagoz. 2013. "Experimental investigation of heat transfer and pressure drop characteristics of Al_2O_3–water nanofluid." *Experimental Thermal and Fluid Science* 50:21–28.

Salman, B.H., H.A. Mohammed, K.M. Munisamy, and A.Sh. Kherbeet. 2013. "Characteristics of heat transfer and fluid flow in microtube and microchannel using conventional fluids and nanofluids: A review." *Renewable and Sustainable Energy Reviews* 28:848–880.

Sarkar, J. 2011. "A critical review on convective heat transfer correlations of nanofluids." *Renewable and Sustainable Energy Reviews* 15(6):3271–3277.

Sparrow, E.M., T.M. Hallman, and R. Siegel. 1957. Turbulent heat transfer in the thermal entrance region of a pipe with uniform heat flux. *Applied Scientific Research, Section A* 7(1):37–52.

Sundar, L.S., and M.K. Singh. 2013. "Convective heat transfer and friction factor correlations of nanofluid in a tube and with inserts: A review." *Renewable and Sustainable Energy Reviews* 20:23–35.

Sureshkumar, R., S.T. Mohideen, and N. Nethaji. 2013. "Heat transfer characteristics of nanofluids in heat pipes: A review." *Renewable and Sustainable Energy Reviews* 20:397–410.

Wu, J.M., and J. Zhao. 2013. "A review of nanofluid heat transfer and critical heat flux enhancement—Research gap to engineering application." *Progress in Nuclear Energy* 66:13–24.

Xu, Z., 2014. "An improved thermal conductivity model for nanofluids with applications to concentration photovoltaic-thermal systems," MS thesis, North Carolina State University, Raleigh, NC.

Xu, Z., and C. Kleinstreuer. 2014a. "Computational analysis of nanofluid cooling of high concentration photovoltaic cells." *ASME Journal of Thermal Science and Engineering Applications* 6:031009-1-9.

Xu, Z., and C. Kleinstreuer. 2014b. "Concentration photovoltaic–thermal energy co-generation system using nanofluids for cooling and heating." *Energy Conversion and Management* 87:504–512.

Yu, W., D.M. France, E.V. Timofeeva, D. Singh, and J.L. Routbort. 2012. "Comparative review of turbulent heat transfer of nanofluids." *International Journal of Heat and Mass Transfer* 55(21–22):5380–5396.

Yu, L., D. Liu, and F. Botz. 2012. "Laminar convective heat transfer of alumina-polyalphaolefin nanofluids containing spherical and non-spherical nanoparticles." *Experimental Thermal and Fluid Science* 37:72–83.

CHAPTER 9

Use of Nanofluids for Heat Transfer Enhancement in Mixed Convection

Hakan F. Oztop, Eiyad Abu-Nada, and Khaled Al-Salem

CONTENTS

9.1 LITERATURE SURVEY

Mixed convection is a combination of natural convection and forced convection, where the Rayleigh number gives an indication to which mode of heat transfer is dominant. Mixed convection heat transfer can be observed in many engineering applications such as solar air collectors, heat exchangers, cooling of electronic equipments, and heating and cooling of buildings.

Over the past two decades, nanofluids have emerged as a promising technology for the enhancement of the intrinsic thermophysical properties of conventional heat transfer fluids. This innovative technology caught the attention of the heat transfer community as a way to improve the poor heat transfer properties of many fluids used in heat transfer applications, for example, water and oils. Many researchers in the community investigated the merits of dispersing nanometer-sized particles into base fluids to enhance heat transfer in mixed convection settings. Abu-Nada and Chamkha [1] conducted a computational study on mixed convection in a lid-driven cavity filled with a nanofluid (water + Al_2O_3). Significant heat transfer enhancement was observed in the study due to the addition of nanoparticles. In another investigation, Chamkha and Abu-Nada [2] tested the effect of using different viscosity models on mixed convection heat transfer in single- and double-lid-driven square cavities filled with the same nanofluid. They found that for small Richardson numbers, the Pak and Cho model gives a reduction in the average Nusselt number for the single-lid-driven cavities compared with other viscosity models. Tiwari and Das [3] studied heat transfer augmentation in a two-sided nanofluid-filled, lid-driven cavity. They solved two-dimensional, single-phase equations using the finite volume method and observed the need for a higher Richardson number to move nanofluids at higher solid volume fraction as a result of the increased viscosity of nanofluids compared to base fluids. Afrouzi and Farhadi [4] investigated mixed convection in a lid-driven cavity filled with Cu–water nanofluid using a cubical heater located inside the cavity in different positions. They found an enhancement in heat transfer with the increase of nanoparticle volume fraction. Ahmad and Pop [5] studied mixed convection in a vertical boundary layer in a porous medium saturated with nanofluids. They used a two-dimensional similarity solution. Billah et al. [6] conducted a computational study for mixed convection in a lid-driven, inclined, triangular enclosure filled with Cu–water nanofluid. They found that the inclination angle affects the dynamics of the flow more than it affects the thermal field. Arani et al. [7] did research on the effects of inclination angle in a partially heated enclosure filled with nanofluid on mixed convection heat transfer and found that the heat transfer is enhanced with increasing the inclination angle. Esfe et al. [8] studied mixed convection in a lid-driven cavity with an obstacle by investigating nanofluid variable properties. They observed that the heat transfer rate is increased with increasing the geometric parameters of the obstacle block and increasing the nanoparticle volume fraction. A study of using different nanofluids on lid-driven porous cavities in mixed convection was performed by Mittal et al. [9]. In addition to one-phase analysis of nanofluids, two-phase mixture model was used in studying mixed convection in nanofluid [10]. The main finding is that increasing the volume fraction of nanoparticles enhances the convective heat transfer coefficient and consequently the Nusselt number, although it has a negligible effect on the wall shear stress and the corresponding skin friction factor. Sebdani et al. [11] studied the mixed convection heat transfer in a two-sided, lid-driven, nanofluid-filled cavity with partial heater. Both the location of the heat source and the nanoparticle volume fraction were found to affect the heat transfer. Arefmanesh and Mahmoodi [12] investigated the effects of uncertainties in different viscosity models for Al_2O_3 nanofluid on mixed convection. They reported a difference between Maiga and Brinkman models. Izadi et al. [13] investigated the effect of Richardson number ratio at the walls on the laminar mixed convection of a nanofluid flowing in an annulus under uniform heating condition. They found that heat transfer coefficient increased with increasing Al_2O_3 nanoparticle volume fraction. Kalteh et al. [14] focused, in their numerical study, on mixed convection in a lid-driven cavity filled with water-based nanofluid and in the presence of a triangular heat source. The results support the claim that dispersing nanoparticles in pure fluids leads to a significant heat transfer enhancement. Mixed convection in nanofluid-filled triangular enclosure was investigated by Ghasemi and Aminossadati [15] using finite volume method and Brinkman model and found that Al_2O_3 nanoparticles enhance the heat transfer. Nemati et al. [16] applied the lattice Boltzmann method to simulate the nanofluid mixed convection heat transfer in a lid-driven cavity and showed a good agreement with the results reported in literature. They indicated that the effects of solid volume fraction grow stronger sequentially for Al_2O_3, CuO, and Cu. Alinia et al. [17] used a two-phase mixture model to simulate mixed convection in nanofluid-filled lid-driven enclosure. Their results, also, suggest that the addition of nanoparticles enhances the heat transfer. Gümgüm and Sezgin [18] applied dual reciprocity boundary element method (DRBEM) solution technique to solve mixed convection of nanofluids in enclosures with moving walls. They disclosed that the average Nusselt number increased with increasing volume fraction of nanoparticles and decreased with increasing Ri number. Other studies can be found in references [19–21].

9.2 AN EXAMPLE SOLUTION

In this section, a case study is presented to illustrate the mathematics involved in analyzing heat transfer in nanofluids. Figure 9.1 shows the geometry of the cavity under investigation. The height and width of the cavity are given as H and L, respectively. The top wall is moving with a constant speed and maintained at a constant temperature (T_H). The bottom wall is stationary and kept at a lower constant temperature (T_L). The fluid in the cavity is a water-based nanofluid containing Al_2O_3 nanoparticles. The flow inside the cavity is assumed laminar, incompressible, and two dimensional. Also, it is assumed that the water and nanoparticles are in thermal equilibrium where no slip occurs between base fluid and particles. The thermophysical properties of the nanofluid are assumed to be constant except for the density variation, which is approximated by the Boussinesq assumption. The thermophysical properties of the nanofluid are listed in Table 9.1.

The governing equations for steady, laminar, lid-driven flow in a cavity filled with a nanofluid in terms of the stream function–vorticity formulation are given as follows:

Kinematics

$$\frac{\partial^2 \psi}{\partial x'^2} + \frac{\partial^2 \psi}{\partial y'^2} = -\omega \qquad (9.1)$$

where:

ψ is the stream function
ω is the vorticity

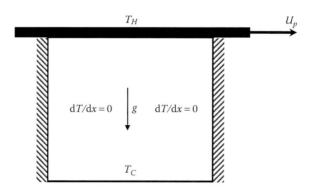

FIGURE 9.1   Schematic of the lid-driven cavity.

TABLE 9.1   Thermophysical Properties of Base Fluid and Nanoparticles

Physical Properties	Fluid Phase (Water)	Al₂O₃	CuO
c_p (J/kg K)	4179	765	540
P (kg/m³)	997.1	3970	6500
K (W/m K)	0.613	25.0	18.0
$B \times 10^{-5}$ (1/K)	21	0.85×10^{-5}	0.85×10^{-5}
d_p (nm)	0.384	47	29

Vorticity

$$\frac{\partial}{\partial x'}\left(\omega\frac{\partial \psi}{\partial y'}\right) - \frac{\partial}{\partial y'}\left(\omega\frac{\partial \psi}{\partial x'}\right) = \frac{1}{\rho_{nf}}\left[\frac{\partial}{\partial x'}\left(\mu_{nf}\frac{\partial \omega}{\partial x'}\right) + \frac{\partial}{\partial y'}\left(\mu_{nf}\frac{\partial \omega}{\partial y'}\right)\right]$$

$$+ \frac{\left[\phi\rho_s\beta_s + (1-\phi)\rho_f\beta_f\right]}{\rho_{nf}}g\left(\frac{\partial T}{\partial x'}\right)$$

(9.2)

Energy

$$\frac{\partial}{\partial x'}\left(T\frac{\partial \psi}{\partial y'}\right) - \frac{\partial}{\partial y'}\left(T\frac{\partial \psi}{\partial x'}\right) = \frac{\partial}{\partial x'}\left(\alpha_{nf}\frac{\partial T}{\partial x'}\right) + \frac{\partial}{\partial y'}\left(\alpha_{nf}\frac{\partial T}{\partial y'}\right)$$

(9.3)

The horizontal and vertical velocities are computed from the stream function relations:

$$u = \frac{\partial \psi}{\partial y'}, \quad v = -\frac{\partial \psi}{\partial x'}$$

(9.4)

In Equations 9.2 and 9.3, the nanofluid thermal diffusivity, effective density, and heat capacitance are given by the following:

$$\alpha_{nf} = \frac{k_{nf}}{\left(\rho c_p\right)_{nf}}$$

(9.5)

$$\rho_{nf} = (1-\phi)\rho_f + \phi\rho_p$$

(9.6)

$$\left(\rho c_p\right)_{nf} = (1-\phi)\left(\rho c_p\right)_f + \phi\left(\rho c_p\right)_p$$

(9.7)

The effective thermal conductivity of the nanofluid is approximated by the Maxwell–Garnett (MG) model [22]:

$$\frac{k_{nf}}{k_f} = \frac{k_s + 2k_f - 2\phi(k_f - k_s)}{k_s + 2k_f + \phi(k_f - k_s)} \tag{9.8}$$

Another effective thermal conductivity model of nanofluids is given by Chon et al. [23] as

$$\frac{k_{nf}}{k_{bf}} = 1 + 64.7\phi^{0.7640} \left(\frac{d_{bf}}{d_p}\right)^{0.3690} \left(\frac{k_{bf}}{k_p}\right)^{0.7476} Pr_T^{0.9955} Re^{1.2321} \tag{9.9}$$

In the model equation, Pr_T and Re are given as

$$Pr_T = \frac{\mu_f}{\rho_f \alpha_f} \tag{9.10}$$

$$Re = \frac{\rho_f k_b T}{3\pi \mu_f^2 l_f} \tag{9.11}$$

where:
   The subscript f stands for the base fluid, which, in this case, is water
   k_b is the Boltzmann constant, 1.3807×10^{-23} J/K
   l_f is the mean path of fluid particles given as 0.17 nm [23]

This model considers the effect of nanoparticle size and temperature on nanofluid thermal conductivity encompassing a wide temperature range between 21°C and 70°C. It is also worth emphasizing that this model is based on experimental measurements of Al_2O_3 nanoparticles in water for volume fractions up to 4%. However, this model was tested experimentally by Angue Minsta et al. [24] for the pair of Al_2O_3 and CuO nanoparticles and found to predict the thermal conductivity of these nanoparticles up to a volume fraction of 9%.

   The MG model is found to be appropriate for studying heat transfer enhancement using nanofluids [25–27]. Abu-Nada et al. [28] reported that the thermal conductivity of nanofluids has lower impact on heat transfer compared to nanofluid viscosity. The difference in the heat transfer predictions between the MG and Chon et al. models is insignificant. Therefore, the MG model is generally found to be a suitable model for nanofluids. The viscosity of the nanofluid can be approximated as the viscosity of the base fluid μ_f containing dilute suspension of fine spherical particles. Some models used in the literature for the viscosity of the nanofluids in mixed convection applications are the Brinkman model [29], the Pak and Cho [30] correlation, and the Nguyen et al. [31] experimental data fitted by Abu-Nada et al. [28] for water–CuO and water–Al$_2$O$_3$ nanofluids. These correlations are given, respectively, as

$$\mu_{nf} = \frac{\mu_f}{(1-\phi)^{2.5}} \tag{9.12}$$

$$\mu_{nf} = \mu_f \left(1 + 39.11\phi + 533.9\phi^2\right) \tag{9.13}$$

$$\mu_{Al_2O_3}(cp) = \exp\left(\begin{array}{c} 3.003 - 0.04203T - 0.5445\phi + 0.0002553T^2 \\ + 0.0524\phi^2 - 1.622\phi^{-1} \end{array}\right) \tag{9.14}$$

$$\mu_{CuO}(cp) = -0.6967 + \frac{15.937}{T} + 1.238\phi + \frac{1356.14}{T^2} - 0.259\phi^2 - 30.88\frac{\phi}{T}$$

$$- \frac{19652.74}{T^3} + 0.01593\phi^3 + 4.38206\frac{\phi^2}{T} + 147.573\frac{\phi}{T^2} \tag{9.15}$$

The viscosity in Equations 9.14 and 9.15 is expressed in centipoise and the temperature in celsius. The following dimensionless variables are introduced:

$$x = \frac{x'}{H}, \quad y = \frac{y'}{H}, \quad \Omega = \frac{\omega}{U_p/H}, \quad \Psi = \frac{\psi}{U_p H}, \quad V = \frac{v}{U_p},$$

$$U = \frac{u}{U_p}, \quad \theta = \frac{T - T_L}{T_H - T_L}, \quad k = \frac{k_{nf}}{k_{fo}}, \quad \alpha = \frac{\alpha_{nf}}{\alpha_{fo}}, \quad \mu = \frac{\mu_{nf}}{\mu_{fo}} \tag{9.16}$$

where the subscript "o" indicates reference conditions. The governing equations in nondimensional form are as follows:

$$\frac{\partial^2 \Psi}{\partial x^2} + \frac{\partial^2 \Psi}{\partial y^2} = -\Omega \tag{9.17}$$

$$\frac{\partial}{\partial x}\left(\Omega\frac{\partial\Psi}{\partial y}\right) - \frac{\partial}{\partial y}\left(\Omega\frac{\partial\Psi}{\partial x}\right) = \frac{Pr}{(1-\phi)+\phi(\rho_p/\rho_f)}\left[\frac{\partial}{\partial x}\left(\mu\frac{\partial\Omega}{\partial x}\right) + \frac{\partial}{\partial y}\left(\mu\frac{\partial\Omega}{\partial y}\right)\right]$$

$$+ Re\,Pr\left\{\frac{1}{\left[(1-\phi)/\phi\right]\left[(\rho_f/\rho_p)+1\right]}\frac{\beta_s}{\beta_f} + \frac{1}{\left[\phi/(1-\phi)\right]\left[(\rho_f/\rho_p)+1\right]}\right\}\left(\frac{\partial\theta}{\partial x}\right) \tag{9.18}$$

$$\frac{\partial}{\partial x}\left(\theta\frac{\partial\Psi}{\partial y}\right)-\frac{\partial}{\partial y}\left(\theta\frac{\partial\Psi}{\partial x}\right)$$

$$=\frac{1}{(1-\phi)+\phi\left[(\rho c_p)_p/(\rho c_p)_f\right]}\left[\frac{\partial}{\partial x}\left(k\frac{\partial\theta}{\partial x}\right)+\frac{\partial}{\partial y}\left(k\frac{\partial\theta}{\partial y}\right)\right] \qquad (9.19)$$

where the dimensionless numbers

$$\mathrm{Ri}=\frac{\mathrm{Gr}}{\mathrm{Re}^2}, \quad \mathrm{Gr}=\frac{g\beta(T_H-T_L)H^3}{v_f^2}, \quad \mathrm{Pr}=\frac{v_f}{\alpha_f}, \quad \mathrm{Re}=\frac{U_pH}{v_f} \qquad (9.20)$$

are the Richardson, Grashof, Prandtl, and Reynolds numbers, respectively. The dimensionless horizontal and vertical velocities are written as

$$U=\frac{\partial\Psi}{\partial y}, \quad V=-\frac{\partial\Psi}{\partial x} \qquad (9.21)$$

The appropriate dimensionless boundary conditions can be written as follows:

On the left wall: $U = V = \Psi = 0, \Omega = -\dfrac{\partial^2\Psi}{\partial x^2}, \dfrac{\partial\theta}{\partial x}=0 \qquad$ (9.22a)

On the right wall: $U = V = \Psi = 0, \Omega = -\dfrac{\partial^2\Psi}{\partial x^2}, \dfrac{\partial\theta}{\partial x}=0 \qquad$ (9.22b)

On the top wall: $U = 1, V = \Psi = 0, \Omega = -\dfrac{\partial^2\Psi}{\partial y^2}, \theta = 1 \qquad$ (9.22c)

On the bottom wall: $U = 0$ (one lid), -1 (double lids),

$$V = \Psi = 0, \Omega = -\frac{\partial^2\Psi}{\partial y^2}, \theta = 0 \qquad (9.22d)$$

9.2.1 Numerical Implementation

Equations 9.17 through 9.19, along with the corresponding boundary conditions given in Equation 9.22a–d, are solved using an efficient finite volume method [32,33]. The diffusion term in the vorticity and energy equations is approximated by a second-order central difference scheme. A second-order upwind differencing scheme is adopted for the convective terms. The algebraic finite volume equations for the vorticity and energy equations are written in the following form:

$$a_P \lambda_P = a_E \lambda_E + a_W \lambda_W + a_N \lambda_N + a_S \lambda_S + b$$

where:
The subscript P denotes the cell location
The subscripts W, E, N, and S denote the west, the east, the north, and the south face of the control volume, respectively
The symbol λ denotes the scalar transport quantities such as Ψ, Ω, and θ
The variables a and b are the coefficients of the transport quantities and the source term respectively

Similar expression is also used for the kinematic equation where only central difference is used for the discritization at the cell P of the control volume. The resulting algebraic equations are solved using successive over/under-relaxation method. Successive under-relaxation was used due to the nonlinear nature of the governing equations especially for the vorticity equation. The convergence criterion is defined by the following expression:

$$\varepsilon = \frac{\sum_{j=1}^{j=M} \sum_{i=1}^{i=N} \left| \lambda^{n+1} - \lambda^{n} \right|}{\sum_{j=1}^{j=M} \sum_{i=1}^{i=N} \left| \lambda^{n+1} \right|} < 10^{-6} \tag{9.23}$$

where:
ε is the tolerance
M and N are the number of grid points in the x and y directions, respectively

An accurate representation of vorticity at the surface is the most critical step in the stream function–vorticity formulation. A second-order accurate formula is used for the vorticity boundary condition. For example, the vorticity at the bottom wall is expressed as

$$\Omega = -\frac{\left(8\Psi_{1,j} - \Psi_{2,j} \right)}{2(\Delta y)^2} \tag{9.24}$$

Similar formula can be written for the left and right walls. However, on the moving lid surface, the velocity of the lid has to be taken into account; therefore, the vorticity on the top wall is given as

$$\Omega = -\frac{\left(8\Psi_{N-1,j} - \Psi_{N-2,j} - 7\Psi_{N,j} + 6\Delta y\right)}{2\left(\Delta y\right)^2} \tag{9.25}$$

Similar expression for the lower lid if it is moving (left) and the last two terms in the numerator vanishes for static lower lid. After the distributions of Ψ, Ω, and θ are found, more useful quantities can easily be obtained. For example, the Nusselt number can be expressed as

$$Nu = \frac{hH}{k_f} \tag{9.26}$$

where the heat transfer coefficient is computed from

$$h = \frac{q_w}{T_H - T_L} \tag{9.27}$$

The thermal conductivity of the nanofluid is expressed as

$$k_{nf} = -\frac{q_w}{\left(\partial T / \partial x\right)} \tag{9.28}$$

Substituting Equations 9.27 and 9.28 into 9.26, and using the dimensionless quantities, the local Nusselt number along the top wall can be written as

$$Nu = -\left(\frac{k_{nf}}{k_f}\right)\frac{\partial \theta}{\partial y} \tag{9.29}$$

where (k_{nf}/k_f) is calculated using Equation 9.8 or 9.9. Finally, the average Nusselt number is determined from

$$Nu_{avg} = \int_0^1 Nu(y)dx \tag{9.30}$$

To evaluate the above equation, a Simpson's 1/3 rule of integration is implemented. For convenience, a normalized average Nusselt number is defined as the ratio of Nusselt number at any volume fraction of nanoparticles to that of pure water:

$$Nu_{avg}^*(\phi) = \frac{Nu_{avg}(\phi)}{Nu_{avg}(\phi = 0)} \tag{9.31}$$

9.2.2 Validation of the Case Study Code

The present results are compared against the previous published work of Iwatsu et al. [34] and Khanafer and Chamkha [35], who studied horizontal lid-driven cavity (see Figure 9.2) using the Navier–Stokes equations. These results were obtained using the same conditions as in the mentioned studies (Gr = 100 and Re = 400). The x components of velocity and temperature profiles of the present code are benchmarked against the results of Iwatsu et al. [34] and Khanafer and Chamkha [35], as shown in Figure 9.3. Figure 9.3a shows the x components of velocity at $X = 0.5$ (i.e., mid width of the cavity) and Figure 9.3b shows another comparison for the temperature profiles at $X = 0.5$. Also, further comparison with the natural convection in differentially heated cavity filled with nanofluid is shown in Figure 9.2. As shown in Figures 9.2 and 9.3, the calculations in the case study are in good agreement with the published data in literature.

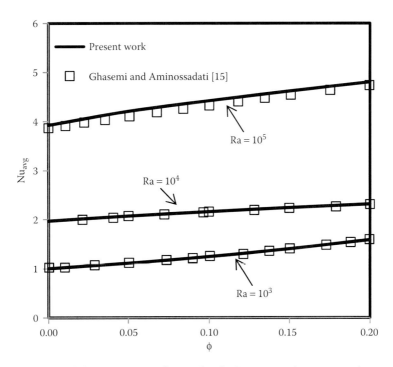

FIGURE 9.2 Validation against the work of Ghasemi and Aminossadati using Cu–water nanofluid. (Data from Ghasemi, B., and Aminossadati, S.M., *Int. Commun. Heat Mass*, 37, 1142–1148, 2010.)

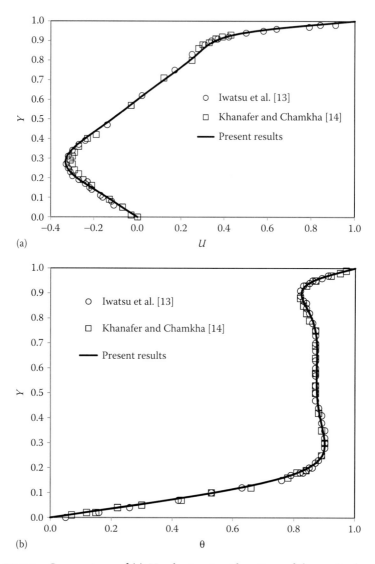

FIGURE 9.3   Comparison of (a) *U*-velocity at mid section of the cavity (*x* = 0.5) and (b) temperature at mid section of the cavity (*x* = 0.5) with those of Iwatsu et al. and Khanafer and Chamkha for Gr = 100 and Re = 400. (Data from Iwatsu, R. et al., *Int. J. Heat Mass Transf.*, 36, 1601–1608, 1993; Khanafer, K.M., Chamkha, A.J., *Int. J. Heat Mass Transf.*, 42, 2465–2481.)

9.3 SUMMARY

We presented in this chapter a brief review of mixed convection in nanofluids. The review was in the form of a literature review and a case study to illustrate the mathematics involved in nanofluid heat transfer analysis. The addition of nanoparticles to base fluids is shown to enhance mixed convection heat transfer. However, the heat transfer enhancement comes at a price, as dispersing nanoparticles into base fluids increases the viscosity of the resulting nanofluid. Also, the heat transfer computations in nanofluids are found to be sensitive to the nanofluid viscosity model employed in the calculations.

REFERENCES

1. Abu-Nada, E., Chamkha, A.J., 2010, Mixed convection flow in a lid-driven inclined square enclosure filled with a nanofluid, *European Journal of Mechanics B/Fluids*, 29, 472–482.
2. Chamkha, A.J., Abu-Nada, E., 2012, Mixed convection flow in single and double-lid driven square cavities filled with water–Al_2O_3 nanofluid: Effect of viscosity models, *European Journal of Mechanics B/Fluids*, 36, 82–96.
3. Tiwari, R.K., Das, M.K., 2007, Heat transfer augmentation in a two-sided lid-driven differentially heated square cavity utilizing nanofluids, *International Journal of Heat and Mass Transfer*, 50, 2002–2018.
4. Afrouzi, H.H., Farhadi, M., 2013, Mixed convection heat transfer in a lid driven enclosure filled by nanofluid, *Iranica Journal of Energy and Environment*, 4, 376–384.
5. Ahmad, S., Pop, I., 2010, Mixed convection boundary layer flow from a vertical flat plate embedded in a porous medium filled with nanofluids, *International Communications in Heat and Mass Transfer*, 37, 987–991.
6. Billah, M.M., Rahman, M.M., Shahabuddin, M., Azad, A.K., 2011, Heat transfer enhancement of copper-water nanofluids in an inclined lid-driven triangular enclosure, *Journal of Scientific Research*, 3, 525–538.
7. Arani, A.A.A., Amani, J., Esfeh, M.H., 2012, Numerical simulation of mixed convection flows in a square double lid-driven cavity partially heated using nanofluid, *Journal of Nanostructures*, 2, 301–311.
8. Esfe, M.H., Ghadi, A.Z., Esforjani, S.S.M., Akbari, M., 2013, Combined convection in a lid-driven cavity with an inside obstacle subjected to Al_2O_3-water nanofluid: Effect of solid volume fraction and nanofluid variable properties, *Acta Physica Polonica A*, 124, 665–672.
9. Mittal, N., Satheesh, A., Kumar, D.S., 2014, Numerical simulation of mixed-convection flow in a lid-driven porous cavity using different nanofluids, *Heat Transfer Asian Research*, 43, 1–16.
10. Goodarzi, M., Safaei, M.R., Vafai, K., Ahmadi, G., Dahari, M., Kazi, S.N., Jomhari, N., 2014, Investigation of nanofluid mixed convection in a shallow cavity using a two-phase mixture model, *International Journal of Thermal Sciences*, 75, 204–220.

11. Sebdani, S.M., Mahmoodi, M., Hashemi, S.M., 2012, Effect of nanofluid variable properties on mixed convection in a square cavity, *International Journal of Thermal Sciences*, 52, 112–126.

12. Arefmanesh, A., Mahmoodi, M., 2011, Effects of uncertainties of viscosity models for Al_2O_3-water nanofluid on mixed convection numerical simulations, *International Journal of Thermal Sciences*, 50, 1706–1719.

13. Izadi, M., Shahmardan, M.M., Behzadmehr, A., 2013, Richardson number ratio effect on laminar mixed convection of a nanofluid flow in an annulus, *International Journal for Computational Methods in Engineering Science and Mechanics*, 14, 304–316.

14. Kalteh, M., Javaherdeh, K., Azarbarzin T., 2014, Numerical solution of nanofluid mixed convection heat transfer in a lid-driven square cavity with a triangular heat source, *Powder Technology*, 253, 780–788.

15. Ghasemi, B., Aminossadati, S.M., 2010, Mixed convection in a lid-driven triangular enclosure filled with nanofluids, *International Communications in Heat and Mass Transfer*, 37, 1142–1148.

16. Nemati, H., Farhadi, M., Sedighi, K., Fattahi, E., Darzi, A.A.R., 2010, Lattice Boltzmann simulation of nanofluid in lid-driven cavity, *International Communications in Heat and Mass Transfer*, 37, 1528–1534.

17. Alinia, M., Ganji, D.D., Gorji-Bandpy, M., 2011, Numerical study of mixed convection in an inclined two sided lid driven cavity filled with nanofluid using two-phase mixture model, *International Communications in Heat and Mass Transfer*, 38, 1428–1435.

18. Gümgüm, S., Sezgin, M.T., 2014, DRBEM solution of mixed convection flow of nanofluids in enclosures with moving walls, *Journal of Computational and Applied Mathematics*, 259, 730–740.

19. Karimipour, A., Esfe, M.H., Safaei, M.R., Semiromi, D.T., Kazi, S.N., 2014, Mixed convection of copper–water nanofluid in a shallow inclined lid driven cavity using the lattice Boltzmann method, *Physica A*, 402, 150–168.

20. Talebi, F., Mahmoudi, A.H., Shahi, M., 2010, Numerical study of mixed convection flows in a square lid-driven cavity utilizing nanofluid, *International Communications in Heat and Mass Transfer*, 37, 79–90.

21. Shahi, M., Mahmoudi, A.H., Talebi, F., 2010, Numerical study of mixed convective cooling in a square cavity ventilated and partially heated from the below utilizing nanofluid, *International Communications in Heat and Mass Transfer*, 37, 201–213.

22. Maxwell-Garnett, J.C. 1904, Colours in metal glasses and in metallic films, *Philosophical Transactions of the Royal Society A*, 203, 385–420.

23. Chon, C.H., Kihm, K.D., Lee, S.P., Choi, S.U.S. 2005, Empirical correlation finding the role of temperature and particle size for nanofluid (Al_2O_3) thermal conductivity enhancement, *Applied Physics Letters*, 87, 153107.

24. Angue Minsta, H., Roy, G., Nguyen, C.T., Doucet, D. 2009, New temperature and conductivity data for water-based nanofluids, *International Journal of Thermal Sciences*, 48, 363–371.

25. Akbarinia, A., Behzadmehr, A. 2007, Numerical study of laminar mixed convection of a nanofluid in horizontal curved tubes, *Applied Thermal Engineering*, 27, 1327–1337.
26. Abu-Nada, E., Oztop, H.F. 2009, Effects of inclination angle on natural convection in enclosures filled with Cu-water nanofluid, *International Journal of Heat and Fluid Flow*, 30, 669–678.
27. Maiga, S.E.B., Palm, S.J., Nguyen, C.T., Roy, G., Galanis, N. 2005, Heat transfer enhancement by using nanofluids in forced convection flows, *International Journal of Heat and Fluid Flow*, 26, 530–546.
28. Abu-Nada, E., Masoud, Z., Oztop, H.F., Campo, A. 2010, Effect of nanofluid variable properties on natural convection in enclosures, *International Journal of Thermal Sciences*, 49, 479–491.
29. Brinkman, H.C. 1952, The viscosity of concentrated suspensions and solution, *Journal of Chemical Physics*, 20, 571–581.
30. Pak, B.C., Cho, Y. 1998, Hydrodynamic and heat transfer study of dispersed fluids with submicron metallic oxide particle, *Experimental Heat Transfer*, 11, 151–170.
31. Nguyen, C.T., Desgranges, F., Roy, G., Galanis, N., Mare, T., Boucher, S., Angue Minsta, H. 2007, Temperature and particle-size dependent viscosity data for water-based nanofluids—Hysteresis phenomenon, *International Journal of Heat and Fluid Flow,* 28, 1492–1506.
32. Patankar, S.V. 1980, *Numerical Heat Transfer and Fluid Flow*, Hemisphere Publishing Corporation, New York.
33. Versteeg, H.K., Malalasekera, W. 1995, *An Introduction to Computational Fluid Dynamic: The Finite Volume Method*, John Wiley, New York.
34. Iwatsu, R., Hyun, J.M., Kuwahara, K., 1993, Mixed convection in a driven cavity with a stable vertical temperature gradient, *International Journal of Heat and Mass Transfer*, 36, 1601–1608.
35. Khanafer, K.M., Chamkha, A.J. 1999, Mixed convection flow in a lid-driven enclosure filled with a fluid-saturated porous medium, *International Journal of Heat and Mass Transfer*, 42, 2465–2481.

CHAPTER 10

Buoyancy-Driven Convection of Enclosed Nanoparticle Suspensions

Massimo Corcione and Alessandro Quintino

CONTENTS

10.1 INTRODUCTION

Effective cooling is one of the top challenges that high-tech manufacturing companies are continuously called to face in order to assure the reliability of their products. In fact, both the heat loads and the heat fluxes of modern devices are growing at an exponential pace as a result of the increasing demand for high performance and reduced size. Typical examples are represented by the microelectronics and automotive industries, just to name a few. In this connection, a considerable research effort has been dedicated to the development of advanced methods for heat transfer enhancement, such as those relying on new geometries and configurations, as well as those based on the use of extended surfaces and/or turbulators.

A further important contribution to the cooling issue may be derived by the replacement of traditional heat transfer fluids, such as water, ethylene glycol, and mineral oils, with nanofluids. These are a new type of heat transfer fluids consisting of colloidal suspensions of nanoparticles, whose effective thermal conductivity has been demonstrated to be higher than that of the corresponding pure base liquid. Actually, since their introduction, nanofluids have attracted the attention of the heat transfer community, as reflected by the large amount of experimental and numerical papers published up to now on forced convection applications, whose common conclusion is that nanoparticle suspensions have a valuable potential for heat transfer enhancement—see, for example, the recent review article compiled by Hussein et al. [1].

Conversely, although buoyancy-driven convection is the heat removal strategy often preferred by thermal engineering designers when a small power consumption, a negligible operating noise, and a high reliability of the system are fundamental concerns, only few experimental works have been executed on natural convection of nanoparticle suspensions in enclosed spaces. This is the case of the studies executed in the last decade on differentially heated cavities by Putra et al. [2], Wen and Ding [3,4], Nnanna [5], Chang et al. [6], Ho et al. [7], and Hu et al. [8], whose main achievement is that the addition of nanoparticles to a pure liquid is basically detrimental. Very small enhancements of the heat transfer rate have been detected only in a limited number of situations at extremely low solid-phase concentrations. In their turn, the numerical studies performed on this topic show contradictory results, as displayed in Table 10.1, in which a survey of selected papers readily available in

TABLE 10.1 Selected Numerical Studies Performed on Natural Convection of Nanofluids in Differentially Heated Square Enclosures

Year	Author(s)	Model	Properties	Nanofluid	Volume Fraction φ (%)	k_n (Correlation or Data)	μ_n (Correlation or Data)	Heat Transfer vs. φ
2003	Khanafer et al. [9]	Single-phase	Constant	Cu (d_p = 10 nm) + H_2O	0–25	Maxwell–Garnett [10] and Amiri and Vafai [11]	Brinkman [12]	Increases
2006	Jou and Tzeng [13]	Single-phase	Constant	Cu (d_p = 10 nm) + H_2O	0–20	Maxwell–Garnett [10] and Amiri and Vafai [11]	Brinkman [12]	Increases
2008	Ho et al. [14]	Single-phase	Constant	Al_2O_3 + H_2O	0–4	Maxwell–Garnett [10] or Maxwell–Garnett [10] and Charuyakorn et al. [15]	Brinkman [12] or Maïga et al. [16]	Increases or decreases
2008	Santra et al. [17]	Single-phase	Constant	Cu (d_p = 100 nm) + H_2O	0–2	Patel et al. [18]	Brinkman [12] or Kwak and Kim [19]	Decreases
2008	Santra et al. [20]	Single-phase	Constant	Cu (d_p = 100 nm) + H_2O	0–5	Patel et al. [18]	Putra et al. [2]	Decreases
2009	Abu-Nada and Oztop [21]	Single-phase	Constant	Cu + H_2O	0–10	Maxwell–Garnett [10]	Brinkman [12]	Increases
2009	Ghasemi and Aminossadati [22]	Single-phase	Constant	CuO (d_p = 10 nm) + H_2O	0–10	Koo and Kleinstreuer [23]	Brinkman [12]	Shows a peak
2010	Abu-Nada and Chamkha [24]	Single-phase	$f(T)$	CuO (d_p = 29 nm) + EG/H_2O	0–6	Namburu et al. [26]	Jang and Choi [25]	Decreases

(Continued)

TABLE 10.1 (*Continued*) Selected Numerical Studies Performed on Natural Convection of Nanofluids in Differentially Heated Square Enclosures

Year	Author(s)	Model	Properties	Nanofluid	Volume Fraction φ (%)	k_n (Correlation or Data)	μ_n (Correlation or Data)	Heat Transfer vs. φ
2010	Abu-Nada et al. [27]	Single-phase	$f(T)$	Al_2O_3 ($d_p = 47$ nm) + H_2O CuO ($d_p = 29$ nm) + H_2O	0–6	Chon et al. [28]	Nguyen et al. [29]	Decreases
2010	Kahveci [30]	Single-phase	Constant	Cu or Ag or Al_2O_3 or CuO or TiO_2 + H_2O	0–20	Yu and Choi [31]	Brinkman [12]	Increases
2010	Lin and Violi [32]	Single-phase	Constant	Al_2O_3 ($d_p = 5 - 250$ nm) + H_2O	0–5	Xu et al. [33]	Jang et al. [34]	Increases or decreases
2011	Esfahani and Bordbar [35]	Two-phase (Brownian diffusion)	Constant	Cu or Ag or Al_2O_3 or TiO_2 + H_2O	0–10	Maxwell–Garnett [10]	Brinkman [12]	Increases
2011	Kefayati et al. [36]	Single-phase	Constant	SiO_2 + H_2O	0–4	Maxwell–Garnett [10]	Brinkman [12]	Increases
2011	Lai and Yang [37]	Single-phase	Constant	Al_2O_3 ($d_p = 47$ nm) + H_2O	0–4	Mintsa et al. [38]	Nguyen et al. [29]	Increases
2011	Oueslati and Bennacer [39]	Two-phase (Brownian diffusion)	Constant	Cu or Al_2O_3 or TiO_2 + H_2O	0–10	Maxwell–Garnett [10]	Maïga et al. [16]	Shows a peak
2011	Saleh et al. [40]	Single-phase	Constant	Cu or Al_2O_3 + H_2O	0–5	Maxwell–Garnett [10]	Brinkman [12]	Increases
2012	Alloui et al. [41]	Single-phase	Constant	Cu or Al_2O_3 or TiO_2 + H_2O	0–20	Maxwell–Garnett [10] or Yu and Choi [31]	Brinkman [12] or Maïga et al. [16] or Pak and Cho [42]	Increases or decreases

(*Continued*)

TABLE 10.1 (Continued)   Selected Numerical Studies Performed on Natural Convection of Nanofluids in Differentially Heated Square Enclosures

Year	Author(s)	Model	Properties	Nanofluid	Volume Fraction φ (%)	k_n (Correlation or Data)	μ_n (Correlation or Data)	Heat Transfer vs. φ
2012	Rahimi et al. [43]	Single-phase	Constant	Cu + H_2O	0–8	Maxwell–Garnett [10] and Amiri and Vafai [11]	Brinkman [12]	Increases
2013	Corcione et al. [44]	Two-phase (Brownian diffusion + thermophoresis)	$f(T)$	Al_2O_3 (d_p = 25–100 nm) + H_2O	0–6	Corcione [45]	Corcione [45]	Shows a peak
2013	Aminfar and Haghgoo [46]	Two-phase (Brownian diffusion + thermophoresis)	$f(T)$	Al_2O_3 (d_p = 33 nm) + H_2O	0–3	Maxwell–Garnett [10]	Ho et al. [7]	Decreases
2013	Pakravan and Yaghoubi [47]	Two-phase (Brownian diffusion + thermophoresis)	$f(T)$	Al_2O_3 (d_p = 150 nm) + H_2O	0–3	Maxwell–Garnett [10]	Brinkman [12]	Decreases
2013	Sheikhzadeh et al. [48]	Two-phase (Brownian diffusion + thermophoresis)	$f(T)$	Al_2O_3 (d_p = 33 nm) + H_2O	0–4	Corcione [45]	Corcione [45]	Decreases
2014	Alipanah et al. [49]	Single-phase	$f(T)$	Cu or Al_2O_3 or TiO_2 + H_2O	0–5	Maxwell–Garnett [10] and Amiri and Vafai [11]	Brinkman [12]	Increases
2014	Wang et al. [50]	Single-phase	Constant	Cu + H_2O	0–20	Yu and Choi [31]	Brinkman [12]	Increases
2014	Choi et al. [51]	Two-phase (Brownian diffusion + thermophoresis)	$f(T)$	CuO + H_2O	0–10	Maxwell–Garnett [10]	Brinkman [12]	Increases

the literature on differentially heated square enclosures is presented. The same type of controversial data can be found for other enclosed configurations, which is, for example, the case of differentially heated horizontal annuli, as shown in Table 10.2.

Several reasons can be invoked to explain such conflicting results. First of all, the use of robust theoretical models or accurate empirical equations, capable of predicting the effective thermal conductivity and dynamic viscosity of nanofluids as more precisely as possible, is crucial for obtaining realistic results. Unfortunately, a wide number of studies miss this requirement for one reason or another. In fact, very often the effective thermal conductivity and dynamic viscosity are calculated using traditional mean-field equations, which, originally developed for composites and mixtures with microsized or millisized inclusions, tend to fail in describing the actual thermomechanical behavior of nanofluids. Misleading conclusions may also derive from the calculation of the effective physical properties by either partly inconsistent semiempirical models or correlations based on experimental data that are in evident contrast with the main body of the literature results or equations whose validity is restricted to situations very different from those investigated. Additionally, too often the physical properties are assumed to be independent of temperature, which does not seem to be an accurate approach.

Second, most studies are based on the so-called single-phase or homogeneous model, in which nanofluids are treated as pure fluids, assuming that the solid and liquid phases are in local thermal and hydrodynamic equilibrium. Actually, such an approach cannot take into account the slip motion occurring between suspended nanoparticles and base liquid, whose consequent nonuniform distribution of the solid-phase concentration may have nonnegligible effects on heat and momentum transfer.

Third, the majority of the works based on the two-phase modeling rely on the McNab–Meisen relation [70] for the calculation of the thermophoretic diffusion coefficient, which brings to underevaluate the role of thermophoresis and, consequently, the amount of heat exchanged across the enclosed space.

Framed in this background, the aim of this chapter is to illustrate a two-phase model based on a double-diffusive approach, which incorporates three empirical correlations for the evaluation of the effective thermal conductivity, the effective dynamic viscosity, and the thermophoretic diffusion coefficient, all based on a high number of experimental

TABLE 10.2 Selected Numerical Studies Performed on Natural Convection of Nanofluids in Differentially Heated Horizontal Annuli

Year	Author(s)	Model	Properties	Nanofluid	Volume Fraction φ (%)	k_n (Correlation or Data)	μ_n (Correlation or Data)	Heat Transfer vs. φ
2008	Abu-Nada et al. [52]	Single-phase	Constant	Cu or Ag or Al_2O_3 or TiO_2 + H_2O	0–10	Maxwell–Garnett [10]	Brinkman [12]	Increases
2009	Abu-Nada [53]	Single-phase	Constant	Al_2O_3 (d_p = 47 nm) + H_2O	0–9	Chon et al. [28]	Nguyen et al. [29]	Decreases
2010	Abu-Nada [54]	Single-phase	Constant	CuO (d_p = 29 nm) + H_2O	0–9	Chon et al. [28]	Nguyen et al. [29]	Decreases
2012	Arefmanesh et al. [55]	Single-phase	Constant	TiO_2 (d_p = 21 nm) + H_2O	0–9	He et al. [56]	He et al. [56]	Increases
2012	Parvin et al. [57]	Single-phase	Constant	Al_2O_3 + H_2O	0–15	Maxwell–Garnett [10] or Chon et al. [28]	Brinkman [12]	Increases
2012	Soleimani et al. [58]	Single-phase	Constant	Cu + H_2O	0–6	Maxwell–Garnett [10]	Brinkman [12]	Increases
2012	Sheikholeslami et al. [59]	Single-phase	Constant	Ag + H_2O	0–6	Maxwell–Garnett [10]	Brinkman [12]	Increases
2013	Ashorynejad et al. [60]	Single-phase	Constant	Ag + H_2O	0–6	Maxwell–Garnett [10]	Brinkman [12]	Increases
2013	Sheikhzadeh et al. [61]	Single-phase	Constant	Cu + H_2O	0–10	Maxwell–Garnett [10]	Brinkman [12]	Increases
2013	Abouei Mehrizi et al. [62]	Single-phase	Constant	Cu + H_2O	0–10	Maxwell–Garnett [10]	Brinkman [12]	Increases
2013	Sheikholeslami et al. [63]	Single-phase	Constant	Al_2O_3 (d_p = 47 nm) + H_2O	0–4	Koo and Kleinstreuer [23]	Koo and Kleinstreuer [64]	Increases

(Continued)

TABLE 10.2 (*Continued*) Selected Numerical Studies Performed on Natural Convection of Nanofluids in Differentially Heated Horizontal Annuli

Year	Author(s)	Model	Properties	Nanofluid	Volume Fraction φ (%)	k_n (Correlation or Data)	μ_n (Correlation or Data)	Heat Transfer vs. φ
2013	Sheikholeslami et al. [65]	Single-phase	Constant	$Cu + H_2O$	0–6	Maxwell–Garnett [10]	Brinkman [12]	Increases
2013	Habibi Matin and Pop [66]	Single-phase	Constant	Cu ($d_p = 100$ nm) $+ H_2O$	0–3	Li et al. [67]	Li et al. [67]	Increases
2013	Corcione et al. [68]	Two-phase	$f(T)$	Al_2O_3 ($d_p = 25–100$ nm) $+ H_2O$	0–6	Corcione [45]	Corcione [45]	Shows a peak
2014	Sheikholeslami et al. [69]	Single-phase	Constant	$Cu + H_2O$	0–6	Maxwell–Garnett [10]	Brinkman [12]	Increases

data available in the literature from diverse sources and validated using relations from other authors and experimental data different from those employed in generating them. Additionally, the effects of thermophoresis on the velocity, temperature, and volume fraction fields will be visualized and discussed.

10.2 TWO-PHASE MODELING

The two-phase or heterogeneous model is the unique approach able to account for the effects of the slip motion occurring between suspended nanoparticles and base liquid, which can actually be identified as the main responsible for the performance degradation of nanofluids observed experimentally in buoyancy-induced enclosed flows.

In the traditional two-phase approach to the study of liquid suspensions, the liquid phase is modeled using the conventional Eulerian approach, whereas the solid phase can be described either as a discrete phase (Eulerian–Lagrangian formulation) or as a continuous phase (Eulerian–Eulerian formulation). In the Eulerian–Lagrangian formulation, the Newton's equation of motion has to be solved for each individual particle, which seems hard to apply to nanofluids, owing to the very large number of suspended nanoparticles. Conversely, in the Eulerian–Eulerian formulation, the average local particle concentration and slip velocity are calculated by solving a second set of mass, momentum, and energy transfer governing equations dedicated to the solid phase. Also in this case, computational difficulties may arise.

Within the limits of the Eulerian description of the solid phase, an easier double-diffusive approach can be followed when the objective is to determine the behavior of the mixture as a whole, rather than the behavior of each phase. Such an approach to the study of nanofluids was originally proposed by Buongiorno [71], who developed a four-equation transport model (the continuity, momentum, and energy equations for the mixture and the continuity equation for the suspended nanoparticles) accounting for the effects of Brownian diffusion and thermophoresis as primary slip mechanisms by which the nanoparticles can develop a significant relative velocity with respect to the base liquid (Brownian motion occurs from high to low nanoparticle concentrations, whereas thermophoresis occurs in the direction from hot to cold). Actually, this model is widely consolidated and it is certainly the most frequently used for two-phase simulations.

The basic assumptions of the model discussed below are (1) laminar and incompressible flow, (2) negligible viscous dissipation and pressure work, (3) suspended nanoparticles and base liquid in local thermal equilibrium, (4) effective properties of the nanofluid dependent on both temperature and solid-phase concentration, (5) negligible heat transfer associated with the nanoparticle motion relative to the base fluid, and (6) negligible radiative heat transfer.

10.2.1 Governing Equations

In the hypotheses stated above, the governing equations of continuity, momentum, and energy for the nanofluid, and the equation of continuity for the nanoparticles reduce to

$$\frac{\partial \rho_n}{\partial t} + \nabla \cdot (\rho_n \mathbf{V}) = 0 \tag{10.1}$$

$$\frac{\partial (\rho_n \mathbf{V})}{\partial t} + \nabla \cdot (\rho_n \mathbf{V}\mathbf{V}) = \nabla \cdot \tau + \rho_n \mathbf{g} \tag{10.2}$$

$$\frac{\partial (\rho_n c_n T)}{\partial t} + \nabla \cdot (\rho_n \mathbf{V} c_n T) = \nabla \cdot (k_n \nabla T) \tag{10.3}$$

$$\frac{\partial (\rho_n m)}{\partial t} + \nabla \cdot (\rho_n \mathbf{V} m) = -\nabla \cdot \mathbf{J}_p \tag{10.4}$$

where:
   t is the time
   $\mathbf{V}$ is the velocity vector
   τ is the stress tensor
   $\mathbf{g}$ is the gravity vector
   $\mathbf{J}_p$ is the nanoparticle diffusion mass flux
   T is the temperature
   m is the mass fraction (also called concentration) of the suspended
       nanoparticles
   ρ_n is the effective mass density
   c_n is the effective specific heat at constant pressure
   k_n is the effective thermal conductivity

Assuming that the nanofluid has a Newtonian behavior, as demonstrated by Das et al. [72], Prasher et al. [73], He et al. [74], Chen et al. [75], Chevalier et al. [76], and Cabaleiro et al. [77], the stress tensor can be expressed as

$$\tau = -\left(p + \frac{2}{3} \, \mu_n \nabla \cdot \mathbf{V} \right) \mathbf{I} + \mu_n [\nabla \mathbf{V} + (\nabla \mathbf{V})^t] \qquad (10.5)$$

where:

p is the pressure

μ_n is the effective dynamic viscosity

$\mathbf{I}$ is the unit tensor

The superscript t indicates the transpose of $\nabla \mathbf{V}$

The nanoparticle diffusion mass flux is calculated as the sum of the Brownian and thermophoretic diffusion terms in the hypothesis of dilute mixture (i.e., low mass fraction), thus obtaining

$$\mathbf{J}_p = -\rho_n \left(D_B \nabla m + D_T \frac{\nabla T}{T} \right) \qquad (10.6)$$

where D_B and D_T are the Brownian and thermophoretic diffusion coefficients, respectively.

It seems worth pointing out that the relationship existing between the nanoparticle mass fractions, m, and the most widely used nanoparticle volume fraction, φ, is

$$\rho_s \varphi = \rho_n m \qquad (10.7)$$

where ρ_s is the mass density of the solid nanoparticles.

10.2.2 Boundary and Initial Conditions

As far as the boundary conditions at the walls are concerned, besides the usual Dirichlet or Neumann boundary conditions for temperature and the no-slip boundary condition for velocity, a zero nanoparticle diffusion mass flux, $\mathbf{J}_p = 0$, must be assumed. In fact, this is the only physically realistic condition applicable at the solid surfaces, because the value of the concentration at the walls can no longer be imposed as frequently done in conventional double-diffusive approaches, in which only the Brownian diffusion effects are typically accounted for.

The initial conditions to be assumed throughout the enclosed space must include the details on the velocity and temperature fields, for example, nanofluid at rest at a uniform average temperature, T_{av}, as well as the details on the concentration field, usually an assigned uniform average mass fraction of the suspended nanoparticles, m_{av}.

10.2.3 Diffusion Coefficients

10.2.3.1 Brownian Diffusion

The Brownian diffusion coefficient, D_B, is given by the Einstein–Stokes equation [78]:

$$D_B = \frac{k_b T}{3\pi\mu_f d_p}$$ (10.8)

where:

$k_b = 1.38066 \times 10^{-23}$ J/K is the Boltzmann constant
μ_f is the dynamic viscosity of the base fluid
d_p is the diameter of the suspended nanoparticles

10.2.3.2 Thermophoresis

The thermophoretic diffusion coefficient, D_T, is expressed as

$$D_T = \beta \frac{\mu_f}{\rho_f} m$$ (10.9)

where:

ρ_f is the mass density of the base fluid
β is the so-called thermophoresis parameter, which, for water-based nanofluids containing metal oxide nanoparticles, can be evaluated using the following algebraic empirical correlation recently proposed by Corcione and Quintino [79]:

$$\beta = \left[\left(1.519\times10^4\right) \left(\frac{k_s}{k_f}\right)^{-3} + 0.95 \right] \cdot \left[-16.32\left(\varphi_{av}\right)^{2.34} + 0.0193\right]$$ (10.10)

where:

k_s and k_f are the thermal conductivities of the solid nanoparticles and the base fluid, whose values have to be calculated at the average temperature of the nanofluid, T_{av}
φ_{av} is the average volume fraction of the suspended nanoparticles

The correlation given in Equation 10.10 was derived through the best fit of a number of natural convection heat transfer measurements performed on square enclosures differentially heated at sides by (1) Ho et al. [7], who used three different cavities with side lengths of 25, 40, and 80 mm containing $Al_2O_3 + H_2O$ with $\varphi_{av} = 0.001, 0.01, 0.02, 0.03$, and 0.04; (2) Putra et al. [2], who used a cavity with a side length of 40 mm containing $CuO + H_2O$ with $\varphi_{av} = 0.01$ and 0.04; and (3) Hu et al. [8], who used a cavity with a side length of 80 mm containing $TiO_2 + H_2O$ with $\varphi_{av} = 0.0093$ and 0.0185.

A number of distributions of β versus φ_{av} based on Equation 10.10 are depicted in Figure 10.1 for water-based nanofluids using (k_s/k_f) as a parameter.

The reliability of Equation 10.10 is assessed by reproducing numerically the experiments performed by (1) Putra et al. [2], who used a laterally heated cavity with a side length of 40 mm containing $Al_2O_3 + H_2O$ with $\varphi_{av} = 0.01$ and 0.04; (2) Wen and Ding [3,4], who used bottom-heated horizontal cavities with a width of 240 mm and a height of 10 mm containing $TiO_2 + H_2O$ with $\varphi_{av} = 0.00356$ and 0.006; and (3) Chang et al. [6], who used bottom-heated horizontal and inclined cavities with a width of 152 mm and heights in the range of 3–14 mm containing $Al_2O_3 + H_2O$ with $\varphi_{av} = 0.0131$. The numerical distributions of the ratio (Nu/Nu_0) between the average Nusselt numbers of the nanofluid and the base liquid versus the Rayleigh number of the pure base liquid, Ra_0, achieved using Equation 10.10 for the computation of β, and the corresponding experimental values, are reported in Figure 10.2. A satisfactory agreement between numerical results and experimental data is apparent.

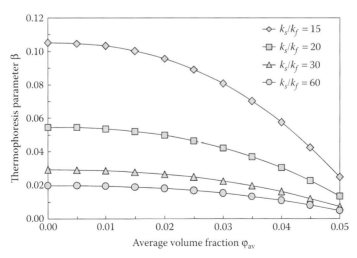

FIGURE 10.1   Distributions of β vs. φ_{av} for water-based nanofluids containing metal oxide suspended nanoparticles, with (k_s/k_f) as a parameter.

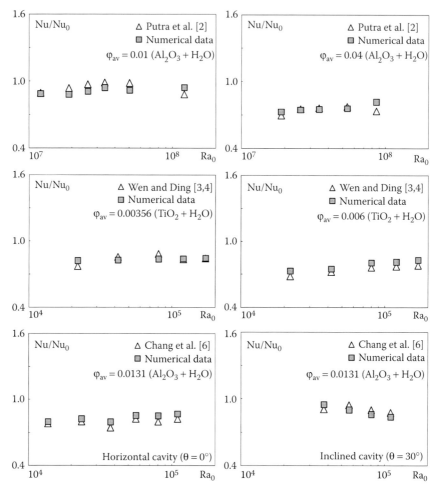

FIGURE 10.2 Comparison between the numerical values of (Nu/Nu$_0$) obtained using Equation 10.10 for the prediction of the thermophoresis parameter β, and the corresponding experimental data of Putra et al. for Al$_2$O$_3$ + H$_2$O in a side-heated cavity, Wen and Ding for TiO$_2$ + H$_2$O in bottom-heated thin cavities, and Chang et al. for Al$_2$O$_3$ + H$_2$O in bottom-heated horizontal and inclined thin cavities. (Data from Putra, N. et al. *Heat Mass Transf.*, 39, 775–784, 2003; Wen, D. and Ding, Y., *Int. J. Heat Fluid Fl.*, 26, 855–864, 2005; Wen, D. and Ding, Y., *IEEE Trans. Nanotechnol.*, 5, 220–227, 2006; Chang, B.H. et al., *Int. J. Heat Mass Transf.*, 51, 1332–1341, 2008.)

10.2.4 Effective Physical Properties

10.2.4.1 Effective Thermal Conductivity

The nanofluid effective thermal conductivity, k_n, can be predicted using an empirical correlation produced by Corcione [45] based on 14 sets of experimental data obtained by eight independent research groups [38,42,80−85]. For any dataset, the details on the nanofluid type, the size of the suspended nanoparticles, the volume fraction range, and the temperature range are listed in Table 10.3. The correlation, whose standard deviation of error is 1.86%, is

$$\frac{k_n}{k_f} = 1 + 4.4\,\mathrm{Re}_p^{0.4}\,\mathrm{Pr}_f^{0.66}\left(\frac{T}{T_{\mathrm{fr}}}\right)^{10}\left(\frac{k_s}{k_f}\right)^{0.03}\varphi^{0.66} \tag{10.11}$$

where:

Re$_p$ is the nanoparticle Reynolds number
Pr$_f$ is the Prandtl number of the base fluid
T_{fr} is the freezing point of the base liquid
φ is the nanoparticle volume fraction

TABLE 10.3   Sources of Thermal Conductivity Data Used in Deriving Equation 10.11

Year	Author(s)	Nanofluid	Nanoparticle Size (nm)	Volume Fraction (%)	Temperature (K)
1993	Masuda et al. [80]	$TiO_2 + H_2O$	27	0−4.3	305−340
1998	Pak and Cho [42]	$TiO_2 + H_2O$	27	0−4.3	300
1999	Lee et al. [81]	$CuO + H_2O$	23.6	0−3.4	294
		$CuO + EG$	23.6	0−4	294
		$Al_2O_3 + H_2O$	38.4	0−4.3	294
		$Al_2O_3 + EG$	38.4	0−5	294
2001	Eastman et al. [82]	$Cu + EG$	10	0−0.6	294
2003	Das et al. [83]	$CuO + H_2O$	28.6	0−4	294−324
		$Al_2O_3 + H_2O$	38.4	0−4	294−324
2005	Chon and Kihm [84]	$Al_2O_3 + H_2O$	47	0−1	294−344
		$Al_2O_3 + H_2O$	150	0−1	294−344
2008	Murshed et al. [85]	$Al_2O_3 + H_2O$	80	0−1	294−334
		$Al_2O_3 + EG$	80	0−5	294−334
2009	Mintsa et al. [38]	$CuO + H_2O$	29	0−10	294−313

EG, ethylene glycol.

The nanoparticle Reynolds number is defined as

$$\mathrm{Re}_p = \frac{\rho_f u_p d_p}{\mu_f} \tag{10.12}$$

where u_p is the nanoparticle Brownian velocity calculated as the ratio between d_p and the time t_D required to cover such a distance, which, according to Keblinski et al. [86], is

$$t_D = \frac{d_p^2}{6 D_B} \tag{10.13}$$

where D_B is the Brownian diffusion coefficient defined in Equation 10.8. Hence,

$$\mathrm{Re}_p = \frac{2 \rho_f k_b T}{\pi \mu_f^2 d_p} \tag{10.14}$$

Notice that in Equations 10.11 through 10.14, all the physical properties are calculated at the nanofluid temperature T.

It is apparent that the thermal conductivity ratio, (k_n/k_f), increases as φ and T increase and d_p decreases. Moreover, (k_n/k_f) depends marginally on the solid–liquid combination, as denoted by the extremely small exponent of (k_s/k_f), at least for metal oxide suspended nanoparticles.

A number of distributions of (k_n/k_f) versus φ based on Equation 10.11 are depicted in Figure 10.3 for alumina–water nanofluids using d_p and T as parameters.

Besides the fact that Equation 10.11 interpolates rather well a wide variety of literature data from different sources, its reliability is tested by a comparative analysis with a number of relations from other authors and experimental data different from those used in generating it [28,87–89]. The results of such a comparison are displayed in Figures 10.4 and 10.5, showing a good degree of agreement.

10.2.4.2 Effective Dynamic Viscosity

The nanofluid effective dynamic viscosity, μ_n, can be predicted using an empirical correlation developed by Corcione [45] based on 14 sets of experimental data reported by 10 independent research groups [42,72–76,80,90–92]. For any dataset, the details are listed in Table 10.4. The correlation, whose standard deviation of error is 1.84%, is

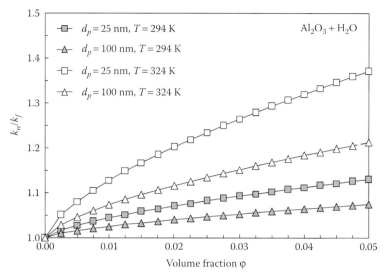

FIGURE 10.3 Distributions of (k_n/k_f) vs. φ for $Al_2O_3 + H_2O$, with d_p and T as parameters.

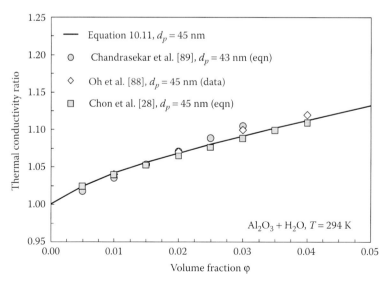

FIGURE 10.4 Comparison between the predictions of Equation 10.11 for Al_2O_3 $(d_p = 45$ nm$) + H_2O$ at temperature $T = 294$ K and some available literature correlations/data. (Data from Chon, C.H. et al., *Appl. Phys. Lett.*, 87, 153107, 2005; Oh, D.W. et al., *Int. J. Heat Fluid Fl.*, 29, 1456–1461, 2008; Chandrasekar, M. et al., *Exp. Therm. Fluid Sci.*, 34, 210–216, 2010.)

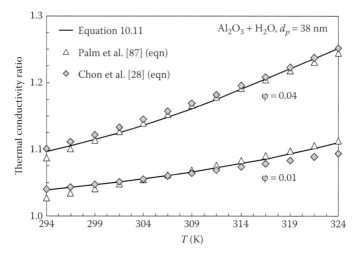

FIGURE 10.5   Comparison between the predictions of Equation 10.11 for Al_2O_3 (d_p = 38 nm) + H_2O at volume fractions φ = 0.01 and 0.04 and some available literature correlations. (Data from Chon, C.H. et al., *Appl. Phys. Lett.*, 87, 153107, 2005; Palm, S.J. et al., *Appl. Therm. Eng.*, 26, 2209–2218, 2006.)

TABLE 10.4   Sources of Dynamic Viscosity Data Used in Deriving Equation 10.15

Year	Author(s)	Nanofluid	Nanoparticle Size (nm)	Volume Fraction (%)	Temperature (K)
1993	Masuda et al. [80]	$TiO_2 + H_2O$	27	0–4.3	305–340
1998	Pak and Cho [42]	$TiO_2 + H_2O$	27	0–4.3	300
1999	Wang et al. [90]	$Al_2O_3 + H_2O$	28	0–6	294
2003	Das et al. [72]	$Al_2O_3 + H_2O$	38	0–4	293–333
2006	Prasher et al. [73]	Al_2O_3 + PG	27	0–3	303–333
		Al_2O_3 + PG	40	0–3	303–333
		Al_2O_3 + PG	50	0–3	303–333
2007	He et al. [74]	$TiO_2 + H_2O$	95	0–1.2	295
2007	Chen et al. [75]	TiO_2 + EG	25	0–1.8	293–333
2007	Chevalier et al. [76]	SiO_2 + ethanol	35	0–5	294
		SiO_2 + ethanol	94	0–7	294
		SiO_2 + ethanol	190	0–5.5	294
2008	Lee et al. [91]	$Al_2O_3 + H_2O$	30	0–0.3	294–312
2008	Garg et al. [92]	Cu + EG	200	0–2	294

EG, ethylene glycol; PG, propylene glycol.

$$\frac{\mu_n}{\mu_f} = \frac{1}{1 - 34.87\left(d_p/d_f\right)^{-0.3} \varphi^{1.03}} \tag{10.15}$$

In the above equation, d_f is the equivalent diameter of a base fluid molecule calculated at the reference temperature $T_0 = 293$ K based on the relation $M = \rho_{f0} V_m N$, where M, ρ_{f0}, and V_m are the molar mass, the mass density at T_0, and the molecular volume of the base fluid, respectively, and $N = 6.022 \cdot 10^{23}$ mol^{-1} is the Avogadro number. If we express V_m as $(4/3)\pi(d_f/2)^3$, we obtain

$$d_f = 0.1\left(\frac{6M}{N\pi\rho_{f0}}\right)^{1/3} \tag{10.16}$$

It may be observed that the dynamic viscosity ratio, (μ_n/μ_f), increases as d_p decreases and φ increases, whereas, within the limits of Equation 10.15, it is independent of both the solid–liquid combination and the temperature.

A set of distributions of (μ_n/μ_f) versus φ that emerge from Equation 10.15 for water-based nanofluids is displayed in Figure 10.6.

As previously done for Equations 10.10 and 10.11, a comparative analysis is conducted to test the strength of Equation 10.15 using relations from other authors and experimental data from sources different from those

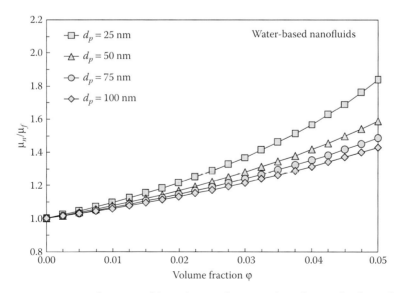

FIGURE 10.6   Distributions of (μ_n/μ_f) vs. φ for water-based nanofluids, with d_p as a parameter.

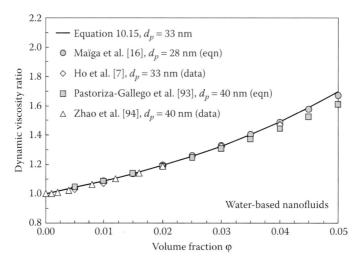

FIGURE 10.7   Comparison between the predictions of Equation 10.15 for water-based nanofluids containing nanoparticles with d_p = 33 nm and some available literature correlations/data. (Data from Ho, C.J. et al., *Int. J. Therm. Sci.*, 49, 1345–1353, 2010; Maïga, S.E.B. et al., *Superlattices Microst.*, 35, 543–557, 2004; Pastoriza-Gallego, M.J. et al., *J. Appl. Phys.*, 106, 064301, 2009; Zhao, J.F. et al., *Chin. Phys. Lett.* 26, 066202, 2009.)

enumerated in Table 10.4 [7,16,93,94]. According to such a comparative analysis, whose results are shown in Figure 10.7, Equation 10.15 seems to be sufficiently reliable to be used for practical applications.

10.2.4.3 Effective Mass Density and Specific Heat at Constant Pressure

The effective mass density and specific heat at constant pressure can be calculated through the customary mixing theory.

The effective mass density of the nanofluid, ρ_n, is given by

$$\rho_n = (1-\varphi)\rho_f + \varphi\rho_s \tag{10.17}$$

whose validity was originally demonstrated by the experimental results of Pak and Cho [42].

The heat capacity at constant pressure per unit volume of the nanofluid, $(\rho c)_n$, is

$$(\rho c)_n = (1-\varphi)(\rho c)_f + \varphi(\rho c)_s \tag{10.18}$$

where $(\rho c)_f$ and $(\rho c)_s$ are the heat capacities at constant pressure per unit volume of the base fluid and the solid nanoparticles, respectively.

Accordingly, the effective specific heat at constant pressure of the nanofluid, c_n, is given by

$$c_n = \frac{(1-\varphi)(\rho c)_f + \varphi(\rho c)_s}{(1-\varphi)\rho_f + \varphi\rho_s} \qquad (10.19)$$

whose predictions were first confirmed experimentally by Zhou and Ni [95].

10.3 A CASE STUDY: THE VERTICAL SQUARE ENCLOSURE DIFFERENTIALLY HEATED AT SIDES

A nanofluid-filled square enclosure of width W is differentially heated at the sidewalls, as shown in Figure 10.8, where the reference Cartesian coordinate system (x, y) is also represented. The heated side is kept at a uniform temperature T_h, whereas the opposite cooled side is maintained at a uniform temperature T_c. The top and bottom walls are assumed to be perfectly insulated.

10.3.1 Computational Procedure

The system of governing Equations 10.1 through 10.4 in conjunction with the proper boundary and initial conditions cited earlier is solved through a control volume formulation of the finite-difference method based on the SIMPLE-C algorithm for pressure–velocity coupling [96]. Convective and diffusive terms are approximated by the QUICK scheme [97], whereas a second-order backward scheme is applied for time integration. The computational spatial

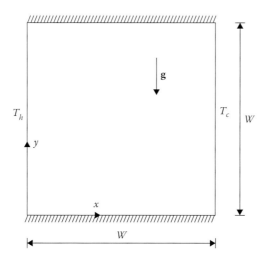

FIGURE 10.8   Sketch of the geometry and coordinate system.

domain is filled with a nonuniform grid, having a higher concentration of grid lines near the boundary walls, and a lower uniform spacing throughout the remainder interior of the cavity. A uniform time discretization is applied. Starting from the assigned initial fields of the dependent variables across the cavity, at each time step the system of discretized algebraic governing equations is solved iteratively by way of a line-by-line application of the Thomas algorithm. A standard under-relaxation technique is enforced in all steps of the computational procedure to ensure adequate convergence. Within each time step, the spatial numerical solution of the velocity, temperature, and concentration fields is considered to be converged when the maximum absolute values of the mass source, as well as the relative changes of the dependent variables at any grid node between two consecutive iterations, are smaller than the prespecified values of 10^{-6} and 10^{-7}, respectively. Time integration is stopped once steady state is reached. This means that the simulation procedure ends when the relative difference between the incoming and outgoing heat transfer rates at the heated and cooled sidewalls, and the relative changes of the time derivatives of the dependent variables at any grid node between two consecutive time steps, is smaller than the preassigned values of 10^{-6} and 10^{-8}, respectively.

Once steady state is reached, the heat fluxes at the heated and cooled sidewalls, q_h and q_c, are obtained using the following expressions:

$$q_h = -(k_n)_h \cdot \frac{\partial T}{\partial x}\bigg|_{x=0} \tag{10.20}$$

$$q_c = -(k_n)_c \cdot \frac{\partial T}{\partial x}\bigg|_{x=W} \tag{10.21}$$

where $(k_n)_h$ and $(k_n)_c$ are the values of the effective thermal conductivity of the nanofluid calculated at temperatures T_h and T_c, respectively. The temperature gradients in Equations 10.20 and 10.21 are evaluated by a second-order temperature profile embracing the wall node and the two subsequent fluid nodes. The heat transfer rates added to the nanofluid by the heated sidewall, Q_h, and withdrawn from the nanofluid by the cooled sidewall, Q_c, are then calculated as

$$Q_h = \int_0^W -(k_n)_h \cdot \frac{\partial T}{\partial x}\bigg|_{x=0} \, dy \tag{10.22}$$

$$Q_c = \int_0^W -(k_n)_c \cdot \frac{\partial T}{\partial x}\bigg|_{x=W} dy \tag{10.23}$$

in which the integrals are computed numerically by means of the trapezoidal rule.

The corresponding average Nusselt numbers for the heated and cooled sidewalls, Nu_h and Nu_c, are as follows:

$$Nu_h = \frac{h_h W}{(k_n)_h} = \frac{Q_h}{(k_n)_h \cdot (T_h - T_c)} \tag{10.24}$$

$$Nu_c = \frac{h_c W}{(k_n)_c} = \frac{Q_c}{(k_n)_c \cdot (T_c - T_h)} \tag{10.25}$$

where h_h and h_c are the average coefficients of convection at the heated and cooled sidewalls, respectively.

Of course, because at steady state the incoming and outgoing heat transfer rates are the same, that is, $Q_h = -Q_c = Q$, the following relationship between Nu_h and Nu_c holds:

$$Nu_h (k_n)_h = Nu_c (k_n)_c \tag{10.26}$$

Numerical tests related to the dependence of the results on the mesh spacing and time stepping have been performed for several combinations of the five controlling parameters, namely, m_{av}, d_p, T_c, T_h, and W. Of course, the nanofluid average temperature, T_{av}, in conjunction with the temperature difference between the cavity sides, ΔT, may be taken as independent variables instead of T_c and T_h. Furthermore, the average nanoparticle volume fraction, φ_{av}, may be used as an independent variable instead of m_{av}, owing to the following relationship derived by combining Equations 10.7 and 10.17:

$$\varphi_{av} = \left[\left(\frac{1}{m_{av}} - 1 \right) \frac{\rho_s}{\rho_f} + 1 \right]^{-1} \tag{10.27}$$

in which the values of the mass densities ρ_s and ρ_f are calculated at the average temperature T_{av}.

The discretization grids and time steps used for computations are chosen in such a way that further refinements do not produce noticeable modifications either in the heat transfer rates or in the flow and volume fraction fields. Specifically, the percentage changes of the heat transfer rates, those of the maximum horizontal and vertical velocity components on the vertical and horizontal midplanes of the enclosure, respectively, as well as those of the maximum and minimum nanoparticle volume fractions on the horizontal midplane of the enclosure, must be smaller than the preestablished accuracy value of 1%. The typical number of nodal points and the time step used for simulations lie in the ranges between 180×180 and 300×300, and between 10^{-3} s and 10^{-2} s, respectively. A number of test runs were also executed with the initial uniform temperature of the nanofluid set to T_c or T_h, rather than T_{av}, with the scope to determine what effect these initial conditions could have on the steady-state flow, temperature, and concentration patterns. Solutions practically identical to those obtained assuming $T = T_{av}$ throughout the enclosure at time $t = 0$ were achieved for all the configurations examined. Finally, three different tests have been executed to validate the numerical code. In the first test, the steady-state solutions obtained for an air-filled square cavity differentially heated at sides, assuming $m_{av} = 0$ and constant physical properties, have been compared with the benchmark results of de Vahl Davis [98] and other authors, that is, Mahdi and Kinney [99], Hortman et al. [100], and Wan et al. [101]. In the second test, the values of the average Nusselt number computed numerically for $Pr = 7$ (which corresponds to water at $T_{av} = 293$ K) and Rayleigh numbers in the range between 10^3 and 5×10^7 (calculated using a fixed $\Delta T = 20$ K) have been compared with the usually recommended Berkovsky–Polevikov correlating equation based on experimental and numerical data of laminar natural convection in a rectangular cavity heated and cooled from the side with an aspect ratio near unity (see, e.g., Bejan [102] and Incropera et al. [103]). In the third test, the solutions obtained for the steady-state double-diffusive convection occurring in a square cavity filled with an air–pollutant mixture having constant physical properties, submitted to simultaneous horizontal temperature and concentration gradients, have been compared with the numerical results of Béghein et al. [104]. An excellent compliance between our results and the literature data was obtained in any validation test carried out.

10.3.2 Selected Results and Discussion

Selected local results are presented in Figures 10.9 and 10.10, in which the streamline, isotherm, and volume fraction contours relative to a square enclosure filled with $Al_2O_3 + H_2O$ or $TiO_2 + H_2O$, respectively, are plotted

FIGURE 10.9 Streamline, isotherm, and volume fraction contour plots for $Al_2O_3 + H_2O$, $W = 40$ mm, $d_p = 33$ nm, $\varphi_{av} = 0.02$, $T_{av} = 300$ K, and $\Delta T = 10$ K. (a) Diagrams obtained by the single-phase approach; (b) diagrams obtained by the two-phase approach with β calculated using the McNab–Meisen relation; (c) diagrams obtained by the two-phase approach with β calculated using Equation 10.10.

for $W = 40$ mm, $d_p = 33$ nm, $\varphi_{av} = 0.02$, $T_{av} = 300$ K, and $\Delta T = 10$ K. In both figures, the corresponding local results obtained using either the single-phase approach, that is, by solving the system of Equations 10.1 through 10.3 with the same boundary and initial conditions of the two-phase model, or a two-phase approach based on the use of the McNab–Meisen relation for the calculation of the thermophoretic diffusion coefficient, are also depicted for comparison.

FIGURE 10.10 Streamline, isotherm, and volume fraction contour plots for $TiO_2 + H_2O$, $W = 40$ mm, $d_p = 33$ nm, $\varphi_{av} = 0.02$, $T_{av} = 300$ K, and $\Delta T = 10$ K. (a) Diagrams obtained by the single-phase approach; (b) diagrams obtained by the two-phase approach with β calculated using the McNab–Meisen relation; (c) diagrams obtained by the two-phase approach with β calculated using Equation 10.10.

As expected, whatever is the model used for simulations, the flow field consists of a single roll cell that derives from the rising of the hot fluid adjacent to the heated side and its descent along the opposite cooled side, which leads to the distinctive temperature distribution featured by a fluid stratification in the core of the cavity. On the contrary, the slip motion occurring between solid and liquid phases markedly affects the shape of such roll cell and the related heat transfer performance. In fact, the combined effects of the nanofluid circulation through the cavity due to the differential heating of the sides, and the thermophoretic diffusion of nanoparticles in the direction from hot to cold, give rise to the formation of low-concentration and high-concentration fluid layers close to the top and bottom walls, respectively. Now, because metal oxides have much higher mass density than water, the top fluid layer is lighter than the bulk fluid, whereas the bottom fluid layer is heavier than the bulk fluid, which means that both layers are stagnant. Notice that for any assigned nanofluid the widths of these layers increase as the value assumed for β is increased (the values of β calculated using the McNab–Meisen relation or Equation 10.10 are 0.0038 or 0.0176 for $Al_2O_3 + H_2O$, and 0.0155 or 0.0990 for $TiO_2 + H_2O$). Indeed, it can be observed that the local solutions obtained through the two-phase model relying on the McNab–Meisen relation are almost the same as those achieved by the single-phase approach, which means that the actual slip motion effects are somehow underestimated. Moreover, it is worth pointing out that, due to the mass continuity for the solid phase, the low-concentration top fluid layer is thicker than the high-concentration bottom fluid layer, which is much more evident for $TiO_2 + H_2O$ than for $Al_2O_3 + H_2O$, owing to the significantly larger value of the thermophoresis parameter, β, as shown in Figure 10.11, which illustrates the profiles of the volume fraction φ along the vertical midplane of the enclosure. Of course, such fluid stagnations result in a limitation of the nanofluid circulation through the cavity, and

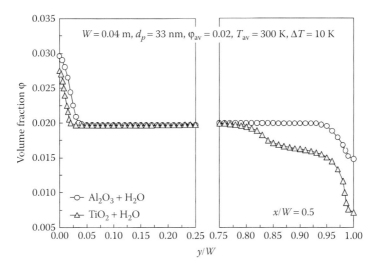

FIGURE 10.11 Distributions of the volume fraction φ along the vertical midplane of the enclosure for both $Al_2O_3 + H_2O$ and $TiO_2 + H_2O$.

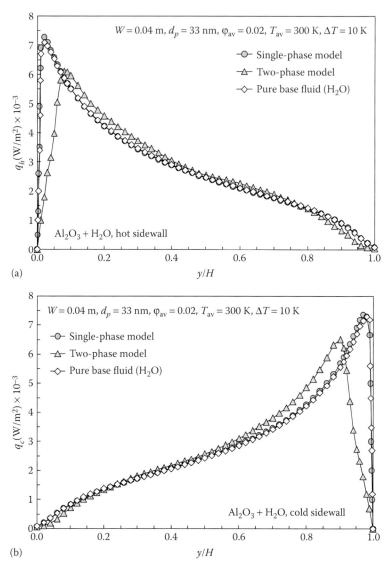

FIGURE 10.12 Heat flux distributions along the thermally active walls of the enclosure filled with $Al_2O_3 + H_2O$: (a) hot sidewall; (b) cold sidewall.

thus in a decrease of the heat transfer rate at the upper and lower portions of the enclosure, as reflected by the distributions of the heat flux along the hot and cold sidewalls displayed in Figure 10.12 for $Al_2O_3 + H_2O$ and in Figure 10.13 for $TiO_2 + H_2O$. In any diagram, the heat flux distribution obtained using the single-phase approach, as well as that for pure water, is also depicted for comparison. It can be seen that the overall heat

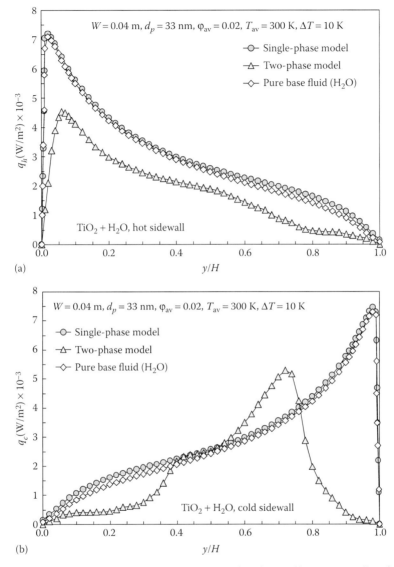

FIGURE 10.13 Heat flux distributions along the thermally active walls of the enclosure filled with $TiO_2 + H_2O$: (a) hot sidewall; (b) cold sidewall.

transfer performance predicted for the nanofluid by our two-phase model is worse than that of the pure base liquid, especially for $TiO_2 + H_2O$, which is in line with the heat transfer degradation detected in almost all the experimental studies available in the specialized literature on this topic. Conversely, according to the single-phase model, an overall slight heat transfer enhancement should occur, which demonstrates that the slip motion effects must necessarily be taken into due account to obtain realistic and reliable results.

Obviously, the fact that in differentially heated square cavities the addition of nanoparticles to a pure base liquid implies an overall thermal performance degradation does not absolutely mean that the use of nanofluids is always detrimental for natural convection applications. Actually, enhancements can be expected for different geometries, boundary conditions, and operating temperatures, which should be the subject for future investigations.

10.4 CONCLUSIONS

The slip motion occurring between suspended nanoparticles and the base fluid has nonnegligible effects on the thermal performance of nanofluids in buoyancy-induced enclosed flows.

This is the reason why the single-phase approach, in which nanofluids are treated as pure fluids in the hypothesis that the solid and liquid phases are in local thermal and hydrodynamic equilibrium, inevitably leads to unreliable results.

On the contrary, the conventional Eulerian–Lagrangian and Eulerian–Eulerian formulations of the two-phase approach may give rise to computational difficulties. In the former case, because the Newton's equation of motion has to be solved for each individual nanoparticle; in the latter case, because a second set of mass, momentum, and energy transfer governing equations dedicated to the solid phase has to be solved in conjunction with the set of governing equations for the liquid phase.

However, because the objective of any study on convective heat transfer of nanofluids is to determine the behavior of the mixture as a whole, rather than the behavior of each phase, an easier double-diffusive approach accounting for the effects of Brownian and thermophoretic diffusion of the suspended nanoparticles can be followed. Of course, the reliability of such an approach is based on the correct evaluation of the effective physical properties of the nanofluid, as well as the nanoparticle coefficients of diffusion.

In this framework, we have illustrated and discussed a two-phase model based on such a double-diffusive approach that incorporates three empirical correlations for the calculation of the effective thermal conductivity, the effective dynamic viscosity, and the thermophoretic diffusion coefficient of the suspended nanoparticles, all based on a high number of experimental data available in the literature from diverse sources, whose application permits to capture the actual thermal performance of nanofluids in buoyancy-induced enclosed flows, as demonstrated by the good degree of agreement between numerical results and experimental data.

LIST OF SYMBOLS

c	specific heat at constant pressure
D_B	Brownian diffusion coefficient
D_T	thermophoretic diffusion coefficient
d_f	equivalent diameter of a base fluid molecule
d_p	nanoparticle diameter
g	gravity vector
h	average coefficient of convection
I	unit tensor
$\mathbf{J}_p$	nanoparticle diffusion mass flux
k	thermal conductivity
k_b	Boltzmann constant $= 1.38066 \times 10^{-23}$ J K^{-1}
m	nanoparticle mass fraction
Nu	Nusselt number
p	pressure
Pr	Prandtl number
Q	heat transfer rate
q	heat flux
Ra	Rayleigh number
Re$_p$	nanoparticle Reynolds number
T	temperature
t	time
u_p	nanoparticle Brownian velocity
V	velocity vector
W	width of the enclosure
x, y	Cartesian coordinates

Greek symbols

β	thermophoresis parameter
φ	nanoparticle volume fraction
μ	dynamic viscosity
ρ	mass density
τ	stress tensor

Subscripts

av	average
c	cooled wall, at the temperature of the cooled wall
f	base fluid
fr	freezing point
h	heated wall, at the temperature of the heated wall
n	nanofluid
s	solid phase

REFERENCES

1. A.M. Hussein, K.V. Sharma, R.A. Bakar, K. Kadirgama, A review of forced convection heat transfer enhancement and hydrodynamic characteristics of a nanofluid, *Renew. Sust. Energ. Rev.* 29, 2014, 734–743.
2. N. Putra, W. Roetzel, S.K. Das, Natural convection of nano-fluids, *Heat Mass Transf.* 39, 2003, 775–784.
3. D. Wen, Y. Ding, Formulation of nanofluids for natural convective heat transfer applications, *Int. J. Heat Fluid Fl.* 26, 2005, 855–864.
4. D. Wen, Y. Ding, Natural convective heat transfer of suspensions of titanium dioxide nanoparticles (nanofluids), *IEEE Trans. Nanotechnol.* 5, 2006, 220–227.
5. A.G.A. Nnanna, Experimental model of temperature-driven nanofluid, *J. Heat Transf.* 129, 2007, 697–704.
6. B.H. Chang, A.F. Mills, E. Hernandez, Natural convection of microparticle suspensions in thin enclosures, *Int. J. Heat Mass Transf.* 51, 2008, 1332–1341.
7. C.J. Ho, W.K. Liu, Y.S. Chang, C.C. Lin, Natural convection heat transfer of alumina-water nanofluid in vertical square enclosures: An experimental study, *Int. J. Therm. Sci.* 49, 2010, 1345–1353.
8. Y. Hu, Y. He, S. Wang, Q. Wang, H.I. Schlaberg, Experimental and numerical investigation on natural convection heat transfer of TiO_2–water nanofluids in a square enclosure, *J. Heat Transf.* 136, 2014, 022502.
9. K. Khanafer, K. Vafai, M. Lightstone, Buoyancy-driven heat transfer enhancement in a two-dimensional enclosure utilizing nanofluids, *Int. J. Heat Mass Transf.* 46, 2003, 3639–3653.
10. J.C. Maxwell, *A Treatise on Electricity and Magnetism*, 3rd ed., Dover, New York, 1954.
11. A. Amiri, K. Vafai, Analysis of dispersion effects and nonthermal equilibrium, non-Darcian, variable porosity, incompressible flow through porous media, *Int. J. Heat Mass Transf.* 37, 1994, 939–954.
12. H.C. Brinkman, The viscosity of concentrated suspensions and solutions, *J. Chem. Phys.* 20, 1952, 571.
13. R.Y. Jou, S.C. Tzeng, Numerical research of nature convective heat transfer enhancement filled with nanofluids in rectangular enclosures, *Int. Commun. Heat Mass* 33, 2006, 727–736.

14. C.J. Ho, M.W. Chen, Z.W. Li, Numerical simulation of natural convection of nanofluid in a square enclosure: Effects due to uncertainties of viscosity and thermal conductivity, *Int. J. Heat Mass Transf.* 51, 2008, 4506–4516.

15. P. Charuyakorn, S. Sengupta, S.K. Roy, Forced convection heat transfer in micro-encapsulated phase change material slurries, *Int. J. Heat Mass Transf.* 34, 1991, 819–833.

16. S.E.B. Maïga, C.T. Nguyen, N. Galanis, G. Roy, Heat transfer behaviours of nanofluids in a uniformly heated tube, *Superlattices Microst.* 35, 2004, 543–557.

17. A.K. Santra, S. Sen, N. Chakraborty, Study of heat transfer characteristics of copper-water nanofluid in a differentially heated square cavity with different viscosity models, *J. Enhanc. Heat Transf.* 15, 2008, 273–287.

18. H.E. Patel, T. Sundararajan, T. Pradeep, A. Dasgupta, N. Dasgupta, S.K. Das, A micro-convection model for the thermal conductivity of nanofluids, *Pramana J. Phys.* 65, 2005, 863–869.

19. K. Kwak, C. Kim, Viscosity and thermal conductivity of copper oxide nanofluid dispersed in ethylene glycol, *Korea-Australia Rheol. J.* 17, 2005, 35–40.

20. A.K. Santra, S. Sen, N. Chakraborty, Study of heat transfer augmentation in a differentially heated square cavity using copper-water nanofluid, *Int. J. Therm. Sci.* 47, 2008, 1113–1122.

21. E. Abu-Nada, H.F. Oztop, Effects of inclination angle on natural convection in enclosures filled with Cu-water nanofluid, *Int. J. Heat Fluid Fl.* 30, 2009, 669–678.

22. B. Ghasemi, S.M. Aminossadati, Natural convection heat transfer in an inclined enclosure filled with a water-CuO nanofluid, *Numer. Heat Tr. A* 55, 2009, 807–823.

23. J. Koo, C. Kleinstreuer, A new thermal conductivity model for nanofluids, *J. Nanopart. Res.* 6, 2004, 577–588.

24. E. Abu-Nada, A.J. Chamkha, Effect of nanofluid variable properties on natural convection in enclosures filled with a CuO-EG-water nanofluid, *Int. J. Therm. Sci.* 49, 2010, 2339–2352.

25. S.P. Jang, S.U.S. Choi, Effects of various parameters on nanofluid thermal conductivity, *J. Heat Transf.* 129, 2007, 617–623.

26. P.K. Namburu, D.P. Kulkarni, D. Misra, D.K. Das, Viscosity of copper oxide nanoparticles dispersed in ethylene glycol and water mixture, *Exp. Therm. Fluid Sci.* 32, 2007, 397–402.

27. E. Abu-Nada, Z. Masoud, H.F. Oztop, A. Campo, Effect of nanofluid variable properties on natural convection in enclosures, *Int. J. Therm. Sci.* 49, 2010, 479–491.

28. C.H. Chon, K.D. Kihm, S.P. Lee, S.U.S. Choi, Empirical correlation finding the role of temperature and particle size for nanofluid (Al_2O_3) thermal conductivity enhancement, *Appl. Phys. Lett.* 87, 2005, 153107.

29. C.T. Nguyen, F. Desgranges, G. Roy, N. Galanis, T. Maré, S. Boucher, H. Angue Mintsa, Temperature and particle-size dependent viscosity data for water-based nanofluids—Hysteresis phenomenon, *Int. J. Heat Fluid Fl.* 28, 2007, 1492–1506.

30. K. Kahveci, Buoyancy driven heat transfer of nanofluids in a tilted enclosure, *J. Heat Transf.* 132, 2010, 062501.
31. W. Yu, S.U.S. Choi, The role of interfacial layers in the enhanced thermal conductivity of nanofluids: A renovated Maxwell model, *J. Nanopart. Res.* 5, 2003, 167–171.
32. K.C. Lin, A. Violi, Natural convection heat transfer of nanofluids in a vertical cavity: Effects of non-uniform particle diameter and temperature on thermal conductivity, *Int. J. Heat Fluid Fl.* 31, 2010, 236–245.
33. J. Xu, B. Yu, M. Zou, P. Xu, A new model for heat conduction of nanofluids based on fractal distributions of nanoparticles, *J. Phys. D* 39, 2006, 4486–4490.
34. S.P. Jang, J.H. Lee, K.S. Hwang, S.U.S. Choi, Particle concentration and tube size dependence of viscosities of Al_2O_3-water nanofluids flowing through micro- and minitubes, *Appl. Phys. Lett.* 91, 2007, 243112.
35. J.A. Esfahani, V. Bordbar, Double diffusive natural convection heat transfer enhancement in a square enclosure using nanofluids, *J. Nanotechnol. Eng. Med.* 2, 2011, 021002.
36. G.H.R. Kefayati, S.F. Hosseinizadeh, M. Gorji, H. Sajjadi, Lattice Boltzmann simulation of natural convection in tall enclosures using water/SiO_2 nanofluid, *Int. Commun. Heat Mass* 38, 2011, 798–805.
37. F.H. Lai, Y.T. Yang, Lattice Boltzmann simulation of natural convection heat transfer of Al_2O_3/water nanofluids in a square enclosure, *Int. J. Therm. Sci.* 50, 2011, 1930–1941.
38. H.A. Mintsa, G. Roy, C.T. Nguyen, D. Doucet, New temperature dependent thermal conductivity data for water-based nanofluids, *Int. J. Therm. Sci.* 48, 2009, 363–371.
39. F.S. Oueslati, R. Bennacer, Heterogeneous nanofluids: Natural convection heat transfer enhancement, *Nanoscale Res. Lett.* 6, 2011, 222.
40. H. Saleh, R. Roslan, I. Hashim, Natural convection heat transfer in a nanofluid-filled trapezoidal enclosure, *Int. J. Heat Mass Transf.* 54, 2011, 194–201.
41. Z. Alloui, J. Guiet, P. Vasseur, M. Reggio, Natural convection of nanofluids in a shallow rectangular enclosure heated from the side, *Can. J. Chem. Eng.* 90, 2012, 69–78.
42. B.C. Pak, Y.I. Cho, Hydrodynamic and heat transfer study of dispersed fluids with submicron metallic oxide particles, *Exp. Heat Transf.* 11, 1998, 151–170.
43. M. Rahimi, A.A. Ranjbar, M.J. Hosseini, M. Abdollahzadeh, Natural convection of nanoparticle–water mixture near its density inversion in a rectangular enclosure, *Int. Commun. Heat Mass Trans.* 39, 2012, 131–137.
44. M. Corcione, M. Cianfrini, A. Quintino, Two-phase mixture modeling of natural convection of nanofluids with temperature-dependent properties, *Int. J. Therm. Sci.* 71, 2013, 182–195.
45. M. Corcione, Empirical correlating equations for predicting the effective thermal conductivity and dynamic viscosity of nanofluids, *Energ. Convers. Manage.* 52, 2011, 789–793.
46. H. Aminfar, M.R. Haghgoo, Brownian motion and thermophoresis effects on natural convection of alumina-water nanofluid, *J. Mech. Eng. Sci.* 227, 2012, 100–110.

47. H.A. Pakravan, M. Yaghoubi, Analysis of nanoparticles migration on natural convective heat transfer of nanofluids, *Int. J. Therm. Sci.* 68, 2013, 79–93.

48. G.A. Sheikhzadeh, M. Dastmalchi, H. Khorasanizadeh, Effects of nanoparticles transport mechanisms on Al_2O_3–water nanofluid natural convection in a square enclosure, *Int. J. Therm. Sci.* 66, 2013, 51–62.

49. M. Alipanah, A.A. Ranjbar, A. Zahmatkesh, Numerical study of natural convection in vertical enclosures utilizing nanofluid, *Adv. Mech. Eng.* 2014, 392610.

50. G. Wang, X. Meng, M. Zeng, H. Ozoe, Q.W. Wang, Natural convection heat transfer of copper–water nanofluid in a square cavity with time-periodic boundary temperature, *Heat Transf. Eng.* 35, 2014, 630–640.

51. S.K. Choi, S.O. Kim, T.H. Lee, D. Hahn, Computation of the natural convection of nanofluid in a square cavity with homogeneous and nonhomogeneous models, *Numer. Heat Tr. A* 65, 2014, 287–301.

52. E. Abu-Nada, Z. Masoud, A. Hijazi, Natural convection heat transfer enhancement in horizontal concentric annuli using nanofluids, *Int. Commun. Heat Mass* 35, 2008, 657–665.

53. E. Abu-Nada, Effects of variable viscosity and thermal conductivity of Al_2O_3-water nanofluid on heat transfer enhancement in natural convection. *Int. J. Heat Fluid Fl.* 30, 2009, 679–690.

54. E. Abu-Nada, Effects of variable viscosity and thermal conductivity of CuO-water nanofluid on heat transfer enhancement in natural convection: Mathematical model and simulation, *J. Heat Transf.* 132, 2010, 052401.

55. A. Arefmanesh, M. Amini, M. Mahmoodi, M. Najafi, Buoyancy-driven heat transfer analysis in two-square duct annuli filled with a nanofluid, *Eur. J. Mech. B Fluids* 33, 2012, 95–104.

56. Y. He, Y. Men, Y. Zhao, H. Lu, Y. Ding, Numerical investigation into the convective heat transfer of TiO_2 nanofluids flowing through a straight tube under the laminar flow conditions, *Appl. Therm. Eng.* 29, 2009, 1965–1972.

57. S. Parvin, R. Nasrin, M.A. Alim, N.F. Hossain, A.J. Chamkha, Thermal conductivity variation on natural convection flow of water-alumina nanofluid in an annulus, *Int. J. Heat Mass Transf.* 55, 2012, 5268–5274.

58. S. Soleimani, M. Sheikholeslami, D.D. Ganji, M. Gorji-Bandpy, Natural convection heat transfer in a nanofluid filled semi-annulus enclosure, *Int. Commun. Heat Mass* 39, 2012, 565–574.

59. M. Sheikholeslami, M. Gorji-Bandpy, D.D. Ganji, Magnetic field effects on natural convection around a horizontal circular cylinder inside a square enclosure filled with nanofluid, *Int. Commun. Heat Mass* 39, 2012, 978–986.

60. H.R. Ashorynejad, A.A. Mohamad, M. Sheikholeslami, Magnetic field effects on natural convection flow of a nanofluid in a horizontal cylindrical annulus using Lattice Boltzmann method, *Int. J. Therm. Sci.* 64, 2013, 240–250.

61. G.A. Sheikhzadeh, M. Arbaban, M.A. Mehrabian, Laminar natural convection of Cu-water nanofluid in concentric annuli with radial fins attached to the inner cylinder, *Heat Mass Transf.* 49, 2013, 391–403.

62. A. Abouei Mehrizi, M. Farhadi, S. Shayamehr, Natural convection flow of Cu–Water nanofluid in horizontal cylindrical annuli with inner triangular cylinder using lattice Boltzmann method, *Int. Commun. Heat Mass* 44, 2013, 147–156.

63. M. Sheikholeslami, M. Gorji-Bandpy, D.D. Ganji, Numerical investigation of MHD effects on Al_2O_3–water nanofluid flow and heat transfer in a semi-annulus enclosure using LBM, *Energy* 60, 2013, 501–510.

64. J. Koo, C. Kleinstreuer, Laminar nanofluid flow in microheat-sinks, *Int. J. Heat Mass Transf.* 48, 2005, 2652–2661.

65. M. Sheikholeslami, M. Gorji-Bandpy, D.D. Ganji, S. Soleimani, Effect of a magnetic field on natural convection in an inclined half-annulus enclosure filled with Cu–water nanofluid using CVFEM, *Adv. Powder Technol.* 24, 2013, 980–991.

66. M. Habibi Matin, I. Pop, Natural convection flow and heat transfer in an eccentric annulus filled by copper nanofluid, *Int. J. Heat Mass Transf.* 61, 2013, 353–364.

67. Q. Li, Y. Xuan, F. Yu, Experimental investigation of submerged single jet impingement using Cu–water nanofluid, *Appl. Therm. Eng.* 36, 2012, 426–433.

68. M. Corcione, E. Habib, A. Quintino, A two-phase numerical study of buoyancy-driven convection of alumina–water nanofluids in differentially-heated horizontal annuli, *Int. J. Heat Mass Transf.* 65, 2013, 327–338.

69. M. Sheikholeslami, M. Gorji-Bandpy, D.D. Ganji, Latticer Boltzmann method for MHD natural convection heat transfer using nanofluid, *Powder Technol.* 254, 2014, 82–93.

70. G.S. McNab, A. Meisen, Thermophoresis in liquids, *J. Colloid Interf. Sci.* 44, 1973, 339–346.

71. J. Buongiorno, Convective transport in nanofluids, *J. Heat Transf.* 128, 2006, 240–250.

72. S.K. Das, N. Putra, W. Roetzel, Pool boiling characteristics of nano-fluids, *Int. J. Heat Mass Transf.* 46, 2003, 851–862.

73. R. Prasher, D. Song, J. Wang, P. Phelan, Measurements of nanofluid viscosity and its implications for thermal applications, *Appl. Phys. Lett.* 89, 2006, 133108.

74. Y. He, Y. Jin, H. Chen, Y. Ding, D. Cang, H. Lu, Heat transfer and flow behaviour of aqueous suspensions of TiO_2 nanoparticles (nanofluids) flowing upward through a vertical pipe, *Int. J. Heat Mass Transf.* 50, 2007, 2272–2281.

75. H. Chen, Y. Ding, Y. He, C. Tan, Rheological behaviour of ethylene glycol based titania nanofluids, *Chem. Phys. Lett.* 444, 2007, 333–337.

76. J. Chevalier, O. Tillement, F. Ayela, Rheological properties of nanofluids flowing through microchannels, *Appl. Phys. Lett.* 91, 2007, 233103.

77. D. Cabaleiro, M.J. Pastoriza-Gallego, M.M. Piñero, L. Lugo, Characterization and measurements of thermal conductivity, density and rheological properties of zinc oxide nanoparticles dispersed in (ethane-1,2-diol + water) mixture, *J. Chem. Thermodyn.* 58, 2013, 405–415.

78. A. Einstein, Über die von der molekularkinetischen Theorie der Wärme geforderte Bewegung von in ruhenden Flüssigkeiten suspendierten Teilchen (in German), *Ann. Phys.* 17, 1905, 549–560.

79. M. Corcione, A. Quintino, A correlation for the prediction of the thermophoretic diffusion effects on natural convection of nanofluids, *Int. J. Therm. Sci.* (in press).

80. H. Masuda, A. Ebata, K. Teramae, N. Hishinuma, Alteration of thermal conductivity and viscosity of liquid by dispersing ultra-fine particles (dispersion of γ-Al$_2$O$_3$, SiO$_2$, and TiO$_2$ ultra-fine particles), *Netsu Bussei* 4, 1993, 227–233.

81. S. Lee, S.U.S. Choi, S. Li, J.A. Eastman, Measuring thermal conductivity of fluids containing oxide nanoparticles, *J. Heat Transf.* 121, 1999, 280–289.

82. J.A. Eastman, S.U.S. Choi, S. Li, W. Yu, L.J. Thompson, Anomalously increased effective thermal conductivity of ethylene glycol-based nanofluids containing copper nanoparticles, *Appl. Phys. Lett.* 78, 2001, 718–720.

83. S.K. Das, N. Putra, P. Thiesen, W. Roetzel, Temperature dependence of thermal conductivity enhancement for nanofluids. *J. Heat Transf.* 125, 2003, 567–574.

84. C.H. Chon, K.D. Kihm, Thermal conductivity enhancement of nanofluids by Brownian motion, *J. Heat Transf.* 127, 2005, 810.

85. S.M.S. Murshed, K.C. Leong, C. Yang, Investigations of thermal conductivity and viscosity of nanofluids, *Int. J. Therm. Sci.* 47, 2008, 560–568.

86. P. Keblinski, S.R. Phillpot, S.U.S. Choi, J.A. Eastman, Mechanisms of heat flow in suspensions of nano-sized particles (nanofluids), *Int. J. Heat Mass Transf.* 45, 2002, 855–863.

87. S.J. Palm, G. Roy, C.T. Nguyen, Heat transfer enhancement with the use of nanofluids in radial flow cooling systems considering temperature-dependent properties, *Appl. Therm. Eng.* 26, 2006, 2209–2218.

88. D.W. Oh, A. Jain, J.K. Eaton, K.E. Goodson, J.S. Lee, Thermal conductivity measurement and sedimentation detection of aluminium oxide nanofluids by using the 3ω method, *Int. J. Heat Fluid Fl.* 29, 2008, 1456–1461.

89. M. Chandrasekar, S. Suresh, A. Chandra Bose, Experimental investigations and theoretical determinations of thermal conductivity and viscosity of Al$_2$O$_3$/water nanofluid, *Exp. Therm. Fluid Sci.* 34, 2010, 210–216.

90. X. Wang, X. Xu, S.U.S. Choi, Thermal conductivity of nanoparticle-fluid mixture, *J. Thermophys. Heat Transf.* 13, 1999, 474–480.

91. J.H. Lee, K.S. Hwang, S.P. Jang, B.H. Lee, J.H. Kim, S.U.S. Choi, C.J. Choi, Effective viscosities and thermal conductivities of aqueous nanofluids containing low volume concentrations of Al$_2$O$_3$ nanoparticles, *Int. J. Heat Mass Transf.* 51, 2008, 2651–2656.

92. J. Garg, B. Poudel, M. Chiesa, J.B. Gordon, J.J. Ma, J.B. Wang, Z.F. Ren, Y.T. Kang, H. Ohtani, J. Nanda, G.H. McKinley, G. Chen, Enhanced thermal conductivity and viscosity of copper nanoparticles in ethylene glycol nanofluid, *J. Appl. Phys.* 103, 2008, 074301.

93. M.J. Pastoriza-Gallego, C. Casanova, R. Páramo, B. Barbés, J.L. Legido, M.M. Piñero, A study on stability and thermophysical properties (density and viscosity) of Al$_2$O$_3$ in water nanofluid, *J. Appl. Phys.* 106, 2009, 064301.

94. J.F. Zhao, Z.Y. Luo, M.J. Ni, K.F. Cen, Dependence of nanofluid viscosity on particle size and pH value, *Chin. Phys. Lett.* 26, 2009, 066202.

95. S.Q. Zhou, R. Ni, Measurement of the specific heat capacity of water-based Al$_2$O$_3$ nanofluid, *Appl. Phys. Lett.* 92, 2008, 093123.

96. J.P. Van Doormaal, G.D. Raithby, Enhancements of the simple method for predicting incompressible fluid flows, *Numer. Heat Tr.* 11, 1984, 147–163.

97. B.P. Leonard, A stable and accurate convective modelling procedure based on quadratic upstream interpolation, *Comput. Methods Appl. Mech. Eng.* 19, 1979, 59–78.

98. G. de Vahl Davis, Natural convection of air in a square cavity: A bench mark numerical solution, *Int. J. Numer. Meth. Fl.* 3, 1983, 249–264.

99. H.S. Mahdi, R.B. Kinney, Time-dependent natural convection in a square cavity: Application of a new finite volume method, *Int. J. Numer. Meth. Fl.* 11, 1990, 57–86.

100. M. Hortmann, M. Peric, G. Scheuerer, Finite volume multigrid prediction of laminar natural convection: Bench-mark solutions, *Int. J. Numer. Meth. Fl.* 11, 1990, 189–207.

101. D.C. Wan, B.S.V. Patnaik, G.W. Wei, A new benchmark quality solution for the buoyancy-driven cavity by discrete singular convolution, *Numer. Heat Tr.* 40, 2001, 199–228.

102. A. Bejan, *Convection Heat Transfer*, 3rd ed., John Wiley & Sons, Hoboken, NJ, 2004.

103. F.P. Incropera, D.P. Dewitt, T.L. Bergman, A.S. Lavine, *Fundamentals of Heat and Mass Transfer*, 6th ed., John Wiley & Sons, Hoboken, NJ, 2007.

104. C. Béghein, F. Haghigat, F. Allard, Numerical study of double-diffusive natural convection in a square cavity, *Int. J. Heat Mass Transf.* 35, 1992, 833–846.

CHAPTER 11

Modeling Convection in Nanofluids

From Clear Fluids to Porous Media

Donald A. Nield and Andrey V. Kuznetsov

CONTENTS

11.1 INTRODUCTION

The term "nanofluid" is now commonly used to refer to a liquid containing a suspension of submicron solid particles (nanoparticles) whose characteristic dimension is on the order of tens or hundreds of nanometers. The fluid can be water or an organic solvent and the particles are commonly metallic oxides. This means that the nanoparticles usually have high thermal conductivity relative to that of the base fluid and that fact favors an increase of heat transfer, but that effect is somewhat offset by the increased viscosity due to the presence of the particles. One of the most interesting features of nanofluids is the enhancement of thermal diffusivity that according to some data may exceed the limits predicted by conventional macroscopic theories of suspensions (Choi, 2009; Choi et al., 2001, 2004; Das et al., 2008; Eastman et al., 2001). The enhancement of effective thermal conductivity was confirmed by experiments conducted by many researchers, including Masuda et al. (1993), although the level of enhancement is still a subject of a debate. For example, in a recent paper by Choi et al. (2014), the authors concluded that, for convection in a laterally heated enclosure with a square cross section, the discrepancy between previously published experimental and theoretical results could be explained by the fact that the respective authors used different definitions of the Nusselt number, one definition being based on the properties of the base fluid and the other on the properties of the suspension. This paper by Choi et al. (2014) is again cited at the end of Section 11.3.

In this chapter, we discuss the mathematical modeling of convection in nanofluids, with emphasis on two problems that are paradigmatic for confined convection and external convection, respectively. The first problem is the onset of convection in a fluid layer heated uniformly from below and the second problem involves a thermal boundary layer adjacent to a vertical wall. We also consider forced convection in a channel.

We first discuss property variations that are the consequence of the fact that nanofluids are suspensions, and then processes that occur due to the smallness of the nanoparticles. (We note in passing that other explanations have been proposed for the increase of thermal conductivity that has been observed in nanofluids. For example, Vadász [2006, 2008] has proposed that thermal lagging between the particles and the fluid could cause such an effect.)

In this chapter, specific attention is given to convection in a porous medium saturated by a nanofluid, a case that has been studied extensively by the present authors.

11.2 PROPERTY VARIATIONS OF SUSPENSIONS

A common approach is to examine the effect of the variation of thermal conductivity and viscosity with nanofluid particle fraction, utilizing expressions obtained using the theory of mixtures, namely,

$$\frac{\mu_{\text{eff}}}{\mu_f} = \frac{1}{\left(1-\phi\right)^{2.5}} \tag{11.1}$$

$$\frac{k_{\text{eff}}}{k_f} = \frac{(k_p + 2k_f) - \phi(k_f - k_p)}{(k_p + 2k_f) + \phi(k_f - k_p)} \tag{11.2a}$$

where:

ϕ denotes the nanoparticle volume fraction

μ and k denote the viscosity and thermal conductivity, respectively

The subscripts f, p, and eff denote the fluid, the particles, and an effective quantity, respectively

Equation 11.1 was obtained by Brinkman (1952) using ideas due to Einstein and Equation 11.2a is the Maxwell–Garnett formula for a suspension of spherical particles that dates back to Maxwell (1881), who considered an electrical conductivity analog to thermal conductivity. An alternative formula for k_{eff}, one based on the effective medium theory, was obtained by Bruggeman (1935). This is obtained by solving the balance equation for k_{eff}:

$$\phi\frac{(k_p - k_{\text{eff}})}{(k_p + k_{\text{eff}})} + (1-\phi)\frac{(k_f - k_{\text{eff}})}{(k_f + k_{\text{eff}})} = 0 \tag{11.2b}$$

The above equation applies for particles of general shape, and it can be readily generalized to the case of more than one type of particle. It yields slightly smaller values of k_{eff} than those given by Equation 11.2a.

For a convection problem, the specific heat c is important, and it is common practice to employ the weighted volumetric average value:

$$c_{\text{eff}} = (1-\phi)c_f + \phi c_p \tag{11.3a}$$

Alternatively, one can consider a weighted volumetric average of the heat capacity:

$$(\rho c)_{\text{eff}} = (1-\phi)(\rho c)_f + \phi(\rho c)_p \tag{11.3b}$$

where ρ is the density.

More precise models have been proposed, but in any case variation of specific heat or heat capacity is normally insignificant in comparison with variation of viscosity and thermal conductivity. For these, Khanafer and Vafai (2011) presented a synthesis of results, and further discussion was presented by Nield and Kuznetsov (2013a).

11.3 PROCESSES ASSOCIATED WITH THE SMALLNESS OF NANOPARTICLES

11.3.1 The Buongiorno Model

An important study of convective transport in nanofluids was made by Buongiorno (2006). He focused on the heat transfer enhancement observed in convective situations. Buongiorno concluded that turbulence is not affected by the presence of the nanoparticles so this cannot explain the observed enhancement. Particle rotation has also been proposed as a cause of heat transfer enhancement, but Buongiorno calculated that this effect is too small to explain the observed results. With dispersion, turbulence, and particle rotation ruled out as significant agencies for heat transfer enhancement, Buongiorno proposed a new model based on the mechanics of the nanoparticle/base fluid relative velocity, which is described in this chapter.

Buongiorno (2006) noted that the nanoparticle absolute velocity can be viewed as the sum of the base fluid velocity and a relative velocity (called the slip velocity). He then considered seven slip mechanisms: inertia, Brownian diffusion, thermophoresis, diffusiophoresis, Magnus effect, fluid drainage, and gravity settling. He concluded that in the absence of turbulent effects, it is the Brownian diffusion and the thermophoresis that will be important. He proceeded to write down conservation equations based on these two effects.

11.3.1.1 Conservation Equations for a Nanofluid

First, we outline the derivation of conservation equations applicable to a nanofluid in the absence of a solid matrix. Later, we modify these equations to the case of a porous medium saturated by the nanofluid. The Buongiorno model treats the nanofluid as a two-component mixture (base fluid and nanoparticles) with the following assumptions:

1. Incompressible flow

2. No chemical reactions

3. Negligible external forces

4. Dilute mixture

5. Negligible viscous dissipation

6. Negligible radiative heat transfer

7. Nanoparticles and base fluid locally in thermal equilibrium

The continuity equation for the nanofluid is

$$\nabla \cdot \mathbf{v} = 0 \tag{11.4}$$

where $\mathbf{v}$ is the nanofluid velocity.

The conservation equation for the nanoparticles in the absence of chemical reactions is

$$\frac{\partial \phi}{\partial t} + \mathbf{v} \cdot \nabla \phi = -\frac{1}{\rho_p} \nabla \cdot \mathbf{j}_p \tag{11.5}$$

where:

t is the time

ϕ is the nanoparticle volume fraction

ρ_p is the nanoparticle mass density

$\mathbf{j}_p$ is the diffusion mass flux for the nanoparticles, given as the sum of two diffusion terms (Brownian diffusion and thermophoresis) by

$$\mathbf{j}_p = \mathbf{j}_{p,B} + \mathbf{j}_{p,T} = -\rho_p D_B \nabla \phi - \rho_p D_T \frac{\nabla T}{T} \tag{11.6}$$

where T is the temperature.

Thermophoresis is analogous to the Soret effect in gaseous or liquid mixtures. It should be noted that Buongiorno departed from the usual tradition by using the notation in which the dependence on temperature of the thermophoretic coefficient is taken into account explicitly. Thus, D_T has the same dimensions as D_B, namely, $m^2 s^{-1}$.

Here, D_B is the Brownian diffusion coefficient given by the Einstein–Stokes equation:

$$D_B = \frac{k_B T}{3\pi \mu d_p} \tag{11.7}$$

where:

k_B is the Boltzmann constant

μ is the viscosity of the fluid

d_p is the nanoparticle diameter

Use has been made of the following expression for the thermophoretic velocity $\mathbf{V}_T$:

$$\mathbf{V}_T = -\tilde{\beta} \frac{\mu}{\rho} \frac{\nabla T}{T} \tag{11.8}$$

where:

ρ is the fluid density

$\tilde{\beta}$ is the proportionality factor, which is given by

$$\tilde{\beta} = 0.26 \frac{k}{2k + k_p} \tag{11.9}$$

where k and k_p are the thermal conductivities of the fluid and the particle material, respectively. Hence, the thermophoretic diffusion flux is given by

$$\mathbf{j}_{p,T} = \rho_p \phi \mathbf{V}_T = -\rho_p D_T \frac{\nabla T}{T} \tag{11.10}$$

where the thermophoretic diffusion coefficient is given by

$$D_T = \tilde{\beta} \frac{\mu}{\rho} \phi \tag{11.11}$$

Equations 11.5 and 11.6 then produce an equation expressing the conservation of nanoparticles in the form:

$$\frac{\partial \phi}{\partial t} + \mathbf{v} \cdot \nabla \phi = \nabla \cdot \left(D_B \nabla \phi + D_T \frac{\nabla T}{T} \right) \tag{11.12}$$

The momentum equation for a nanofluid takes the same form as for a pure fluid, but it should be remembered that μ is a strong function of ϕ. If one introduces a buoyancy force and adopts the Boussinesq approximation, then the momentum equation can be written as

$$\rho \left(\frac{\partial \mathbf{v}}{\partial t} + \mathbf{v} \cdot \nabla \mathbf{v} \right) = -\nabla p + \mu \nabla^2 \mathbf{v} + \rho \mathbf{g} \tag{11.13}$$

where:

$\mathbf{g}$ is the gravitational acceleration vector

$$\rho = \phi \rho_p + (1 - \phi) \rho_f \tag{11.14}$$

The nanofluid density ρ can be approximated by the base fluid density ρ_f when ϕ is small. Then, when the Boussinesq approximation is adopted, the buoyancy term is approximated by

$$\rho \mathbf{g} \cong \left[\phi \rho_p + (1 - \phi) \{ \rho [1 - \beta(T - T_0)] \} \right] \mathbf{g} \tag{11.15}$$

The thermal energy equation for a nanofluid can be written as

$$\rho c\left(\frac{\partial T}{\partial t}+\mathbf{v}\cdot\nabla T\right)=-\nabla\cdot\mathbf{q}+h_p\nabla\cdot\mathbf{j}_p \tag{11.16}$$

where:

    c is the nanofluid specific heat

    T is the nanofluid temperature

    h_p is the specific enthalpy of the nanoparticle material

    $\mathbf{q}$ is the energy flux, relative to a frame moving with the nanofluid velocity $\mathbf{v}$, given by

$$\mathbf{q}=-k\nabla T+h_p\mathbf{j}_p \tag{11.17}$$

where k is the nanofluid thermal conductivity. Substituting Equation 11.17 in 11.16 yields

$$\rho c\left(\frac{\partial T}{\partial t}+\mathbf{v}\cdot\nabla T\right)=\nabla\cdot(k\nabla T)-c_p\mathbf{j}_p\cdot\nabla T \tag{11.18}$$

In deriving this equation, use has been made of a vector identity and the fact that $\nabla h_p=c_p\nabla T$, where c_p is the nanoparticle specific heat of the material constituting the nanoparticles, whereas c is the specific heat (at constant pressure) of the fluid. Then, substitution of Equation 11.6 in 11.18 gives the final form:

$$\rho c\left(\frac{\partial T}{\partial t}+\mathbf{v}\cdot\nabla T\right)=\nabla\cdot(k\nabla T)+\rho_p c_p\left(D_B\nabla\phi\cdot\nabla T+D_T\frac{\nabla T\cdot\nabla T}{T}\right) \tag{11.19}$$

Equations 11.12 and 11.19 constitute a coupled pair of equations for T and ϕ. One observes that the nanofluid terms are similar to the Soret and Dufour cross-diffusion terms that arise in the case of double diffusion in a binary fluid.

We now draw attention to a subtlety that has previously been overlooked. Equations 11.12 and 11.19 are in the form of which was presented by Buongiorno (2006). As they stand, they are adequate for application to natural convection problems. However, in the case of forced convection and mixed convection problems, an additional contribution to the particle flux (a convective term) must be taken into account. Equation 11.12 needs no change, because a convective term is already properly included in that equation. However, in Equation 11.19, only the convection of heat is already incorporated and an additional thermophoresis term needs to be inserted. An externally applied pressure gradient leads to a through-flow of the base fluid and it is the particle flux that results from this that is involved. We believe that Equation 11.19 should then be modified to read

$$\rho c \left(\frac{\partial T}{\partial t} + \mathbf{v} \cdot \nabla T \right) = \nabla \cdot (k \nabla T) + \rho_p c_p \left(\begin{array}{c} D_B \nabla \phi \cdot \nabla T + D_T \dfrac{\nabla T \cdot \nabla T}{T} \\ \\ - \phi_0 \mathbf{v}_0 \cdot \nabla T \end{array} \right) \quad (11.20)$$

where:

ϕ_0 is the mean particle fraction

$\mathbf{v}_0$ is the mean velocity

It is being assumed here that the nanofluid is dilute and so ϕ_0 is small compared with unity. If the velocity is small, then the new term (the last term within parentheses on the left-hand side) is small in comparison with the other terms within parentheses on the right-hand side of Equation 11.20. The effect of the new term is to reduce the change in temperature in the direction of the velocity.

11.3.1.2 Conservation Equations for a Porous Medium Saturated by a Nanofluid

We consider a porous medium whose porosity is denoted by ε and permeability by K. The Darcy velocity is denoted by $\mathbf{v}_D$. This is related to $\mathbf{v}$ by $\mathbf{v}_D = \varepsilon \mathbf{v}$. We now have to deal with the following four field equations (corresponding to Equations 11.4, 11.13, 11.20, 11.12, respectively), for the total mass, momentum, thermal energy, and nanoparticles, respectively:

$$\nabla \cdot \mathbf{v}_D = 0 \quad (11.21)$$

$$\rho \left(\frac{1}{\varepsilon} \frac{\partial \mathbf{v}_D}{\partial t} + \frac{1}{\varepsilon^2} \mathbf{v}_D \cdot \nabla \mathbf{v}_D \right) = -\nabla p + \tilde{\mu} \nabla^2 \mathbf{v}_D - \frac{\mu}{K} \mathbf{v}_D$$
$$+ \left[\phi \rho_p + (1-\phi) \{ \rho [1 - \beta (T - T_0)] \} \right] \mathbf{g} \quad (11.22)$$

$$(\rho c)_m \frac{\partial T}{\partial t} + (\rho c)_f \mathbf{v}_D \cdot \nabla T = \nabla \cdot (k_m \nabla T) + \varepsilon(\rho c)_p$$

$$\times \left(D_B \nabla \phi \cdot \nabla T + D_T \frac{\nabla T \cdot \nabla T}{T} - \frac{1}{\varepsilon} \phi_0 \mathbf{v}_0 \cdot \nabla T \right) \tag{11.23}$$

$$\frac{\partial \phi}{\partial t} + \frac{1}{\varepsilon} \mathbf{v}_D \cdot \nabla \phi = \nabla \cdot \left(D_B \nabla \phi + D_T \frac{\nabla T}{T} \right) \tag{11.24}$$

Here we have introduced the effective viscosity $\tilde{\mu}$, the effective heat capacity $(\rho c)_m$, and the effective thermal conductivity k_m of the porous medium.

In writing Equations 11.21 through 11.24, we have assumed that the Brownian motion and thermophoretic processes remain coherent, although volume averages over a representative elementary volume are taken. We have noted that thermophoresis and Brownian motion are intrinsic processes in the sense that the appropriate averages are taken over the nanofluid only and the solid matrix is not involved. We assume that the nanoparticles are small compared with the pores.

The extent to which the Buongiorno model is adequate is open to debate. Esalamian and Saghir (2014) discuss the extent to which reliable expressions exist for the estimation of thermophoretic data. They also discuss the similarity between thermophoresis in nanofluids and thermodiffusion in binary mixtures.

Choi et al. (2014) refer to Buongiorno's model as a "nonhomogeneous model" in contrast to the "homogeneous model" where the nanofluid properties and the volume fraction do not vary within the solution domain. They cite a large number of numerical studies in which the homogeneous model was used and in which Rayleigh and Nusselt numbers were defined in terms of the conductivity of the base fluid.

It seems that it should be feasible to perform numerical computations using the combination of the models discussed in Sections 11.2 and 11.3. This could be a model in which Equations 11.21 through 11.24 were employed, but now with the viscosity and thermal conductivity allowed to vary according to Equations 11.1 and 11.2a. A start in this direction was made by Nield and Kuznetsov (2012), who obtained an approximate analytical solution for the Horton–Rogers–Lapwood problem.

11.4 FORCED CONVECTION

Nanofluid-cooled microchannel heat sinks were studied by Ghazvini and Shokouhmand (2009) and Ghazvini et al. (2009) using fin and porous media approaches, but we are aware of just half a dozen papers on forced convection in a porous medium with a nanofluid. Thermally developing forced convection of a nanofluid in a parallel-plate channel was studied numerically by Maghrebi et al. (2012), who employed the Buongiorno model with thermophoresis and Brownian motion. They found that the local Nusselt number is decreased when the Lewis number Le is increased and the Schmidt Sc number is increased, these parameters being defined by

$$Le = \frac{\mu}{\rho D_B}, \quad Sc = \frac{\mu}{D_B \phi_0} \tag{11.25}$$

where:

D_B is the Brownian diffusion coefficient
D_T is the thermophoretic diffusion coefficient
ϕ_0 is the particle fraction at the channel inlet
μ is the nanofluid viscosity
ρ is the nanofluid density

Armaghani, Chamkha et al. (2014) extended this study to include the effect of local thermal nonequilibrium (LTNE). Further numerical work with LTNE, first taking into account particle migration and then using a model in which the heat flux in each of the phases is considered, was carried out by Armaghani, Chamkha et al. (2014) and Armaghani, Maghrebi et al. (2014).

On the other hand, in his analytical study of flow in microchannels, Hung (2010) considered just the variation of thermal conductivity, viscosity, and heat capacity. A more general analytical study using the Buongiorno model was made by Nield and Kuznetsov (2014c). They examined flow in a Darcy porous medium occupying a parallel-plane channel with uniform heat flux on the boundaries. They found that the combined effect of Brownian motion and thermophoresis is to reduce the Nusselt number. The reduction increases as $N_A N_B / \varepsilon$ increases, where N_A and N_B are defined by

$$N_A = \frac{D_T}{D_B}, \quad N_B = \frac{\varepsilon (\rho c)_p \phi_0^*}{(\rho c)_f} \tag{11.26}$$

and so

$$\frac{N_A N_B}{\varepsilon} = \frac{D_T}{D_B} \frac{(\rho c)_p \phi_0^{\star}}{(\rho c)_f} \tag{11.27}$$

which is the product of a diffusivity ratio and a heat capacity ratio. Nield and Kuznetsov (2014c) noted that this reduction in heat transfer due to a modification of the temperature profile by Brownian motion and thermophoresis would oppose any increase due to the thermal conductivity of the nanofluid, which is higher than that of a regular fluid. The predicted reduction is not intuitive. In the case of natural convection (see below), the effect of thermophoresis and Brownian motion is generally to increase the heat transfer. The results of Nield and Kuznetsov (2014c) apply only to the case where the Péclet number based on the thermophoretic diffusivity is small compared with unity. It was pointed out by Nield and Kuznetsov (2014a, 2014b) that net throughflow produces an extra contribution to the nanoparticle flux and hence an additional term into the thermal energy equation (the last term in Equation 11.20).

By contrast, natural convection has been extensively studied.

11.5 NATURAL CONVECTION

11.5.1 Internal Natural Convection

The paradigmatic problem for natural convection is the Rayleigh–Bénard problem, the onset of instability in a horizontal layer of fluid uniformly heated from below. In the case of a porous medium, this is commonly called the Horton–Rogers–Lapwood problem.

The application of the Buongiorno model to the case where there is no solid matrix (in this case for convenience, we refer to a clear fluid), which is the regular Rayleigh–Bénard problem, was first investigated by Tzou (2008a, 2008b). The problem was reexamined by Nield and Kuznetsov (2010a, 2010b, 2011c) and modified to the case of a porous medium, using the Darcy model by Nield and Kuznetsov (2009b). They considered a problem analogous to the double diffusion in a binary fluid, a problem studied by Nield (1968), and accordingly they treated the case of impermeable boundaries held at constant temperatures and constant particle fraction. They recognized that it would be more realistic physically to impose zero particle flux on the boundaries (setting to zero the normal component of the vector $\mathbf{j}_p$ that appears on the left-hand side of Equation 11.6), but in order to make analytical progress, they found it necessary to freeze the basic profile of the particle fraction.

The analysis of Nield and Kuznetsov (2009b) has been extended in several ways. Kuznetsov and Nield (2010a, 2011a) included the effect of LTNE. Kuznetsov and Nield (2010c) employed the Brinkman model. Kuznetsov and Nield (2010d) studied double diffusion. Nield and Kuznetsov (2011b) examined the effect of vertical throughflow. A number of other authors have included additional effects, such as rotation, visco-elasticity, and bioconvection, and this work is surveyed in Section 9.7 in the book of Nield and Bejan (2013). An alternative model, incorporating the effects of conductivity and viscosity variation and with cross-diffusion also included, was examined by Nield and Kuznetsov (2012). The above studies involved bottom heating. The case of uniform volumetric heating was investigated by Nield and Kuznetsov (2013b). In this paper, zero particle-flux boundary conditions were employed. The Horton–Rogers–Lapwood problem, and the corresponding problem in a fluid without a solid matrix, has been revisited by Nield and Kuznetsov (2014d, 2014e). This time they treated the case of zero particle-flux boundary conditions. They showed that in this case oscillatory instability was ruled out. They obtained an approximate expression for the nonoscillatory instability boundary in the form:

$$\text{Ra} = 40 - \left(N_A + \frac{\text{Le}}{\varepsilon} \right) \text{Rn} \tag{11.28}$$

This boundary is attained with a dimensionless wave number $\alpha = 3.16$. In the above equation, ε is the porosity, Ra is the usual Rayleigh–Darcy number, Rn is a nanofluid Rayleigh number, Le is a Lewis number, and N_A is a modified diffusivity ratio, which are now defined as follows:

$$\text{Ra} = \frac{\rho g \beta K H (T_h^* - T_c^*)}{\mu \alpha_m} \tag{11.29}$$

$$\text{Rn} = \frac{(\rho_p - \rho)\phi_0^* g K H}{\mu \alpha_m} \tag{11.30}$$

$$\text{Le} = \frac{\alpha_m}{D_B} \tag{11.31}$$

$$N_A = \frac{D_T (T_h^* - T_c^*)}{D_B T_c^* \phi_0^*} \tag{11.32}$$

where:

T_h^* and T_c^* are the temperatures at the bottom and top boundaries, respectively

ϕ_0^* is a reference nanoparticle volume fraction

K is the permeability of the porous medium

H is the layer depth

α_m is the effective thermal diffusivity of the porous medium

g is the gravitational acceleration

11.5.2 External Natural Convection

The problem of boundary layer flow over a vertical plate held at a constant temperature (a classical problem associated with Pohlhausen and Kukien) was treated by Nield and Kuznetsov (2009a), Kuznetsov and Nield (2010b, 2011b), and (with a revised model, employing zero particle-flux boundary conditions) by Kuznetsov and Nield (2013). The corresponding problem of boundary layer flow in a porous medium over a vertical plate at a constant temperature (the Cheng–Minkowycz problem) was studied by Nield and Kuznetsov (2009a, 2011a) and (with the revised model) by Kuznetsov and Nield (2013).

A very large number of other authors have studied variations of this work. Papers involving a porous medium have been surveyed in Section 9.7 of Nield and Bejan (2013).

11.6 CONCLUSIONS

We surveyed the applications of nanofluid theory to forced and natural convection (and by implication for mixed convection). We tried to present the review from a unifying standpoint, which emphasizes the similarity between modeling convection in nanofluids clear of porous obstacles as well as in nanofluid-saturated porous media. We discussed the modeling of the effects associated with the smallness of nanoparticles, such as thermophoresis and Brownian motion, as well as modifications in the governing equations required for describing nanofluid convection in a porous medium. In nanofluid convection, the amount of theoretical and modeling work currently exceeds the amount of experimental validation of these models. We envision that in the future a large emphasis should be placed on experimental verification of existing theories and modeling results.

REFERENCES

Armaghani, T., Chamkha, A.J., Maghrebi, M.J., and Nazari, M., Numerical analysis of a nanofluid forced convection in a porous channel: A new heat flux model in LTNE condition. *J. Porous Media*,17, 637–646, 2014.

Armaghani, T., Maghrebi, M.J., Chamkha, A.J., and Nazari, M., Effects of particle migration on nanofluid forced convection heat transfer in a local thermal non-equilibrium porous channel. *J. Nanofluids* 3, 51–59, 2014.

Brinkman, H.C., The viscosity of concentrated suspensions and solutions. *J. Chem. Phys.* 20, 571–581, 1952.

Bruggeman, D.A.G., Berechnung verschiedener physikalischer konstanten von heterogenen substanzen. I. Dielektrizitätskonstahnten und leitfähigkeiten der mischkörper aus isotropen substanzen (Calculation of different physical constants of heterogeneous substances. I. Dielectric constants and conductivities of mixtures from isotropic substances). *Ann. Phys. Lpz.* 24, 636–679, 1935.

Buongiorno, J., Convective heat transfer in nanofluids. *ASME J. Heat Transf.* 128, 240–250, 2006.

Choi, S.U.S., Nanofluids: From vision to reality through research. *ASME J. Heat Transf.* 131, 033106, 2009.

Choi, S.K., Kim, S.O., Lee, T.H., and Dohee-Hahn, Computation of the natural convection of nanofluid in a square cavity with homogeneous and nonhomogeneous models. *Numer. Heat Tr. A* 65, 287–301, 2014.

Choi, S.U.S., Zhang, Z., and Keblinski, P., Nanofluids. In: *Encyclopedia of Nanoscience and Nanotechnology*, H. Nalwa, ed., American Scientific Publishers, New York, pp. 757–773, 2004.

Choi, S.U.S., Zhang, Z.G., Yu, W., Lockwood, F.E., and Grulke, E.A., Anomalous thermal conductivity enhancement in nanotube suspensions. *Appl. Phys. Lett.* 79, 2252–2254, 2001.

Das, S., Choi, S.U.S., Yu, W., and Pradeep, T., *Nanofluids Science and Technology*, Wiley, Hoboken, NJ, 2008.

Eastman, J.A., Choi, S.U.S., Li, S., Yu, W., and Thompson, L.J., Anomalously increased effective thermal conductivities of ethylene glycol-based nanofluids containing copper nanoparticles. *Appl. Phys. Lett.* 78, 718–720, 2001.

Ghazvini, M., Akhavan-Behabadi, M.A., and Esmaeili, M., The effect of viscous dissipation on laminar nanofluid flow in a microchannel heat sink. *IME J. Mech. Eng. Sci.* 263, 2697–2706, 2009.

Ghazvini, M., and Shokouhmand, H., Investigation of nanofluid-cooled microchannel heat sink using fin and porous media approaches. *Energ. Convers. Manag.* 50, 2373–2380, 2009.

Hung, Y.M., Analytical study on forced convection of nanofluids with viscous dissipation in microchannels. *Heat Transf. Eng.* 31, 1184–1192, 2010.

Khanafer, K., and Vafai, K., A critical analysis of thermophysical characteristics of nanofluids. *Int. J. Heat Mass Transf.* 54, 4410–4428, 2011.

Kuznetsov, A.V., and Nield, D.A., Effect of thermal non-equilibrium on the onset of convection in a porous medium layer saturated by a nanofluid. *Transport Porous Med.* 83, 425–436, 2010a.

Kuznetsov, A.V., and Nield, D.A., Natural convective boundary-layer flow of a nanofluid past a vertical plate. *Int. J. Therm. Sci.* 49, 243–247, 2010b.

Kuznetsov, A.V., and Nield, D.A., The onset of double-diffusive nanofluid convection in a layer of a saturated porous medium. *Transport Porous Med.* 85, 941–951, 2010c.

Kuznetsov, A.V., and Nield, D.A., Thermal instability in a porous medium layer saturated by a nanofluid: Brinkman model. *Transport Porous Med.* 81, 409–422, 2010d.

Kuznetsov, A.V., and Nield, D.A., The effect of local thermal nonequilibrium on the onset of convection in a porous medium layer saturated by a nanofluid: Brinkman model. *J. Porous Media* 14, 285–293, 2011a.

Kuznetsov, A.V., and Nield, D.A., Natural convective boundary-layer flow of a nanofluid past a vertical plate. *Int. J. Therm. Sci.* 50, 712–717, 2011b.

Kuznetsov, A.V., and Nield, D.A., The Cheng-Minkowycz problem for natural convective boundary layer flow in a porous medium saturated by a nanofluid: A revised model. *Int. J. Heat Mass Transf.* 65, 682–685, 2013.

Maghrebi, M.J., Nazari, M., and Armaghansi, T., Forced convection heat transfer of nanofluids in a porous channel. *Transport Porous Med.* 93, 401–413, 2012.

Masuda, H., Ebata, A., Teramae, K., and Hishinuma, N., Alteration of thermal conductivity and viscosity of liquid by dispersing ultra-fine particles. *Netsu Bussei* 7, 227–233, 1993.

Maxwell, J.C., *A Treatise on Electricity and Magnetism*, 2nd edn., Clarendon Press, Oxford, 1881.

Nield, D.A., Onset of thermohaline convection in a porous medium. *Water Resour. Res.* 11, 553–560, 1968.

Nield, D.A., and Bejan, A., *Convection in Porous Media*, 4th edn., Springer, New York, 2013.

Nield, D.A., and Kuznetsov, A.V., The Cheng-Minkowycz problem for natural convective boundary layer flow in a porous medium saturated by a nanofluid. *Int. J. Heat Mass Transf.* 52, 5792–5795, 2009a.

Nield, D.A., and Kuznetsov, A.V., Thermal instability in a porous medium layer saturated by a nanofluid. *Int. J. Heat Mass Transf.* 52, 5796–5801, 2009b.

Nield, D.A., and Kuznetsov, A.V., The effect of local thermal non-equilibrium on the onset of convection in a nanofluid. *ASME J. Heat Transf.* 132, 052405, 2010a.

Nield, D.A., and Kuznetsov, A.V., The onset of convection in a nanofluid layer. *Eur. J. Mech. B. Fluids* 29, 217–223, 2010b.

Nield, D.A., and Kuznetsov, A.V., The Cheng-Minkowycz problem for the double-diffusive natural convection boundary layer flow in a porous medium saturated by a nanofluid. *Int. J. Heat Mass Transf.* 54, 374–378, 2011a.

Nield, D.A., and Kuznetsov, A.V., Effect of vertical throughflow on thermal instability in a porous medium layer saturated by a nanofluid. *Transport Porous Med.* 87, 765–775, 2011b.

Nield, D.A., and Kuznetsov, A.V., The onset of double-diffusive convection in a nanofluid layer. *Int. J. Heat Fluid Fl.* 32, 771–776, 2011c.

Nield, D.A., and Kuznetsov, A.V., The onset of convection in a layer of a porous medium saturated by a nanofluid: Effects of conductivity and viscosity variation and cross-diffusion. *Transport Porous Med.* 92, 837–846, 2012.

Nield, D.A., and Kuznetsov, A.V., A note on the variation of nanofluid viscosity with temperature. *Int. Commun. Heat Mass* 41, 17–18, 2013a.

Nield, D.A., and Kuznetsov, A.V., Onset of convection with internal heating in a porous medium saturated by a nanofluid. *Transport Porous Med.* 99, 73–83, 2013b.

Nield, D.A., and Kuznetsov, A.V., Corrigendum to "Forced convection in a parallel-plate channel occupied by a nanofluid or a porous medium saturated by a nanofluid" [*International Journal of Heat and Mass Transfer* 70 (2014) 430–433]. 76, 534–534, 2014a.

Nield, D.A., and Kuznetsov, A.V., The effect of vertical throughflow on thermal instability in a porous medium layer saturated by a nanofluid: A revised model. *ASME J. Heat Mass Transf.* 2014b.

Nield, D.A., and Kuznetsov, A.V., Forced convection in a parallel-plate channel occupied by a nanofluid or a porous medium saturated by a nanofluid. *Int. J. Heat Mass Transf.* 70, 430–433 2014c.

Nield, D.A., and Kuznetsov, A.V., The onset of convection in a horizontal nanofluid layer of finite depth: A revised model. *Int. J. Heat Mass Transf.* 77, 915–918, 2014d.

Nield, D.A., and Kuznetsov, A.V., Thermal instability in a porous medium layer saturated by a nanofluid: A revised model. *Int. J. Heat Mass Transf.* 68, 211–214, 2014e.

Tzou, D.Y., Instability of nanofluids in natural convection. *ASME J. Heat Transf.* 130, 072401, 2008a.

Tzou, D.Y., Thermal instability of nanofluids in natural convection. *Int. J. Heat Mass Transf.* 51, 2967–2979, 2008b.

Vadász, P., Heat conduction in nanofluid suspensions. *ASME J. Heat Transf.* 128, 465–477, 2006.

Vadász, P., (ed.), Nanofluids suspensions and bi-composite media as derivatives of interface heat transfer modeling in porous media. In: *Emerging Topics in Heat and Mass Transfer in Porous Media*, Springer, New York, pp. 283–326, 2008.

CHAPTER 12

Convection and Instability Phenomena in Nanofluid-Saturated Porous Media

Antonio Barletta, Eugenia Rossi
di Schio, and Michele Celli

CONTENTS

12.1 INTRODUCTION

In the last decade, a wide research effort from the heat transfer community has been oriented to nanofluid-saturated porous media. The interest in this subject is mainly motivated by the technological value of an improved design of heat exchangers with high thermal performances and a relatively small size. With this aim in mind, an optimal fluid-to-solid heat exchange is devised through a metal foam, and the usual fluid is replaced by a nanofluid. Hence, when nanofluid-saturated porous media are studied, the porous medium is preferably considered as one with a high thermal conductivity. The argument supporting the use of a nanofluid, instead of an ordinary fluid, is grounded on its enhanced features relative to the high thermal conductivity and the high heat transfer rates. These specific features, widely studied in the existing literature, are likely to go beyond what one could expect from reasonings based on a mere evaluation of the volumetric concentration of (metal or carbon) nanoparticles or nanotubes. It is what has been called the "magic" power of nanoparticles by Rao (2010). What Rao calls magic is the not yet convincingly explained anomalous thermal conductivity, heat transfer rates, and viscosity experimentally observed for nanofluids.

This chapter is organized as follows: first a short review of the properties and modeling of nanofluids is provided. The momentum and energy transfer in saturated porous media are described. Then, the analysis of the literature on heat and fluid flow in nanofluid-saturated porous media is surveyed, by focusing on three main topics: forced convection, natural or mixed convection, and onset of thermal instability.

12.2 NANOFLUID MODELS FOR HEAT TRANSFER AND MASS DIFFUSION

A nanofluid is a suspension of particulate of the nanometer dimension in a base fluid. The nanofluids were born as answer to the need of more conductive fluids to employ in the heat transfer devices. A model that describes the heat and mass transfer behavior of these nanofluids is of key importance for an exhaustive analysis of the systems where nanofluids are involved. Two different approaches can be distinguished at this stage: homogeneous or nonhomogeneous. These two different approaches lead to different sets of governing equations: the homogeneous model is totally equivalent to the model employed for clear fluids except for a suitable rescaling of the governing parameters due to the presence of nanoparticles inside the base fluid (Magyari, 2011). Among all possible homogeneous models, an example is that defined by Tiwari and Das (2007). A nonhomogeneous model, however, adds one or more equations to the set of governing equations in order to describe the nanoparticle distribution and/or the nanoparticle velocity. Among the possible nonhomogeneous models, the most widely employed is the model by Buongiorno (2006), a model suitable for the heat transfer analysis and for the investigation of the nonhomogeneous distributions of nanoparticles.

12.2.1 Buongiorno's Model

Buongiorno's model responds to the need of investigating the spatial distribution of the nanoparticles and the thermal behavior of the nanofluid. Buongiorno (2006), in order to proper modeling the nanofluid heat transfer and the nanoparticle distribution, adds a dedicated diffusion equation for the nanoparticles to the set of governing equations, already constituted by a continuity equation, a momentum balance equation, and an energy balance equation. Moreover, inside the other governing balance equations, he assumes the thermophysical properties to be a combination of the properties of the base fluid and the nanoparticles. In order to develop a realistic two-component model for transport phenomena in nanofluids, Buongiorno compares the mechanisms allowing the nanoparticles to develop a slip velocity with respect to the base fluid. Seven are the leading slip mechanisms that the literature identifies: inertia, Brownian diffusion, thermophoresis, diffusiophoresis, Magnus effect, fluid drainage, and gravity settling. The analysis of Buongiorno concludes that the most relevant effects that have to be considered are only the thermophoresis and the Brownian diffusion, and he adds these effects to the energy balance equation and the diffusion equation of the nanoparticles. On assuming the Oberbeck–Boussinesq approximation, negligible viscous dissipation, negligible radiative heat transfer, relatively low concentration of nanoparticles, and local thermal equilibrium (LTE) between nanoparticles and base fluid, the set of governing equations of Buongiorno's model are as follows:

$$\nabla \cdot \mathbf{u} = 0$$

$$\rho_P \left(\frac{\partial \varphi}{\partial t} + \mathbf{u} \cdot \nabla \varphi \right) = -\nabla \cdot \mathbf{j}_p$$

$$\rho_{nf} \left(\frac{\partial \mathbf{u}}{\partial t} + \mathbf{u} \cdot \nabla \mathbf{u} \right) = -\nabla p + \mu_{nf} \nabla^2 \mathbf{u} + \rho_{nf} \beta_{nf} (T_{nf} - T_0) \mathbf{g} \qquad (12.1)$$

$$(\rho c)_{nf} \left(\frac{\partial T_{nf}}{\partial t} + \mathbf{u} \cdot \nabla T_{nf} \right) = k_{nf} \nabla^2 T_{nf} - c_p \mathbf{j}_p \cdot \nabla T_{nf}$$

where:

$\mathbf{j}_p$ is the nanoparticle mass flux given by

$$\mathbf{j}_p = -\rho_P \left(D_B \nabla \varphi + D_T \frac{\nabla T}{T} \right), \quad D_B = \frac{k_B T_0}{3 \pi \mu a}, \quad D_T = \frac{B \mu_f \varphi_0}{\rho_f} \qquad (12.2)$$

u represents the nanofluid velocity field

ρ is the density

φ is the volumetric fraction of nanoparticles present inside the base fluid

t is the time variable

p is the pressure field

μ is the dynamic viscosity

β is the thermal expansion coefficient

T_{nf} is the nanofluid temperature field

g is the acceleration due to gravity

c is the specify heat

k is the thermal conductivity

D_B is the Brownian diffusion coefficient

k_B is the Boltzmann constant

T_0 is the reference temperature

a is the diameter of the nanoparticles

D_T is the thermophoretic coefficient

B is a proportionality factor (see McNab and Meisen, 1973)

φ_0 is the average volume fraction of nanoparticles dispersed inside the nanofluid

The subscript p refers to the properties of the nanoparticles

The subscript nf refers to the properties of the nanofluid

The nanofluid properties have to be expressed as functions of the properties of the base fluid and the nanoparticles. The thermophysical properties such as the density, the specific heat, and the thermal expansion coefficient can be expressed analytically as combination of the components properties:

$$\rho_{nf} = (1-\varphi_0)\rho_f + \varphi_0\rho_p$$

$$(\rho c)_{nf} = (1-\varphi_0)(\rho c)_f + \varphi_0(\rho c)_p \qquad (12.3)$$

$$(\rho\beta)_{nf} = (1-\varphi_0)(\rho\beta)_f + \varphi_0(\rho\beta)_p$$

where f stands for fluid. On the contrary, the nanofluid dynamic viscosity μ_{nf} and the nanofluid thermal conductivity k_{nf} cannot be defined as simple combination of the known parameters, but empirical correlations are needed. The results of the investigations are sensitive to the choice of μ_{nf} and k_{nf}, and the expressions of μ_{nf} and k_{nf} available in the literature have nonnegligible differences. For instance, the Brinkman model for μ_{nf} and the Maxwell model for k_{nf} may be employed:

$$\mu_{nf} = \frac{\mu_f}{(1-\varphi_0)^{2.5}}$$

$$k_{nf} = k_f \left[\frac{k_p + 2k_f - 2\varphi_0(k_f - k_p)}{k_p + 2k_f + \varphi_0(k_f - k_p)} \right]$$

(12.4)

Maxwell's model for the thermal conductivity works for liquid–solid mixtures characterized by relatively big spherical particles, but it seems to be accurate enough also for nanoparticles with low values of φ_0. An alternative to Maxwell's model could be Bruggeman's (1935) model or Hamilton–Crosser's model (Hamilton and Crosser, 1962). When low values of φ_0 are assumed, Bruggeman's model gives the same values of Maxwell's model. Bruggeman's model fits better the experimental data with respect to Maxwell's model for larger values of φ_0. Hamilton–Crosser's model takes into account the possible nonspherical shape of the nanoparticles, and it is reliable only for $k_p/k_f > 100$.

Brinkman's model, however, is just one of the possible models that can be employed in order to define the dynamic viscosity of the nanofluid: Einstein (1956), Saitô (1950), and others. Each model is reliable for a particular kind of nanoparticle shape rather than for particular ranges of volumetric concentrations of nanoparticles.

A fairly important simplification can be performed concerning the nanoparticle flux by assuming that $\Delta T/T_0 \ll 1$, where ΔT refers to the reference temperature difference. This assumption allows one to simplify Equation 12.2 to

$$\mathbf{j}_p = -\rho_p \left(D_B \nabla \varphi + \frac{D_T \Delta T}{T_0} \nabla T_{nf} \right)$$

(12.5)

12.3 DESCRIPTION OF THE POROUS MEDIUM

A porous medium is a solid material with void inner structures entirely saturated by a moving fluid, or a nanofluid, such as, for instance, sand, pebbles, and metallic foam. Referring to Figure 12.1, one can consider a representative volume V to be chosen as small on a macroscopic scale even if it is large on the scale of the single structure present in the porous medium. If V_f is the void part of V, then let us call *porosity*, ε, the dimensionless quantity strictly smaller than unity:

$$\varepsilon = \frac{V_f}{V}$$

(12.6)

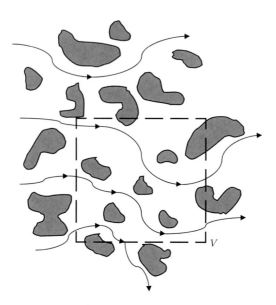

FIGURE 12.1   Representative elementary volume.

In order to study the convection in a porous medium, let us describe it as a continuum. One can define both the *intrinsic velocity*, as an average value of the local fluid velocity $\mathbf{u}^*$ performed in the void part V_f of the representative elementary volume V (REV):

$$\mathbf{U} = \frac{1}{V_f} \int_{V_f} \mathbf{u}^* \, dV \tag{12.7}$$

and the *seepage velocity* (also known as *Darcy's velocity*), defined as an average value performed in the REV V:

$$\mathbf{u} = \frac{1}{V} \int_{V} \mathbf{u}^* \, dV \tag{12.8}$$

Because $\mathbf{u}^* = 0$ in the part of V not included in V_f, one obtains the well-known *Dupuit–Forchheimer's relationship*:

$$\mathbf{u} = \varepsilon U \tag{12.9}$$

The mathematical model used for the description of a fluid-saturated porous medium is given by local mass and momentum equations. By following the procedure described in detail by Bejan (2004), one takes the average over a REV of the local mass balance for the fluid phase, under the conditions of incompressible flow, so that the mass balance equations of a fluid-saturated porous medium can be expressed as

$$\varepsilon \frac{\partial \rho}{\partial t} + \nabla \cdot (\rho \mathbf{u}) = 0 \tag{12.10}$$

The oldest, simplest, and most widely employed constitutive equation to express the local momentum balance is *Darcy's law* (ca. 1856):

$$\frac{\mu}{K} \mathbf{u} = -\nabla p + \mathbf{f} \tag{12.11}$$

where:
  K is the so-called *permeability*, a property of the system that depends on the number of pores per unit area present in a cross section transverse to the fluid flow, the shape of the pores, and their size
  μ is the dynamic viscosity of the fluid
  p is the fluid pressure
  $\mathbf{f}$ is the external body force per unit volume applied to the fluid (in the simplest case, the gravitational body force $\rho \, \mathbf{g}$)

If the hypothesis of laminar fully developed flow in each pore cannot be applied, Darcy's law should be replaced by *Darcy–Forchheimer's model* (ca. 1901):

$$\frac{\mu}{K} \left(1 + \frac{\rho c_f \sqrt{K}}{\mu} |\mathbf{u}| \right) \mathbf{u} = -\nabla p + \mathbf{f} \tag{12.12}$$

where:
  $|\mathbf{u}|$ is the modulus of $\mathbf{u}$
  ρ is the fluid mass density
  c_f is a dimensionless property of the porous medium called *form-drag coefficient*, which depends on the porous material

Obviously, Darcy–Forchheimer's model includes Darcy's law as a special case, that is, in the limit $c_f \to 0$. On the contrary, this model allows one to consider a gradual transition toward a hydraulic regime where acting forces are proportional to the square of the fluid velocity in each pore, and the hydraulic regime for the fluid flow inside the pores is reached whenever $\rho c_f |\mathbf{u}| \sqrt{K} / \mu \gg 1$. A widely accepted criterion to establish whether Darcy's law or Darcy–Forchheimer's model has to be employed is formulated by means of the permeability-based Reynolds number:

$$\mathrm{Re}_K = \frac{\rho|\mathbf{u}|\sqrt{K}}{\mu} \qquad (12.13)$$

According to this definition, Darcy's law gradually loses its validity when $\mathrm{Re}_K \sim 10^2$ or greater, and a clever way to apply the criterion is to take $|\mathbf{u}|$ as the maximum value in the domain.

A common feature of Darcy's law and Forchheimer's extension of this law is that they refer to a tight-packed solid with a fluid flowing in very small pores. Indeed, this is a circumstance very far from a free flowing fluid. The boundary condition for the seepage velocity can be, for instance, impermeability ($\mathbf{u}\cdot\mathbf{n} = 0$, where $\mathbf{n}$ is the unit vector normal to the surface). However, one cannot allow also a no-slip condition on the same surface, as the problem would be overconditioned. This feature is similar to that arising in perfect clear fluids (Euler's equation). The impossibility of prescribing no-slip conditions at the boundary walls creates a sharp distinction between the Navier–Stokes fluid model and the models of fluid-saturated porous media based on either Darcy's law or Forchheimer's extension of this law.

A continuous transition from the momentum balance equation of a clear fluid (Navier–Stokes equation) to Darcy's law is provided by the so-called *Brinkman's model* (ca. 1948):

$$\frac{\mu}{K}\mathbf{u} - \mu'\nabla^2\mathbf{u} = -\nabla p + \mathbf{f} \qquad (12.14)$$

where the quantity μ' is called the *effective viscosity*: it depends on the fluid viscosity μ and the permeability K of the medium. A commonly employed correlation for the effective viscosity is *Einstein's formula* for dilute suspensions:

$$\mu' = \mu\left[1 + 2.5\left(1 - \varepsilon\right)\right] \qquad (12.15)$$

Brinkman's model allows one to prescribe no-slip wall conditions as for a Navier–Stokes clear fluid. If the porosity is equal to 1, one has a clear fluid and Equation 12.15 implies that $\mu' = \mu$. Moreover, if $\varepsilon = 1$, Equation 12.14 reduces to the Navier–Stokes equation without the inertial contribution (negligible acceleration), provided that the limit of infinite permeability is also taken ($K \to \infty$). On the contrary, in the limit of a very small permeability ($K \to 0$), Brinkman's model reduces to Darcy's law, Equation 12.11.

Referring to the flow in a plane parallel porous channel having distance $2L$ between the plates, as sketched in Figure 12.2, Figure 12.3

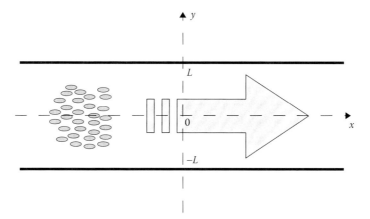

FIGURE 12.2   Flow in a porous channel.

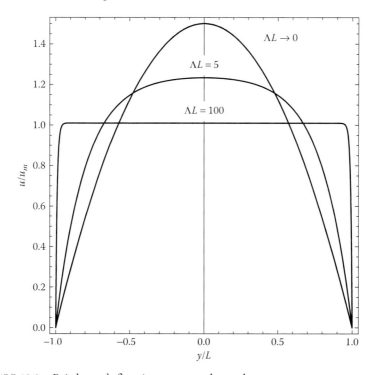

FIGURE 12.3   Brinkman's flow in a porous channel.

shows the dimensionless velocity profiles for Brinkman's flow in a cross section as a function of the parameter $\Lambda = \sqrt{\mu / (K\mu')}$. This figure reveals the main features of Brinkman's model, and in particular its bridging role to the clear fluid. In fact, in the limit $\Lambda \to \infty$, the Brinkman model tends to coincide with Darcy's law, whereas the limit $\Lambda \to 0$ is achieved when the permeability K tends to infinity, thus corresponding to the case of clear fluid.

An extended form of Equation 12.14 including inertial effects was introduced by Vafai and Tien (1982) and Hsu and Cheng (1990):

$$\rho\left[\frac{1}{\varepsilon}\frac{\partial\mathbf{u}}{\partial t}+\frac{1}{\varepsilon}(\mathbf{u}\cdot\nabla)\left(\frac{\mathbf{u}}{\varepsilon}\right)\right]=-\nabla p-\frac{\mu}{K}\mathbf{u}+\frac{\mu}{\varepsilon}\nabla^2\mathbf{u}-\frac{\rho c_f}{\sqrt{K}}|\mathbf{u}|\mathbf{u}+\mathbf{f} \quad (12.16)$$

An extensive treatment of Darcy's law and its extensions as well as a detailed description of the local volume-averaging procedure can be found in the books by Nield and Bejan (2013) and Kaviany (1995).

12.3.1 Local Energy Balance

By taking the average over a REV of the local energy balance for the solid and fluid phases, respectively, under the conditions of incompressible flow, and by denoting the local temperatures of the solid and fluid phases as T_s and T_f, one obtains

$$\varepsilon\rho c\frac{\partial T_f}{\partial t}+\rho c\mathbf{u}\cdot\nabla T_f=\varepsilon\nabla\cdot(k\nabla T_f)+h(T_s-T_f)+q_{gf}+\Phi \quad (12.17)$$

$$(1-\varepsilon)\rho_s c_s\frac{\partial T_s}{\partial t}=(1-\varepsilon)\nabla\cdot(k_s\nabla T_s)+h(T_f-T_s)+q_{gs} \quad (12.18)$$

where q_{gs} and q_{gf} are the power per unit volume generated within the solid phase and the power per unit volume generated within the fluid phase, respectively. These source terms are due, for instance, to Joule heating or chemical reactions. In Equations 12.17 and 12.18, the properties ρ, c, and k refer to the fluid, whereas ρ_s, c_s, and k_s refer to the solid matrix. The two equations are coupled by the volumetric interphase heat transfer coefficient h, and they represent the most utilized local thermal nonequilibrium (LTNE) model, nowadays expressed according to the form proposed by Nield and Bejan (2013).

The last term on the right-hand side of Equation 12.17, Φ, is the power per unit volume generated by viscous dissipation, and its expression is specific to the momentum balance model employed. As pointed out by Nield (2007), the general rule to evaluate Φ is

$$\Phi = F_d \cdot \mathbf{u} \quad (12.19)$$

where F_d is the drag force:

$$F_d = -\nabla p + \mathbf{f} \quad (12.20)$$

The drag force has an expression that depends on the model adopted:

$$\text{Darcy's law} \rightarrow F_d = \frac{\mu}{K}\mathbf{u} \tag{12.21}$$

$$\text{Darcy–Forchheimer's model} \rightarrow F_d = \frac{\mu}{K}\left(1 + \frac{\rho c_f \sqrt{K}}{\mu}|\mathbf{u}|\right)\mathbf{u} \tag{12.22}$$

$$\text{Brinkman's model} \rightarrow F_d = \frac{\mu}{K}\mathbf{u} - \mu'\nabla^2\mathbf{u} \tag{12.23}$$

There are, however, some controversies relative to Nield's rule expressed by Equation 12.19, especially with reference to its application in the case of Brinkman's model.

If the local average temperature of the solid phase coincides with the local average temperature of the fluid phase, $T_s = T_f = T$ (*LTE hypothesis*), the sum of Equations 12.17 and 12.18 yields

$$\rho c\left(\sigma\frac{\partial T}{\partial t} + \mathbf{u}\cdot\nabla T\right) = k'\nabla^2 T + q_g + \Phi \tag{12.24}$$

where σ is the *heat capacity ratio* defined as

$$\sigma = \frac{\varepsilon\rho c + (1-\varepsilon)\rho_s c_s}{\rho c} \tag{12.25}$$

whereas k', assumed to be constant, is the effective thermal conductivity of the fluid-saturated porous medium, given by

$$k' = \varepsilon k + (1-\varepsilon)k_s \tag{12.26}$$

Equation 12.24 is the local energy balance for the fluid-saturated porous medium.

The expressions of heat capacity ratio σ and the effective thermal conductivity k' given by Equations 12.25 and 12.26 are widely employed. However, Nield (2002) points out that these expressions correspond to the case where the thermal resistances of the solid matrix and the liquid phase are taken in parallel, that is, assuming that the conduction phenomenon occurs in parallel between the two phases. Otherwise, if we consider a porous medium consisting of stacked parallel layers, this assumption loses its validity because in such a medium the conduction proceeds in parallel in the longitudinal direction and in series in the direction transverse to the layers, thus leading to an anisotropy of the medium. In particular, the effective conductivity of a series of stacked parallel layers of solid and fluid is given by the expression:

$$k'' = \frac{k\,k_s}{\varepsilon k_s + (1-\varepsilon)k} \tag{12.27}$$

Let us fix a system of Cartesian axes such that the x-axis corresponds to the longitudinal direction and the y-axis to the transverse direction to the layers. The LTE equation to be considered, in order to model the anisotropy, by neglecting the heat source and viscous dissipation terms, is given by

$$\rho c \left(\sigma \frac{\partial T}{\partial t} + \mathbf{u} \cdot \nabla T \right) = \frac{\partial}{\partial x}\left(k' \frac{\partial T}{\partial x} \right) + \frac{\partial}{\partial y}\left(k'' \frac{\partial T}{\partial y} \right) \tag{12.28}$$

where k' and k'' are given by Equations 12.26 and 12.27, respectively. In the LTNE case, one has to formulate two equations, valid for the solid and fluid phases such that, in the limit $T_f = T_s$, Equation 12.28 is obtained:

$$\varepsilon \rho c \frac{\partial T_f}{\partial t} + \rho c \mathbf{u} \cdot \nabla T_f = \varepsilon \left[\frac{\partial}{\partial x}\left(k \frac{\partial T_f}{\partial x} \right) + \frac{\partial}{\partial y}\left(k'' \frac{\partial T_f}{\partial y} \right) \right] + h(T_s - T_f) \tag{12.29}$$

$$(1-\varepsilon)\rho_s c_s \frac{\partial T_s}{\partial t} = (1-\varepsilon)\left[\frac{\partial}{\partial x}\left(k_s \frac{\partial T_s}{\partial x} \right) + \frac{\partial}{\partial y}\left(k'' \frac{\partial T_s}{\partial y} \right) \right] + h(T_f - T_s) \tag{12.30}$$

The above equations show that a close interplay between the two phases and their thermophysical properties arises and, in cases of strong anisotropy, the thermophysical properties of the two phases appear to be coupled in both balance equations.

Other LTNE models with two temperatures have been introduced in the literature. For instance, Alazmi and Vafai (2000) use the following equations for the steady energy balance:

$$\rho c \mathbf{u} \cdot \nabla T_f = \nabla \cdot (k_{f,\text{eff}} \nabla T_f) + h_{\text{sf}} a_{\text{sf}} (T_s - T_f) \tag{12.31}$$

$$\nabla \cdot (k_{s,\text{eff}} \nabla T_s) - h_{\text{sf}} a_{\text{sf}} (T_s - T_f) = 0 \tag{12.32}$$

where:

$$k_{f,\text{eff}} = \varepsilon k \tag{12.33}$$

$$k_{s,\text{eff}} = (1-\varepsilon)k_s \tag{12.34}$$

For the steady case with constant porosity ε, Equations 12.31 and 12.32 reduce to Equations 12.17 and 12.18, provided that heat source and viscous dissipation are negligible. The volumetric heat transfer coefficient h is replaced by the product $h_{sf}a_{sf}$, where h_{sf} is the usual heat transfer coefficient per unit area and a_{sf} is the solid–fluid interface area per unit volume. Reported correlations for h_{sf} and a_{sf} refer to a packed bed of spherical particles with diameter d_p:

$$h_{sf} = \frac{k_f(2+1.1\mathrm{Pr}^{1/3}\mathrm{Re}^{0.6})}{d_p}, \qquad a_{sf} = \frac{6(1-\varepsilon)}{d_p} \qquad (12.35)$$

The correlation for h_{sf} was obtained experimentally by Wakao et al. (1979) and the correlation for a_{sf} by Vafai and Sŏzen (1990). Concerning Equation 12.35, Pr is the Prandtl number and $\mathrm{Re} = \rho\, ud_p/\mu$ is the Reynolds number.

If a first kind or Dirichlet temperature condition must be prescribed on a boundary wall and the LTNE is assumed, this boundary condition can be written as $T_f = T_s = T_0$ on the boundary surface. This condition implies that the LTE is prescribed at the boundary wall.

However, in the case of a second kind or Neumann boundary condition, two alternatives are possible (see Amiri et al. [1995]). According to the first option, a prescribed heat flux is divided between the two phases of the porous medium depending on the porosity and the thermal conductivities. Then, another boundary condition has to be added to determine uniquely the solution: usually, an LTE condition on the boundary surface is invoked, that is, $T_f = T_s$. In the second approach, one considers equal wall heat fluxes in the two phases. The porous medium is ideally bounded by impermeable walls having a negligible thickness. However, real impermeable walls display a finite thickness and the uniform heat flux prescribed on the external side of the wall may not yield, at the interface with the porous medium, a heat flux uniform and equally divided between the two phases. For this reason, in practical cases, the second kind boundary conditions may differ.

In order to define the applicability of the two approaches, as a general rule, the first approach is reliable in most practical cases, whereas the second approach should be used only when the impermeable boundary wall has a very small thickness. Several possibilities concerning the conducting walls were compared by Alazmi and Vafai (2002).

The main features of the LTNE model applied to a Darcy–Graetz problem are reported in Figures 12.4 through 12.6. Reference is made to flow in a Darcy porous channel having a constant velocity u_0, as sketched in Figure 12.2, with uniform and equal boundary temperatures $T = T_w$

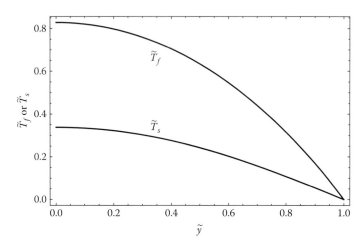

FIGURE 12.4   Dimensionless temperatures of the fluid and solid phases in the thermally developed region.

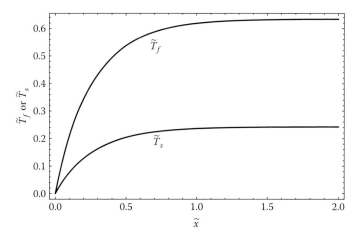

FIGURE 12.5   Dimensionless temperatures of the solid and fluid phases in the thermal entrance region.

prescribed after an adiabatic preparation region. The following dimensionless quantities are introduced:

$$\tilde{x} = \frac{x}{L\mathrm{Pe}}, \quad \tilde{y} = \frac{y}{L}, \quad \tilde{T}_{s,f} = \frac{k\,K}{\mu\,u_0^2\,L^2}(T_{s,f} - T_w),$$

$$\mathrm{Pe} = \frac{\rho\,c\,u_o\,L}{k\varepsilon}, \quad H = \frac{h\,L^2}{k\varepsilon}, \quad \gamma = \frac{k}{k_s}\frac{\varepsilon}{1-\varepsilon}$$

(12.36)

where Pe refers to the Péclet number.

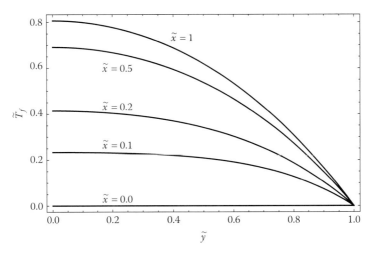

FIGURE 12.6 Dimensionless temperatures of the fluid phase in the thermal entrance region at different longitudinal positions.

In Figure 12.4, the fully developed temperature profiles $T_f(\tilde{y})$ and $T_{s0}(\tilde{y})$ versus $\tilde{y}$ for $H = 2$, $\gamma = 0.8$, and $\varepsilon = 0.4$ are plotted. Figures 12.5 and 12.6 show the thermal entrance region. In particular, in Figure 12.5, the dimensionless temperatures of the solid and fluid phases versus the dimensionless longitudinal coordinate $\tilde{x}$ for $\tilde{y} = 0.5$ and for $H = 2$, $\gamma = 0.8$, and $\varepsilon = 0.4$ are reported.

The longitudinal evolution of the fluid temperature field is presented in Figure 12.6, where the dimensionless temperature of the fluid phase versus $\tilde{y}$ at different dimensionless longitudinal positions is reported for $H = 2$, $\gamma = 0.8$ and $\varepsilon = 0.4$.

12.4 FORCED CONVECTION IN A NANOFLUID-SATURATED POROUS MEDIUM

The heat transfer performances of forced convection regime in a nanofluid-saturated porous medium is summarized by Nield and Kuznetsov (2014) and Hung (2011). The problem can be treated by means of Buongiorno's model, as Nield and Kuznetsov do, and also by means of the alternative approach of considering the porous medium as composed by an array of microchannels, as Hung does.

Nield and Kuznetsov investigate the temperature behavior of a thermally developed nanofluid flow inside a porous medium confined by two horizontal plates placed in $z = \pm L$. Darcy' law is assumed as a momentum balance equation and a uniform heat flux on the boundaries is imposed. As Darcy's law is employed, the velocity profile is constant and directed along the x-axis: $\mathbf{u} = (u_0,0,0)$. A vanishing flux of nanoparticles at the boundaries is assumed. An analytical solution of the problem is found by means of the ansatz that the variables can be expressed as $\{T_{nf}, \varphi\} = \{C_1 x + f(z), C_2 x + h(z)\}$. The approximate dimensionless analytical solution for the temperature and nanoparticle concentration fields that Nield and Kuznetsov obtain is

$$T_{nf} \approx \frac{x}{Pe} + \frac{1}{2}(z^2 - 1) + \frac{\beta^2}{2}(z^4 - 1)$$

$$\varphi \approx -N_A T_{nf}$$

(12.37)

where:

Pe $= L u_0 (\rho c)_f / k_{nf}$ is the Péclet number

k_{nf} is the thermal diffusivity of the nanofluid

$\beta^2 = N_A N_B / \varepsilon$, where $N_A = D_T / (D_B T_0)$, $N_B = \varepsilon (\rho c)_p \varphi_0 / (\rho c)_f$, and ε is the porous medium porosity

In order to analyze the thermal performances of the system, the definition of the Nusselt number is needed:

$$Nu = \frac{2 L q_w}{k_s (T_w - \overline{T}_{nf})}$$

(12.38)

where:

q_w is the wall heat flux

T_w is the wall temperature

$\overline{T}_{nf}$ is the bulk average temperature

The solution of Equation 12.37, combined with the definition of Equation 12.38, yields a Nusselt number given by

$$Nu = 6 \left(1 - \frac{2}{5} \beta^2 \right)$$

(12.39)

The combination of Brownian motion and thermophoresis has the effect of reducing the Nusselt number as the quantity β increases, as shown in Figure 12.7. Instead of modeling the porous medium by means of Darcy's law balances, an alternative approach may be found

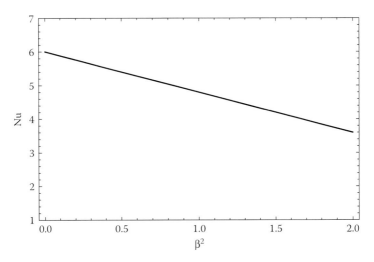

FIGURE 12.7   Nusselt number as a function of the parameter β.

by Hung (2011). The porous medium may, in fact, be considered as composed by an array of microchannels of radius r_0. The analysis of Hung is focused on a thermally developed flow inside a single cylindrical microchannel thermally stressed by a uniform wall heat flux q_w. The pressure gradient is assumed to be constant and directed along the x-axis, so that the nondimensional velocity profile obtained is the Hagen–Poiseuille profile:

$$\mathbf{u} = \left\{ \frac{2}{\mu_n}\left(1-r^2\right),0,0 \right\} \tag{12.40}$$

where $\mu_n = \mu_{nf}/\mu_f$ from Equation 12.4. Hung chooses to employ both Brinkman's model for μ_{nf} and Maxwell's model for k_{nf}. The temperature profile over the microchannel is obtained by solving the energy balance equation for a nanofluid including the effect of viscous dissipation;

$$T_{nf} = -\frac{1}{8k_n\mu_n}\left[(\mu_n+16Br)r^4 - 4(\mu_n+8Br)r^2 + 3\mu_n + 16Br\right] \tag{12.41}$$

where the Brinkman number Br and k_n are defined as

$$Br = \frac{\mu_f u_m^2}{2q_w r_0}, \quad k_n = \frac{k_{nf}}{k_f} \tag{12.42}$$

where u_m is the average velocity over the microchannel.

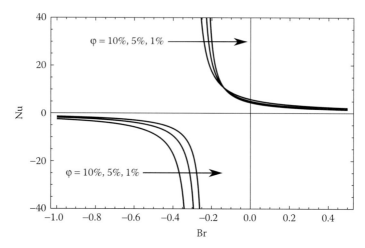

FIGURE 12.8   Nusselt number as a function of the Brinkman number for different values of nanoparticle concentration and $k_n = 10^3$.

In order to complete the analysis of the thermal behavior of the system, the Nusselt number is evaluated and given by

$$\text{Nu} = \frac{2\, r_0\, q_w}{k_f(T_w - \bar{T}_{\text{nf}})} = \frac{48 k_n \mu_n}{48 \text{Br} + 11 \mu_n} \tag{12.43}$$

where T_w is the temperature at r_0. The value of Nu becomes singular if $\text{Br} = -11\mu_n/48$. The behavior of the Nusselt number as a function of the Brinkman number is shown in Figure 12.8.

12.5 MIXED AND NATURAL CONVECTION

In this section, the papers available in the literature dealing with the effect of the buoyancy on convection of a nanofluid-saturated porous medium are presented.

The Cheng–Minkowycz problem of natural convection past a vertical plate, in a porous medium saturated by a nanofluid, is studied analytically by Nield and Kuznetsov (2009a). Darcy's model is employed and the description of the nanofluid incorporates the effects of Brownian motion and thermophoresis. A similarity solution is presented, depending on the Lewis number, the buoyancy ratio number, the Brownian motion number, and the thermophoresis number. The Nusselt number is evaluated and its dependence on these four parameters is investigated.

In the work of Gorla et al. (2011), a numerical study of the natural convective boundary layer flow over a vertical wedge embedded in a porous medium saturated with a nanofluid is presented. The velocity, temperature, and nanoparticle volume fraction profiles, as well as the friction factor, surface heat and mass transfer rates, are presented for parametric variations of the buoyancy ratio parameter, Brownian motion parameter, thermophoresis parameter, and Lewis number.

Ahmad and Pop (2010) study the steady mixed convection boundary layer flow past a vertical flat plate embedded in a porous medium filled with nanofluids, referring to different types of nanoparticles. The authors employ the homogeneous nanofluid model proposed by Tiwari and Das (2007).

The same nanofluid model is considered by Nazar et al. (2011) to investigate the steady mixed convection boundary layer flow from an isothermal horizontal circular cylinder embedded in a porous medium filled with a nanofluid. Both cases of a heated and a cooled cylinder are studied numerically, by means of an implicit finite-difference scheme. Three different types of nanoparticles are considered, and it is found that for each particular nanoparticle, as the nanoparticle volume fraction increases, the magnitude of the skin friction coefficient decreases, and this leads to an increase in the value of the mixed convection parameter, which first produces no separation. Moreover, it is also found that heating the cylinder delays separation of the boundary layer, and if the cylinder is hot enough, then it is suppressed completely. On the contrary, cooling the cylinder brings the boundary layer separation point nearer to the lower stagnation point, and for a sufficiently cold cylinder, there will not be a boundary layer on the cylinder.

The effect of the inclination angle to the vertical is studied by Rana, Bhargava et al. (2012). This topic deserves particular interest because of applications such as solar film collectors or geothermal energy storage. The authors investigate numerically the steady mixed convection boundary layer flow of an incompressible nanofluid along an inclined plate in a porous medium. The Nusselt number is found to decrease with increasing Brownian motion number or thermophoresis number, whereas it increases with increasing tilt angle. The effects of Lewis number, buoyancy ratio, and mixed convection parameter on temperature and concentration distributions are also examined in detail.

12.6 STABILITY ANALYSES

In the literature on nanofluid-saturated porous media, many authors pay attention to the conditions for the onset of convection.

With reference to a horizontal porous layer, the Horton–Rogers–Lapwood problem is revisited in the case of a nanofluid by Nield and Kuznetsov (2009b) by employing Darcy's model and by considering the model for the nanofluid behavior described by Buongiorno (2006).

Kuznetsov and Nield (2010c) extend the above-mentioned analysis to the case of Brinkman's model and consider the nanoparticle concentration at the lower boundary different from the upper boundary, that is, top-heavy and bottom-heavy arrangements. The paper shows that the critical Rayleigh number can be reduced or increased by a substantial amount, depending on whether the basic nanoparticle distribution is top heavy or bottom heavy.

A theory of double-diffusive nanofluid convection in porous media is developed by Kuznetsov and Nield (2010b) to investigate the onset of nanofluid convection in a horizontal layer for the case where the base fluid of the nanofluid is itself a binary fluid such as salty water. The model used for the nanofluid incorporates the effects of Brownian motion and thermophoresis, whereas Darcy's model is used for the porous medium. In addition, the thermal energy equations include regular diffusion and cross-diffusion terms.

In the work of Nield and Kuznetsov (2010), a combined conductive-convective–radiative problem, in which radiative heat transfer is treated as a diffusion process, is considered to investigate the onset of convection in a horizontal layer of a cellular porous material heated from below. The problem, as described by the authors, is relevant to cellular foams formed from plastics, ceramics, and metals. The effective thermal conductivity of the porous medium is assumed to be a function of temperature. It is shown that the variation of conductivity with temperature above that of the cold boundary leads to an increase in the critical Rayleigh number and an increase in the critical wave number. However, the critical Rayleigh number based on the conductivity at the mean temperature decreases with an increase in the thermal variation parameter if the radiative contribution to the effective conductivity is sufficiently large compared with the nonradiative component.

In the work of Nield and Kuznetsov (2011), the effect of vertical throughflow on the onset of convection in a horizontal layer is studied analytically. The model used for the nanofluid incorporates the effects of Brownian motion and thermophoresis. The authors show that in the absence of throughflow the critical Rayleigh number is independent of both the thermophoresis and Brownian motion parameters, whereas the critical frequency depends on the thermophoresis effect. On the contrary, if vertical throughflow is considered, there are additional terms in the critical stability criterion values that depend on both the thermophoresis and Brownian motion parameters. Because the Lewis number is large, the effect of each additional term is destabilizing.

Oscillatory convection in a horizontal layer of nanofluid in a Darcy's porous medium is studied by Chand and Rana (2012b).

The thermal instability in a rotating porous layer saturated by a nanofluid is studied by Agarwal et al. (2011) considering Darcy's model and for top-heavy and bottom-heavy suspensions, and by Chand and Rana (2012a) for a Darcy–Brinkman porous medium. In the work of Chand and Rana (2012a), a linear stability analysis based on normal modes is used to find a solution for the layer confined between two free boundaries.

All the above-mentioned papers assume that the LTE hypothesis holds. On the contrary, LTNE models have been employed by Kuznetsov and Nield (2010a) and Bhadauria and Agarwal (2011). Both papers consider a three-temperature LTNE model, assuming one-temperature field for the solid matrix, one for the base fluid, and one for the suspended nanoparticles. In particular, Kuznetsov and Nield (2010a) employ Darcy's model for the porous medium and consider in the model used for the nanofluid the effects of Brownian motion and thermophoresis. The analysis reveals that in some circumstances the effect of LTNE can be significant, but remains insignificant for a typical dilute nanofluid (with large Lewis number and small particle-to-fluid heat capacity ratio). Bhadauria and Agarwal (2011) point out that convection sets in earlier for LTNE compared to LTE. The authors consider a horizontal porous layer described through Brinkman's model, although nanofluid incorporates the effect of Brownian motion along with thermophoresis.

Moreover, few papers aim to investigate rheologic effects, assuming that the base fluid displays a non-Newtonian behavior. Nield (2011) points out that the Horton–Rogers–Lapwood problem becomes singular when a Newtonian fluid is replaced by a standard power-law fluid. It is shown how this singularity can be removed. When this is done, the nanofluid effects due to thermophoresis and Brownian motion become independent of the power-law index.

Sheu (2011) studies the onset of convection in a horizontal layer of a porous medium saturated with a viscoelastic nanofluid. More in detail, an Oldroyd-B type constitutive equation is used to describe the rheologic behavior of viscoelastic nanofluids, and both Brownian motion and thermophoresis are taken into account. The onset criterion for stationary and oscillatory convection is analytically derived. The authors show that oscillatory instability is possible in both bottom-heavy and top-heavy nanoparticle distributions, and that there is competition among the processes of thermophoresis, Brownian diffusion, and viscoelasticity, which cause the convection to be set in through oscillatory rather than stationary modes.

Other rheologies are investigated by Chand (2011), Chand and Rana (2012c), Rana and Takur (2012), and Rana, Takur et al. (2012).

Finally, the effect of buoyancy on the onset of convection in a nanofluid-saturated porous medium, considering variable gravity effects, is investigated by Chand et al. (2013).

REFERENCES

Agarwal, S., B. Bhadauria, and P. Siddheshwar (2011). Thermal instability of a nanofluid saturating a rotating anisotropic porous medium. *Special Topics and Reviews in Porous Media 2*, 53–64.

Ahmad, S. and I. Pop (2010). Mixed convection boundary layer flow from a vertical flat plate embedded in a porous medium filled with nanofluids. *International Communications in Heat and Mass Transfer 37*, 987–991.

Alazmi, B. and K. Vafai (2000). Analysis of variants within the porous media transport models. *Journal of Heat Transfer 122*, 303–326.

Alazmi, B. and K. Vafai (2002). Constant wall heat flux boundary conditions in porous media under local thermal non-equilibrium conditions. *International Journal of Heat and Mass Transfer 45*, 3071–3087.

Amiri, A., K. Vafai, and T. Kuzay (1995). Effects of boundary conditions on non-Darcian heat transfer through porous media and experimental comparisons. *Numerical Heat Transfer, Part A 27*, 651–664.

Bejan, A. (2004). *Convection Heat Transfer*, 3rd ed. Wiley, New York.

Bhadauria, B. and S. Agarwal (2011). Convective transport in a nanofluid saturated porous layer with thermal non equilibrium model. *Transport in Porous Media 88*, 107–131.

Bruggeman, D.A.G. (1935). Berechnung verschiedener physikalischer konstanten von heterogenen substanzen. i. dielektrizitätskonstanten und leitfhigkeiten der mischkörper aus isotropen substanzen. *Annalen der Physik 416*(7), 636–664.

Buongiorno, J. (2006). Convective transport in nanofluids. *ASME Journal of Heat Transfer 128*, 240–250.

Chand, R. (2011). Effect of suspended particles on thermal instability of Maxwell visco-elastic fluid with variable gravity in porous medium. *Antarctica Journal of Mathematics 8*, 487–497.

Chand, R. and G. Rana (2012a). On the onset of thermal convection in rotating nanofluid layer saturating a Darcy-Brinkman porous medium. *International Journal of Heat and Mass Transfer 55*, 5417–5424.

Chand, R. and G. Rana (2012b). Oscillating convection of nanofluid in porous medium. *Transport in Porous Media 95*, 269–284.

Chand, R. and G. Rana (2012c). Thermal instability of Rivlin-Ericksen elastico-viscous nanofluid saturated by a porous medium. *Journal of Fluid Engineering 134*, 121203.

Chand, R., G. Rana, and S. Kumar (2013). Variable gravity effects on thermal instability of nanofluid in anisotropic porous medium. *International Journal of Applied Mechanics and Engineering 18*, 631–642.

Einstein, A. (1956). *Investigations on the Theory of the Brownian Movement*. Dover Publications, New York.

Gorla, R., A. Chamkha, and A. Rashad (2011). Mixed convective boundary layer flow over a vertical wedge embedded in a porous medium saturated with a nanofluid: Natural convection dominated regime. *Nanoscale Research Letters 6*, 207–215.

Hamilton, R.L. and O.K. Crosser (1962). Thermal conductivity of heterogeneous two-component systems. *Industrial and Engineering Chemistry Fundamentals 1*(3), 187–191.

Hsu, C. and P. Cheng (1990). Thermal dispersion in a porous medium. *International Journal of Heat and Mass Transfer 33*, 1587–1597.

Hung, Y.M. (2011). Analytical study on forced convection of nanofluids with viscous dissipation in microchannels. *Heat Transfer Engineering 31*, 1184–1192.

Kaviany, M. (1995). *Principles of Heat Transfer in Porous Media*. Springer, New York.

Kuznetsov, A. and D. Nield (2010a). Effect of local thermal non-equilibrium on the onset of convection in a porous medium layer saturated by a nanofluid. *Transport in Porous Media 83*, 425–436.

Kuznetsov, A. and D. Nield (2010b). The onset of double-diffusive nanofluid convection in a layer of a saturated porous medium. *Transport in Porous Media 85*, 941–951.

Kuznetsov, A. and D. Nield (2010c). Thermal instability in a porous medium layer saturated by a nanofluid: Brinkman model. *Transport in Porous Media 81*, 409–422.

Magyari, E. (2011). Comment on the homogeneous nanofluid models applied to convective heat transfer problems. *Acta Mechanica 222*, 381–385.

McNab, G.S. and A. Meisen (1973). Thermophoresis in liquids. *Journal of Colloid and Interface Science 44*, 339–346.

Nazar, R., L. Tham, I. Pop, and D. Ingham (2011). Mixed convection boundary layer flow from a horizontal circular cylinder embedded in a porous medium filled with a nanofluid. *Transport in Porous Media 86*, 517–536.

Nield, D. (2002). A note on the modeling of local thermal non-equilibrium in a structured porous medium. *International Journal of Heat and Mass Transfer 45*, 4367–4368.

Nield, D. (2007). The modeling of viscous dissipation in a saturated porous medium. *ASME Journal of Heat Transfer 129*, 1459–1463.

Nield, D. (2011). A note on the onset of convection in a layer of a porous medium saturated by a non-Newtonian nanofluid of power-law type. *Transport in Porous Media 87*, 121–123.

Nield, D. and A. Bejan (2013). *Convection in Porous Media*, 4th ed. Springer, New York.

Nield, D. and A. Kuznetsov (2009a). The Cheng-Minkowycz problem for natural convective boundary-layer flow in a porous medium saturated by a nanofluid. *International Journal of Heat and Mass Transfer 52*, 5792–5795.

Nield, D. and A. Kuznetsov (2009b). Thermal instability in a porous medium layer saturated by a nanofluid. *International Journal of Heat and Mass Transfer 52*, 5796–5801.

Nield, D. and A. Kuznetsov (2010). The onset of convection in a layer of cellular porous material: Effect of temperature-dependent conductivity arising from radiative transfer. *Journal of Heat Transfer 132*, 074503.

Nield, D. and A. Kuznetsov (2011). The effect of vertical throughflow on thermal instability in a porous medium layer saturated by a nanofluid. *Transport in Porous Media 87*, 765–775.

Nield, D.A. and A.V. Kuznetsov (2014). Forced convection in a parallel-plate channel occupied by a nanofluid or a porous medium saturated by a nanofluid. *International Journal of Heat and Mass Transfer 70*, 430–433.

Rana, P., R. Bhargava, and O. Bég (2012). Numerical solution for mixed convection boundary layer flow of a nanofluid along an inclined plate embedded in a porous medium. *International Journal of Heat and Mass Transfer 64*, 2816–2832.

Rana, G. and R. Takur (2012). Effect of suspended particles on thermal convection in Rivlin-Ericksen elastico-viscous fluid in a Brinkman porous medium. *Journal of Mechanical Engineering and Sciences 2*, 162–171.

Rana, G., R. Takur, and K. Kumar (2012). Thermosolutal convection in compressible Walters (model b) fluid permeated with suspended particles in a Brinkman porous medium. *Journal of Computational Multiphase Flows 4*, 211–224.

Rao, Y. (2010). Nanofluids: Stability, phase diagram, rheology and applications. *Particuology 8*, 549–555.

Saitô, N. (1950). Concentration dependence of the viscosity of high polymer solutions. i. *Journal of the Physical Society of Japan 5*(1), 4–8.

Sheu, L. (2011). Thermal instability in a porous medium layer saturated with a viscoelastic nanofluid. *Transport in Porous Media 88*, 461–477.

Tiwari, R. and M. Das (2007). Heat transfer augmentation in a two-sided lid-driven differentially heated square cavity utilizing nanofluids. *International Journal of Heat and Mass Transfer 50*, 2002–2018.

Vafai, K. and M. Sözen (1990). Analysis of energy and momentum transport for fluid flow through a porous bed. *Journal of Heat Transfer 112*, 690–699.

Vafai, K. and E. Tien (1982). Boundary and inertia effects on convective mass transfer in porous media. *International Journal of Heat and Mass Transfer 25*, 1183–1190.

Wakao, N., S. Kaguei, and T. Funazkri (1979). Effect of fluid dispersion coefficients on particle-to-fluid heat transfer coefficients in packed beds: Correlation of Nusselt numbers. *Chemical Engineering Science 34*, 325–336.

CHAPTER 13

Nanofluid Two-Phase Flow and Heat Transfer

Lixin Cheng

CONTENTS

13.1 INTRODUCTION

Nanofluids are engineered colloids made of a base fluid and nanoparticles (1–100 nm). Common base fluids include water, organic liquids (e.g., ethylene, triethylene glycols, refrigerants), oils and lubricants, biofluids, polymeric solutions, and other common liquids. Materials commonly used as nanoparticles include chemically stable metals (e.g., gold, copper), metal oxides (e.g., alumina, silica, zirconia, titania), oxide ceramics (e.g., Al_2O_3, CuO), metal carbides (e.g., SiC), metal nitrides (e.g., aluminum nitride, SiN), carbon in various forms (e.g., diamond, graphite, carbon nanotubes [CNTs], fullerene), and functionalized nanoparticles. As a new research frontier, nanofluid two-phase flow and heat transfer have the potential to improve heat transfer and energy efficiency in many applications, such as microelectronics, power electronics, transportation, nuclear engineering, heat pipes, refrigeration, air-conditioning, and heat pump systems [1–3]. Nanofluids can have significantly better heat transfer characteristics than base fluids, with the following features:

1. They have larger thermal conductivities compared to conventional fluids.

2. They have a strongly nonlinear temperature dependency on the effective thermal conductivity.

3. They enhance or diminish heat transfer in single-phase flow.

4. They enhance or reduce nucleate pool boiling heat transfer.

5. They yield higher critical heat fluxes (CHFs) under pool boiling conditions.

More attention to the thermal conductivities of nanofluids than to the resulting heat transfer characteristics has been paid over the past few years. The use of nanofluids appears promising in several aspects of two-phase flow and heat transfer, but still faces several challenges: (1) the lack of agreement between experimental results from different research groups and (2) the lack of theoretical understanding of the underlying mechanisms with respect to nanoparticles. Thus, it is essential to review the state-of-the art review on nanofluid two-phase flow and heat transfer, including nucleate pool boiling, flow boiling, and CHF, and to identify the future research needs, which are the purposes of this chapter.

13.2 THERMAL PHYSICAL PROPERTIES OF NANOFLUIDS

Most studies on nanofluid thermal properties are focused on thermal conductivity and viscosity. However, two-phase flow and heat transfer characteristics also depend on other thermal physical properties, such as specific heat, latent heat, density, and surface tension, but they are less investigated. Thus, it is essential to conduct relevant research to provide a basis for nanofluid two-phase flow and heat transfer research.

By suspending nanoparticles in conventional heat transfer fluids, thermal conductivity of the base fluid can be significantly improved [4–11]. As shown in Table 13.1, solids have thermal conductivities that are orders of magnitude larger than those of conventional heat transfer fluids.

TABLE 13.1   Thermal Conductivities of Various Solids and Liquids at Room Temperature

Material	Form	Thermal Conductivity (W/mK)
Carbon	Nanotubes	1800–6600
	Diamond	2300
	Graphite	110–190
	Fullerene films	0.4
Metallic solids (pure)	Silver	429
	Copper	401
	Nickel	237
Nonmetallic solids	Silicon	148
Metallic liquids	Aluminum	40
	Sodium at 644 K	72.3
Others	Water	0.613
	Ethylene glycol	0.253
	Engine oil	0.145
	R134a	0.0811

Substantially increased thermal conductivities of nanofluids containing a small amount of metal, such as Cu or Fe, or metal oxide, such as SiO_2, Al_2O_3, WO_3, TiO_2, or CuO, have been reported. Experimental results show that these nanofluids have substantially higher thermal conductivities than the same liquids without nanoparticles. The nanoparticle thermal conductivity increases with the nanoparticle volume fraction. In general, metallic nanofluids show much more dramatic enhancements than metallic oxide nanofluids. Furthermore, nanofluid thermal conductivities are also strongly dependent on temperature. Particle size, shape, and volume concentration also influence the thermal conductivity of nanofluids [1–3]. The enhancement of thermal conductivities by nanoparticles may be explained as follows: (1) The suspended nanoparticles increase the surface area and the heat capacity of the fluid; (2) the suspended nanoparticles increase the effective (or apparent) thermal conductivity of the fluid; (3) the interactions and collisions among particles, fluid, and the flow passage surface are intensified; (4) the mixing and turbulence of the fluid are intensified; and (5) the dispersion of nanoparticles enable a more uniform temperature distribution in the fluid.

As a type of new structure material, CNTs have unusually high thermal conductivity up to 6000 W/mK compared to 0.08 W/mK of a liquid refrigerant [12–19]. Figure 13.1 shows the microstructures of single-walled

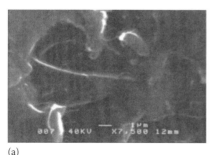
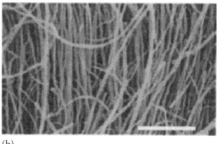

(a)   (b)

FIGURE 13.1   Scanning electron micrographs of CNT samples. (a) Single-walled nanotubes obtained by arc discharge. (Reprinted with permission from Bieruck, M.J. et al., *Appl. Phys. Lett.*, 80, 15, 2767, 2002, Copyright, American Institute of Physics.) (b) Multiwalled CNTs obtained by chemical vapor deposition. (Reprinted with permission from Choi, S.U.S. et al., *Appl. Phys. Lett.*, 79, 2252, 2001, Copyright, American Institute of Physics.)

and multiwalled CNTs. CNTs have a very high aspect ratio. CNTs from a highly entangled fiber network are not very mobile, as demonstrated by viscosity measurements, and thus, their effect on the thermal transport in fluid suspensions is expected to be similar to that of polymer composites. The first reported work on a single-walled CNT–polymer epoxy composite by Biercuk et al. [12] (Figure 13.1a) demonstrated a 70% increase in thermal conductivity at 40 K, rising to 125% at room temperature with 1 wt.% nanotube loading. They also observed that thermal conductivity increased with increasing temperature. Berber et al. [14] reported an unusually high thermal conductivity of CNTs, reaching 6600 W/mK at room temperature. Kim et al. [15] reported that the thermal conductivity of individual multiwalled nanotubes reached 3000 W/mK at room temperature. Dispersion of a small amount of nanotubes produces a remarkable change in thermal conductivity of the base fluid (up to 259% at 1 vol.% in Figure 13.2 [16]). The thermal conductivity of nanotube suspensions (solid circles) is 1 order of magnitude greater than predicted by the existing models (dotted lines). The measured thermal conductivity of nanotube suspensions is nonlinear with nanotube volume fraction, whereas theoretical predictions show a linear relationship (inset), which is thus significantly contradictive to what is expected.

Due to the absence of a theory for thermal conductivities of nanofluids, the existing models developed for conventional solid/liquid systems have been used to estimate the effective conductivities of nanofluids. For example, the Hamilton and Crosser [20] model has been applied to nanofluids.

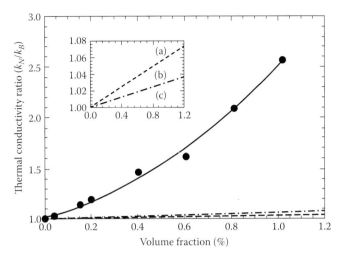

FIGURE 13.2 Effective thermal conductivity of CNT suspensions with respect to the pure fluid as a function of nanotube volume ratio and the prediction of Hamilton–Crosser equation (a), the Boonecaze and Brady equation (b), and Maxwell's equation (c). (Reprinted with permission from Choi, S.U.S. et al., *Appl. Phys. Lett.*, 79, 2252, 2001, Copyright, American Institute of Physics.)

However, measured thermal conductivities are substantially greater than theoretical predictions. Furthermore, a number of investigations have been conducted to identify the possible mechanisms that contribute to the enhanced effective thermal conductivity of nanoparticle suspensions [21]. The Brownian motion of the nanoparticles in these suspensions is one of the potential contributors to this enhancement. A number of theoretical studies have accounted for the higher thermal conductivity considering other various factors such as concentration, temperature, and nanoparticle size as well. However, the fundamental mechanisms are not yet well understood and no concrete conclusions have been reached that prove which is/are the controlling mechanisms. Further research efforts are needed to develop a suitable model to predict the thermal conductivity of nanofluids and should take into account the important molecular and nanomechanisms that are responsible for enhancing the thermal conductivity of nanofluids [1,2].

In general, the viscosity of nanofluids is much higher than that of their base fluids. The viscosity is a strong function of temperature and the volumetric concentration. Furthermore, a particle size effect seems to be important only for sufficiently high particle fractions. Kulkarni et al. [22] studied the rheological property of CuO–water nanofluids with volumetric concentrations of 5%–15% at temperatures from 278 to 323 K. Their nanofluids behave as time-independent, shear-thinning, pseudoplastic fluids. They also proposed a new correlation to predict the viscosity of these nanofluids as a function of temperature and volumetric concentration based on their own data. Kulkarni et al. [23] also conducted an experimental investigation on the rheological behavior of copper oxide nanoparticles dispersed in a 60:40 propylene glycol and water mixture, with particle volumetric concentrations from 0% to 6% at temperatures from –35°C to 50°C. Their results showed that these nanofluids exhibited a Newtonian fluid behavior. Nguyen et al. [24] studied the effect of temperature and particle volume concentration on the dynamic viscosity for water–Al_2O_3 nanofluids at temperatures from 22°C to 75°C. They found a hysteresis phenomenon on the viscosity of nanofluids. For a given particle volume concentration, there is a critical temperature beyond which nanofluid viscous behavior becomes drastically altered. If a fluid sample is heated beyond such a critical temperature, a striking increase of viscosity occurs. If it is cooled after being heated beyond this critical temperature, then a hysteresis phenomenon can occur as shown in Figure 13.3. Such an intriguing hysteresis phenomenon still remains poorly understood. Furthermore, the critical temperature was found to be strongly dependent on both particle fraction and size. So far, studies

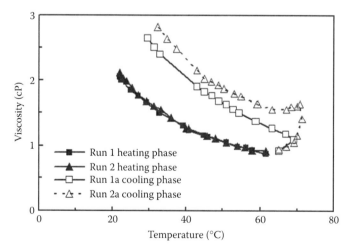

FIGURE 13.3　Hysteresis observed for water–Al_2O_3 (47 nm, 7% particle volume fraction). (Reprinted from *Int. J. Therm. Sci.*, 47, Nguyen, C.T. et al., Viscosity data for Al_2O_3–water nanofluid—Hysteresis: Is heat transfer enhancement using nanofluids reliable? 103, Copyright 2008, with permission from Elsevier Limited.)

on the viscosity of nanofluids have been still quite limited. Furthermore, there is at present no systematic theory or model to predict the viscosity of nanofluids. Further experimental study is needed to expand the database, although fundamental investigations on fluid/particle surface interactions should be made as a prerequisite to development of theoretical models.

Various mixing rules are generally applied to specific heat and density of nanofluids but not validated. Further measurements should be done to verify their applicability to nanofluids. Furthermore, surface tension is a very important parameter in boiling and two-phase flow heat transfer phenomena. However, such study is very limited [25–28].

13.3 POOL BOILING HEAT TRANSFER AND CHF OF NANOFLUIDS

Researchers on pool boiling heat transfer and CHF have extensively been conducted over the past few years. The measured pool boiling heat transfer results are conflicting with some studies showing a decrease or no change in nucleate boiling heat transfer with nanofluids, whereas some show an increase. However, all the studies showed an enhancement in CHF at pool boiling conditions. It should be noticed that nucleate pool boiling data are often measured with about ±10% errors and experimental data from independent studies on the same pure fluid often disagree by 30%–50% or more. Furthermore, an enhanced boiling surface typically increases heat transfer coefficients significantly and the deposition of nanoparticles on the heat transfer surfaces may be another factor but not yet well understood as pointed out by Cheng et al. [1,2].

13.3.1 Nucleate Pool Boiling Heat Transfer

Witharana [29] measured heat transfer coefficients of Au (unspecified size)–water, SiO_2 (30 nm)–water, and SiO_2–ethylene glycol nanofluids under pool boiling conditions. Their experimental results for Au–water nanofluids showed the nanofluid heat transfer coefficients were higher than those of pure water and also increased with increasing gold particle concentration. The enhancement of heat transfer was only about 11% at the intermediate heat fluxes (3 W/cm²) and 21% at a higher heat flux (4 W/cm²), whereas SiO_2–water and SiO_2–ethylene glycol nanofluids depicted a decrease in their heat transfer coefficients. These contradictory behaviors were not explained in their study. Li et al. [30] investigated pool boiling heat transfer characteristics of a CuO/water nanofluid and found that heat transfer deteriorated. They attributed this to the decrease in active nucleation sites caused by nanoparticle sedimentation on the boiling surface based on observations. Das et al. [31,32] carried out an experimental study on pool boiling characteristics of Al_2O_3–water nanofluids under atmospheric conditions and found that the nanoparticles degraded the boiling heat transfer performance. They speculated that the deterioration in boiling performance was not due to a change in fluid property but due to the change in surface wetting characteristics because of entrapment of nanoparticles in the surface cavities. You et al. [33] also reported deterioration in nucleate pool boiling heat transfer for Al_2O_3–water nanofluids. Kim et al. [34] found that heat transfer coefficients of Al_2O_3–water nanofluids remained unchanged compared to those of water. Bang and Chang [35] studied the boiling heat transfer using Al_2O_3–water nanofluids on a horizontal smooth surface. Their results showed lower heat transfer coefficients compared to pure water in natural convection as well as in nucleate pool boiling. Vassallo et al. [36] compared the heat transfer performance of pure water, silica nano-solutions, and silica micro-solutions under atmospheric pressure boiling on a 0.4 mm NiCr wire submerged in nanofluids. Their results showed no appreciable differences in nucleate boiling heat transfer. A thick (0.15–0.2 mm) silica coating was observed to form on their wire heater. They speculated that the roughness of the solid substrate might be responsible for the observed results. The additional thermal resistance of the silica could also have played a role. Prakash Narayan et al. [37] studied the effect of surface orientation on nucleate pool boiling of Al_2O_3–water nanofluids and found that heat transfer deteriorated for all test cases. A significant effect of surface orientation on heat transfer was found, where a horizontal orientation gave the best heat transfer and the heater surface at an inclination of 45° gave the worst heat transfer.

On the contrary, several studies have shown that nanoparticles can enhance nucleate pool boiling heat transfer. Tu et al. [38] found a significant enhancement in pool boiling heat transfer for an Al_2O_3–water nanofluid, up to 64% for a small fraction of nanoparticles. Wen and Ding [39] conducted experiments on nucleate pool boiling using γ-Al_2O_3–water nanofluids and reported that the presence of alumina in the nanofluid enhanced heat transfer significantly, by up to 40% for a 1.25 wt.% concentration of the nanoparticles. Wen et al. [40] conducted experiments on pool boiling of TiO_2–water nanofluids and showed that heat transfer increased by up to 50% at a concentration of 0.7 vol.%. Chopkar et al. [41] reported that ZrO_2–water can enhance nucleate boiling heat transfer at low particle volumetric concentrations, but heat transfer decreases with a further increase in concentration. They also mentioned that an addition of a surfactant to the nanofluids drastically decreased heat transfer, whereas surfactants often increase nucleate boiling heat transfer. Park and Jung [42] studied the effect of CNTs on nucleate boiling heat transfer of two refrigerants (R123 and R134a). In their studies, 1 vol.% of CNTs was added to the refrigerants and they found that CNTs increased nucleate boiling heat transfer coefficients for both refrigerants. Figure 13.4 shows their heat transfer coefficients of R134a with and without CNTs [42]. In particular, enhancements up to 36.6% were observed at low heat fluxes. With increasing heat flux, however, the enhancement diminished due to more vigorous bubble generation according to their visual observations. In addition, no deposit of the particles on their heat transfer surface was observed in their study.

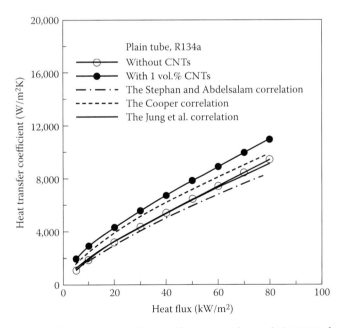

FIGURE 13.4 Boiling heat transfer coefficients with 1 vol.% CNTs for R134a. (Reprinted from *Energ. Buildings*, 39, Park, K.J. and Jung, D., Boiling heat transfer enhancement with carbon nanotubes for refrigerants used in building air-conditioning, 1061, Copyright 2007, with permission from Elsevier Limited.)

13.3.2 CHF Phenomena of Nanofluids in Pool Boiling

All the reported experimental studies mentioned on CHF have shown some enhancement of CHF. The maximum increase in CHF with a nanofluid can reach 3 times compared to that of the base fluid. You et al. [33] found drastic enhancement in the CHF for Al_2O_3–water nanofluids. Figure 13.5 shows their experimental results for CHF in pool boiling, where up to threefold enhancement was achieved. They also performed a visualization study and found that the average size of departing bubbles increased, but the frequency of departing bubble decreased as shown in Figure 13.6. They concluded that the unusual CHF enhancement of nanofluids could not be explained by any existing CHF model, that is, no pool boiling CHF model includes thermal conductivity or liquid viscosity and hence cannot explain this phenomenon, and the enhancement on liquid-to-vapor phase change was not related to the creased thermal conductivity. They also reported enhancement in CHF for both horizontal and vertical surface orientations with the nanofluids. Noting a change of the roughness of

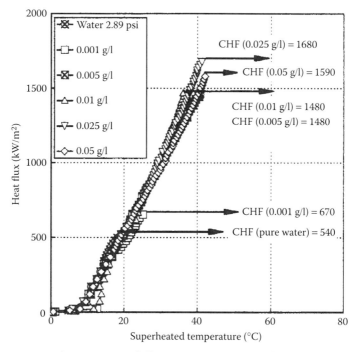

FIGURE 13.5 Boiling curves at different concentrations of Al_2O_3–water nanofluids. (Reprinted with permission from You, S.M. et al., *Appl. Phys. Lett.*, 83, 3374, 2003, Copyright 2003, American Institute of Physics.)

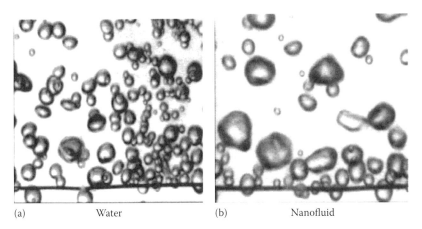

(a) Water   (b) Nanofluid

FIGURE 13.6 Sample pictures of bubbles growing on a heated wire (300 kW/m^2). (Reprinted with permission from You, S.M. et al., *Appl. Phys. Lett.*, 83, 3374, 2003, Copyright 2003, American Institute of Physics.)

the heater surface before and after their experiments, they hypothesized that the reason for the increase in CHF might be due to a surface coating formed on the heater with nanoparticles. They also observed that the size of departing bubbles increased and the bubble frequency decreased significantly in nanofluids compared to those in pure water. The orientation of the heater surface had a great effect on CHF. Kim et al. [43,44] also reported that a significant enhancement in CHF was achieved for Al_2O_3–water, ZrO_2–water, and SiO_2–water nanofluids at concentrations less than 0.1 vol.%. They noted that a porous layer of nanoparticles formed on the heat transfer surface and this layer apparently significantly improved the surface wettability, which may explain a plausible mechanism of CHF enhancement. Kim et al. [45,46] and Kim and Kim [47] found that CHF enhancement was achieved for Al_2O_3–water and TiO_2–water nanofluids at concentrations from 0.005 to 0.1 vol.% with a maximum increase of about 100%. They also found that nanoparticles coated their heat transfer surfaces and thus apparently enhanced the surface wettability. Hence, if the augmentation is the result of the coating, one would conclude that it is more effective to directly coat the surface rather than using nanoparticles in the fluid to do it. Tu et al. [38] also found that CHF enhancement was achieved for Al_2O_3–water nanofluids. Vassallo et al. [36] reported a marked increase in CHF for SiO_2–water nanoparticles with a maximum value of 3 times compared to that of pure water. Milanova and Kumar [48] found that CHF of ionic solutions with SiO_2 nanoparticles was enhanced up to 3 times compared to that of conventional fluids. Nanofluids in a strong electrolyte, that is, in a high ionic concentration, yielded higher CHF than buffer solutions. Xue et al. [49] conducted experiments on CNT–water nanofluids and found that CHF, transition boiling, and the minimum heat flux in film boiling were enhanced. Park et al. [50] also reported CHF enhancement of CNT nanofluids. As there is no deposition, a reasonable explanation of CHF mechanisms is still unclear.

13.4 FLOW BOILING AND CHF PHENOMENA OF NANOFLUIDS

Studies on flow boiling of nanofluids are limited and the available studies show contradictory experimental results. Xue et al. [25] studied the thermal performance of a CNT–water nanofluid in a closed two-phase thermosyphon and found that the nanofluid deteriorated the heat transfer performance. Liu et al. [52] reported that the boiling heat transfer in their thermosyphon was greatly enhanced using a Cu–water nanofluid in a miniature thermosyphon. Ma et al. [53] reported that the heat transport capacity of an oscillating heat pipe was significantly increased using a diamond nanoparticle–water nanofluid. Liu et al. [54] studied in a flat heat pipe evaporator and found that the heat transfer coefficient and CHFs of CuO–water nanofluids were enhanced by about 25% and 50%, respectively, at atmospheric pressure, and about 100% and 150%, respectively, at a pressure of 7.4 kPa, and also found that there was an optimum mass concentration for attaining a maximum heat transfer enhancement. Furthermore, Liu and Qiu [55] studied the boiling heat transfer and CHF of jet impingement with CuO–water nanofluids on a large flat surface and found that the boiling heat transfer was poorer, whereas CHF was enhanced compared to that of pure water. Lee and Mudawar [56] conducted flow boiling experiments in a microchannel heat sink using pure water and a 1% Al_2O_3 nanofluid solution. They suggested that nanofluids should not be used in microchannels due to the deposition of the nanoparticles on the channel surface. In fact, no heat transfer data were presented in their paper. Park et al. [57] studied the flow boiling of nanofluids in a horizontal plain tube with an inside diameter of 8 mm, and a noticeable decrease in the heat transfer coefficient was observed and a liquid film of high particle concentration may be formed on the tube surface.

In recent years, several researchers have shown the heat transfer and CHF enhancement in flow boiling of nanofluids [58–72]. However, understanding the fundamentals and mechanisms of CHF still needs to be investigated. Especially, no relevant research, such as the influence on flow patterns and their transitions, pressure drops, and CHF, has been available so far. Furthermore, investigation on the nanoparticle size effect on flow boiling is not well understood. Apparently, the corresponding mechanisms and theoretical modeling are not available either. Therefore, more experiments should be conducted and a new theoretical study is needed as well to explain and predict the results. Furthermore, two-phase flow and heat transfer in microscale channels are becoming important in many applications [71–75], but understanding the mechanisms of flow boiling of nanofluids in microscale channels is necessary but has not yet been investigated so far. Especially, one could also note that some nanofluids coat the heat transfer surface, and hence this may significantly influence the results. The surface effects need to be clearly separated from the fluidic effects in order to deduce the actual trends in the nanofluid data and thus build new models. No systematic knowledge on the nanoparticle size effects on two-phase flow patterns, pressure drop, heat transfer, and CHF in flow boiling of nanofluids is available and the coated surface effect on flow boiling in microscale channels. Thus, future research should be aimed at developing new fabrication technology for stable nanofluids at first, characterizing the nanofluids, modeling their physical properties, and conducting experimental and theoretical investigation on flow boiling of nanofluids in single microscale channels with various nanoparticle sizes. Surface coat effect will also be considered in the modeling aspect. Furthermore, based on the planned experimental results, new theoretical work is needed to achieve an advanced knowledge in modeling the properties of nanofluids and the evaporation of nanofluids in both macroscale and microscale channels.

13.5 BOILING HEAT TRANSFER AND CHF MECHANISMS OF NANOFLUIDS

Explanation of the mechanisms of deterioration or enhancement of nucleate pool boiling heat transfer with nanofluids has been tried by several researchers. These include the decreasing of active nucleation sites from nanoparticle sedimentation on the boiling surface, the change of wettability of the surface, and nanoparticle coatings on the surface. Furthermore, bubble dynamics has also been studied. Bang and Chang [35] and Bang et al. [51] conducted visualization on nucleate pool boiling and the liquid film separating a vapor bubble from a heated surface, which was used to explain the deterioration of nucleate boiling heat transfer. You et al. [33] observed that the average size of departing bubbles increased significantly and the bubble frequency decreased significantly in nanofluids compared to those in pure water. Tu et al. [38] reported that there were smaller bubbles with no obvious changes of bubble departure frequency compared to pure water. The different observed bubble behaviors thus apparently account for the deterioration or enhancement of nucleate boiling heat transfer. However, the various contradictory results make it difficult to explain the phenomena utilizing methods for pure fluids.

All the available studies seem to clearly conclude that the primary reason of CHF enhancement in pool boiling of nanofluids was the change of surface microstructure of the boiling surface due to a nanoparticle layer coating formed on the surface during pool boiling of nanofluids. If this is the case, it would be easier to use an enhanced surface with a porous coating from a practical viewpoint. However, further studies are still needed to clarify the heat transfer enhancement and CHF mechanism.

Examining widely quoted correlations for nucleate pool boiling heat transfer, it is not evident as to how a nanofluid will have an influence. For example, the Cooper [76] correlation (Equation 13.1) is based on the reduced pressure p_r, but nothing is known about the effect of nanofluids on the critical pressure or vapor pressure curve:

$$h_{nb} = 55 p_r^{0.12-0.2\log_{10} R_p} \left(-\log_{10} p_r\right)^{-0.55} M^{-0.5} q^{0.67} C \tag{13.1}$$

where:

h_{nb} is the nucleate boiling heat transfer coefficient
R_p is the surface roughness (μm)
M is the molecular weight
q is the heat flux
C is a constant, which is 1 for horizontal plane surfaces and 1.7 for horizontal copper tubes according to Cooper's original paper

However, comparison with experimental data suggests that better agreement is achieved if a value of 1 is used also for horizontal tubes. Note that the heat transfer coefficient is a fairly weak function of the surface roughness parameter R_p, which is seldom well known. A value of $R_p = 1$ is suggested for technically smooth surfaces. Thus, a nanocoating may have an effect but would be very small.

Taking the Forster and Zuber [77] correlation,

$$h_{nb} = 0.00122 \left(\frac{k_L^{0.79} c_{pL}^{0.45} \rho_L^{0.49}}{\sigma^{0.5} \mu_L^{0.29} h_{LV}^{0.24} \rho_V^{0.24}} \right) \Delta T_{sat}^{0.24} \Delta p_{sat}^{0.75} \tag{13.2}$$

it would predict an increase in heat transfer coefficients through the increase in liquid thermal conductivity and a decrease in heat transfer coefficients through the increase in liquid viscosity and surface tension. In Equation 13.2, k_L is the liquid thermal conductivity; c_{pL} is the liquid specific heat; ρ_L and ρ_V are the liquid and vapor densities, respectively; σ is the surface tension; μ_L is the liquid dynamic viscosity; h_{LV} is the latent heat; and ΔT_{sat} and Δp_{sat} are the superheated temperature and pressure, respectively.

Taking the Stephan and Abdelsalam [78] correlation for water derived by multiple regression,

$$h_{nb} = 0.0546 \left[\left(\frac{\rho_V}{\rho_L} \right)^{1/2} \left(\frac{q D_{bub}}{k_L T_{sat}} \right) \right]^{0.67} \left(\frac{h_{LV} D_{bub}^2}{a_L^2} \right)^{0.248} \left(1 - \frac{\rho_V}{\rho_L} \right)^{-4.33} \frac{k_L}{D_{bub}} \tag{13.3}$$

$$D_{bub} = 0.0146 \beta \left[\frac{2\sigma}{g(\rho_L - \rho_V)} \right]^{1/2} \tag{13.4}$$

where:

D_{bub} is the bubble departure diameter

The contact angle β is assigned a fixed value of 35° irrespective of the fluid

T_{sat} is the saturation temperature of the fluid (K)

a_L is the liquid thermal diffusivity

g is the gravity constant

It can be summarized that the dependency of heat transfer on the liquid thermal conductivity, density, and viscosity is as follows:

$$h_{nb} \propto k_L^{-0.166}$$

$$h_{nb} \propto \sigma^{0.083}$$

Thus, it would predict a decrease in heat transfer coefficients through the increase in liquid thermal conductivity and an increase in heat transfer coefficients through the increase in surface tension, although no liquid viscosity effect is concerned.

On the contrary, neither liquid thermal conductivity nor liquid viscosity is found in the CHF model of Lienhard and Dhir [79] for pool boiling:

$$q_{crit} = 0.149 h_{LV} \rho_V \left[\frac{\sigma g (\rho_L - \rho_V)}{\rho_V^2} \right]^{1/4} \tag{13.5}$$

where q_{crit} is the CHF. According to this correlation, CHF increases with increasing surface tension and liquid density. On the contrary, q_{crit} is only proportional to $\sigma^{1/4}$, so its effect is rather weak.

With respect to flow boiling heat transfer models, the nanofluid effect on the nucleate boiling contribution would be the same as a forementioned for nucleate pool boiling, utilizing the convective heat transfer correlation for annular flow of Kattan et al. [80]:

$$h_{cb} = 0.0133 \mathrm{Re}_L^{0.69} \mathrm{Pr}_L^{0.4} \frac{k_L}{\delta} \tag{13.6}$$

$$\mathrm{Re}_L = \frac{4 \rho_L u_L \delta}{\mu_L} \tag{13.7}$$

$$\mathrm{Pr}_L = \frac{c_{pL} \mu_L}{k_L} \tag{13.8}$$

where:
   h_{cb} is the convective heat transfer coefficient
   Re_L is the liquid film Reynolds number
   Pr_L is the liquid Prandtl number
   δ is the liquid film thickness

It can be summarized that the dependency of heat transfer on the liquid thermal conductivity, density, and viscosity are as follows:

$$h_{cb} \propto k_L^{0.6}$$

$$h_{cb} \propto \mu_L^{-0.29}$$

Thus, this predicts an increase in heat transfer coefficient through the increase in the liquid thermal conductivity but a decrease in heat transfer coefficient through the increase in the liquid viscosity, although no surface tension effect is concerned.

According to the analysis, it is clearly shown that the physical properties such as surface tension, liquid density, and viscosity have an effect on nucleate pool boiling heat transfer, convective flow boiling, and CHF in both pool and flow boiling processes. So far, the lack of knowledge of these physical properties of nanofluids has greatly limited an evaluation of the possible effect. This also poses a serious question: which physical properties should we use to reduce experimental data for nanofluids? The data reduction methods used might be one of the reasons why the available experimental are contradictory. Furthermore, nucleation density site, bubble dynamics, thin-film evaporation, dryout, liquid–vapor interfacial force, and boiling surface structures are the main factors that affect nucleate boiling heat transfer and CHF. For nanofluids, the size and type of nanoparticles could be important, but it is still unclear that they would affect the underlying mechanisms. Considering the controversies from the previous studies, the aggregation of nanofluids could be an important factor affecting the boiling performance, which needs to be clarified quantitatively. Furthermore, the mechanisms that explain the substantial increase in CHF still need to be verified.

13.6 FUTURE RESEARCH NEEDS

The available experimental data on nucleate pool boiling heat transfer of nanofluids are quite limited and many conflicts still exist between different studies on the heat transfer characteristics. The inconsistencies indicate that the understanding of the thermal behaviors of nanofluids related to nucleate pool boiling heat transfer is still poor. The results on CHF enhancement by nanofluids are consistent with each other in the literature; however, the mechanism responsible for this is not yet clear. Advanced physical models are required to explain and predict the influence of nanoparticles on nucleate pool boiling and CHF.

Explanations and new theories are also needed to take into account all important effects on the nanofluid flow boiling characteristics. No systematic knowledge of their effects has been achieved yet. Two-phase flow and flow boiling of other nanofluids should also be investigated in the future. Based on the experimental results, new theoretical work should be developed to achieve an advanced knowledge in modeling of the two-phase flow and heat transfer of nanofluids.

Flow boiling compared to nucleate pool boiling can greatly enhance the cooling performance of a microchannel heat sink by increasing the heat transfer coefficient. Furthermore, because flow boiling relies to a great degree on latent heat transfer, better temperature axial uniformity is realized in both the coolant and the wall compared to a single-phase heat sink. The question here is whether nanoparticles could further enhance an already superior heat transfer performance. Furthermore, it would be very valuable indeed to see if the increase in nucleate pool boiling CHF also occurs in flow boiling CHF, for which there may be numerous high heat dissipation applications in micro devices and the new generation of CPU chips, just to name a few. On the contrary, nanoparticles should only be used for processes whose exit quality is less than that of the onset of dryout, in order to minimize the deposition of the nanoparticles on the channel wall. Furthermore, other two-phase flow characteristics such as flow patterns and two-phase pressure drops should be investigated. The flow patterns should be related to the corresponding heat transfer and pressure drop characteristics. Flow pattern-based heat transfer and pressure drop prediction methods should also be developed based on a wide range of experimental data. Apparently, there are many challenges in nanofluid two-phase flow and heat transfer, which is a new frontier research involving multidisciplinary subjects. Much work is needed to achieve the fundamental knowledge and practical applications.

13.7 CONCLUDING REMARKS

There are still too many unresolved problems with respect to our knowledge of two-phase and heat transfer with nanofluids. Many controversies exist with numerous conflicting experimental results and trends of nanofluid two-phase flow and heat transfer. In general, nanofluids were found to increase, decrease, or have no effect on nucleate pool boiling but consistently to increase CHF. The following future research needs have been identified according to this review:

1. Physical properties such as liquid density and viscosity, surface tension, and specific heat have a significant effect on nanofluid two-phase flow and thermal physics such as nucleate pool boiling, convective flow boiling, and CHF in both pool and flow boiling processes. To properly present the experimental results and to understand the physical mechanisms related to the two-phase and heat

transfer phenomena, the nanofluid physical properties should be systematically investigated to set up a consistent database of physical properties.

2. Nucleate pool boiling heat transfer and its mechanisms should be further investigated. The inconsistencies between different studies should be clarified. Furthermore, the effect of nanoparticle size and type on heat transfer should be studied. The heat transfer mechanisms responsible for these trends should be identified and be able to explain why nucleate heat transfer may be enhanced or decreased. Data should also be segregated by fluids that deposit on the boiling surface and those that do not in order to prove if the fluid alone can enhance performance.

3. CHF phenomena in pool boiling should be systematically investigated and the mechanisms responsible for its delay to higher heat fluxes definitively identified. Furthermore, a new model for CHF should be developed based on the experimental nanofluid data and the CHF mechanisms.

4. More experiments on nanofluid two-phase flow and flow boiling should be conducted in both macro- and microchannels to evaluate the potential benefits of nanofluids. These should also include heat transfer performance, CHF, two-phase flow patterns, and pressure drop in various types of channels. Especially, the two-phase flow and heat transfer characteristics should be related to the corresponding flow patterns.

5. Nanofluid two-phase flow and heat transfer phenomena in microscale channels should be understood through systematic experimental and theoretical studies.

6. The sediment or coating of nanoparticles on the heat transfer surface is a big question that needs to be resolved. For example, if such a coating is beneficial, then it could be applied more easily using a coating process rather than nanofluid deposition. If such a nanoparticle layer has adverse effects, then ways to prevent it are needed or the correct nanofluids should be found.

7. Models and prediction methods that include nanoparticle effects should be developed based on accurate measurements and observations of two-phase flow and heat transfer with nanofluids.

REFERENCES

1. Cheng, L., and Liu, L., "Boiling and two phase flow phenomena of refrigerant-based nanofluids: Fundamentals applications and challenges." *International Journal of Refrigeration* 36, 2013: 421–446.

2. Cheng, L., Filho, E.P.B., and Thome, J.R., "Nanofluid two-phase flow and thermal physics: A new research frontier of nanotechnology and its challenges." *Journal of Nanoscience and Nanotechnology* 8, 2008: 3315–3332.

3. Cheng, L., "Nanofluid heat transfer technologies." *Recent Patents Engineering* 3 no. 1, 2009: 1–7.

4. Eastman, J.A., Choi, S.U.S., Li, S., Yu W., and Thompson, L.J., "Anormalously increased effective thermal conductivities of ethylene glycol-based nanofluids containing copper nanoparticles." *Applied Physics Letters* 78, 2001: 718–720.

5. Das, S.K., Putra, N., Thiesen, P., and Roetzel, W., "Temperature dependence of thermal conductivity enhancement for nanofluids." *Journal of Heat Transfer* 125, 2003: 567–574.

6. Lee, S., Choi, S.U.S., Li, S., and Eastman, J.A., "Measuring thermal conductivity of fluids containing oxide nanoparticles." *Journal of Heat Transfer* 121, 1999: 280–289.

7. Jang, S.P., and Choi, S.U.S., "Role of Brownian motion in the enhanced thermal conductivity of nanofluids." *Applied Physics Letters* 84, 2004: 219–246.

8. Hong, T.K., Yang, H.-S., and Choi, C.J., "Study of the enhanced thermal conductivity of Fe nanofluids." *Journal of Applied Physics* 97 no. 6, 2005: 1–4.

9. S.P. Jang, and S.U.S. Choi, "Effects of various parameters on nanofluid thermal conductivity." *Journal of Heat Transfer* 129, 2007: 617–623.

10. Murshed, S.M.S., Leong K.C., and Yang, C., "Enhanced thermal conductivity of TiO_2–water based nanofluids." *International Journal of Thermal Science* 44, 2005: 367–373.

11. Murshed, S.M.S., Leong K.C., and Yang, C., "Investigations of thermal conductivity and viscosity of nanofluids." *International Journal of Thermal Science* 47, 2008: 560–568.

12. Biercuk, M.J., Llaguno, M.C., Radosavljevic, M., Hyun, J.K., and Johnson, A.T., "Carbon nanotube composites for thermal management." *Applied Physics Letters* 80, 2002: 2767–2769.

13. Hone, J., Whitney M., and Zettl, A., "Thermal conductivity of single-walled carbon nanotubes." *Synthetic Metals* 103, 1999: 2498–2499.

14. Berber, S., Kwon, Y.K., and Tomanek, D., "Unusually high thermal conductivity of carbon nanotubes." *Physics Review Letter* 84, 2000: 4613–4616.

15. Kim, P., Shi, L., Majumdar, A., and Mceuen, P.L., "Thermal transport measurements of individual multiwalled nanotubes." *Physics Review Letter* 87, 2001: 215502.

16. Choi, S.U.S., Zhang, Z.G., Yu, W., Lockwood, F.E., and Grulke, E.A., "Anomalous thermal conductivity enhancement in nanotube suspensions, *Applied Physics Letters* 79, 2001: 2252–2254.

17. Xie, H., Lee, H., Youn, W., and Choi, M., "Nanofluids containing multiwalled carbon nanotubes and their enhanced thermal conductivities." *Journal of Applied Physics* 94, 2003: 4967–4971.

18. Liu, M.S., Lin, M.C.C., Huang, I.T., and Wang, C.C., "Enhancement of thermal conductivity with carbon nanotube for nanofluids." *International Communication in Heat and Mass Transfer* 32, 2005: 1202–1210.

19. Assael, M.J., Metaxa, I.N., Arvanitidis, J., Christofilos, D., and Lioutas, C., "Thermal conductivity enhancement in aqueous suspensions of carbon multi-walled and double-walled nanotubes in the presence of two different dispersants." *International Journal of Thermophysics* 26, 2005: 647–664.

20. Hamilton, R.L., and Crosser, O.K., "Thermal conductivity of heterogeneous two component systems." *Industrial Engineering Chemistry Fundamentals* 1 no. 3, 1962: 187–191.

21. Keblinski, P., Phillpot, S.R., Choi, S.U.S., and Eastman, J.A., "Mechanisms of heat flow in suspensions of nano-sized particles (nanofluids)." *International Journal of Heat and Mass Transfer* 45, 2002: 855–863.

22. Kulkarni, D.P., Das, D.K., and Patil, S.L., "Effect of temperature on rheological properties of copper oxide nanoparticles dispersed in propylene glycol and water mixture." *Journal of Nanoscience and Nanotechnology* 7, 2007: 2318–2322.

23. Kulkarni, D.P., Das, D.K., and Chukwu, G.A., "Temperature dependent rheological property of copper oxide nanoparticles suspension (nanofluid)." *Journal of Nanoscience and Nanotechnology* 6, 2006: 1150–1154.

24. Nguyen, C.T., Desgranges, F., Galanis, N. et al. "Viscosity data for Al_2O_3–water nanofluid—Hysteresis: Is heat transfer enhancement using nanofluids reliable?" *International Journal of Thermal Science* 47, 2008: 103–111.

25. Xue, H.S., Fan, J.R., Hu, Y.C., Hong, R.H., and Cen, K.F., "The interface effect of carbon nanotube suspension on thermal performance of a two-phase closed thermosyphon." *Journal of Applied Physics* 100, 2006: 104909.

26. Shin, D., and Banerjee, D., "Enhanced specific heat of silica nanofluid." *Journal of Heat Transfer* 133, 2010: 024510.

27. Shin, D., and Banerjee, D., "Enhancement of specific heat capacity of high-temperature silica-nanofluids synthesized in alkali chloride salt eutectics for solar thermal-energy storage applications." *International Journal of Heat and Mass Transfer* 54, 2011: 1064–1070.

28. Tiznobaik, H., and Shin, D., "Enhanced specific heat capacity of high-temperature molten salt-based nanofluids." *International Journal of Heat and Mass Transfer* 54, 2011: 1064–1070.

29. Witharana, S., Boiling of Refrigerants on Enhanced Surfaces and Boiling of Nanofluids, PhD thesis, The Royal Institute of Technology, Stockholm, Sweden, 2003.

30. Li, C.H., Wang, B.X., and Peng, X.F., "Experimental investigations on boiling of nano-particle suspensions." *2003 Boiling Heat Transfer Conference*, May 4–8, Montego Bay, Jamaica, NY, 2003.

31. Das, S.K., Putra, N., and Roetzel, W., "Pool boiling characteristics of nanofluids." *International Journal of Heat and Mass Transfer* 46, 2003: 851–862.

32. Das, S.K., Putra, N., and Roetzel, W., "Pool boiling characteristics of nanofluids on horizontal narrow tubes." *International Journal of Multiphase Flow* 29, 2003: 1237–1247.

33. You, S.M., Kim, J.H., and Kim, K.H., "Effect of nanoparticles on critical heat flux of water in pool boiling heat transfer." *Applied Physics Letters* 83, 2003: 3374–3376.

34. Kim, J.H., Kim, K.H., and You, S.M., "Pool boiling heat transfer in saturated nanofluids." *2004 ASME International Mechanical Engineering Congress & Exhibition*, November 13–30, ASME, Anaheim, CA, IMECE2004-61108, p. 621, 2004.

35. Bang, I.C., and Chang, S.H., "Boiling heat transfer performance and phenomena of Al_2O_3-water nano-fluids from a plain surface in a pool." *International Journal of Heat and Mass Transfer* 48, 2005: 2407–2419.

36. Vassallo, P., Kumar, R., and D'Amico, S., "Pool boiling heat transfer experiments in silica-water nano-fluids." *International Journal of Heat and Mass Transfer* 47, 2004: 407–411.

37. Prakash Narayan, G., Sateesh A.K.B.G., and Das, S.K., "Effect of surface orientation on pool boiling heat transfer of nanoparticle suspensions." *International Journal of Multiphase Flow* 34, 2008: 145–160.

38. Tu, J.P., Dinh, N., and Theofanous, T., "An experimental study of nanofluid boiling heat transfer." *Proceedings of the 6th International Symposium on Heat Transfer*, June 15–19, Beijing, China, 2004.

39. Wen, D., and Ding, Y., "Experimental investigation into pool boiling heat transfer of aqueous based γ-alumina nanofluids." *Journal of Nanoparticle Research* 7, 2005: 265–274.

40. Wen, D., Ding, Y., and Williams, R.A., "Pool boiling heat transfer of aqueous based TiO_2 nanofluids." *Journal of Enhanced Heat Transfer* 13, 2006: 231–244.

41. Chopkar, M., Das, A.K., Manna, I., and Das, P.K., "Pool boiling heat transfer characteristics of ZrO_2–water nanofluids from a flat surface in a pool." *Heat and Mass Transfer* 44, 2008: 999–1004.

42. Park, K.J., and Jung, D., "Boiling heat transfer enhancement with carbon nanotubes for refrigerants used in building air-conditioning." *Energy and Buildings* 39, 2007: 1061–1064.

43. Kim, S.J., Bang, I.C., Buongiorno, J., and Hu, L.W., "Surface wettability change during pool boiling of nanofluids and its effect on critical heat flux." *International Journal of Heat and Mass Transfer* 50, 2007: 4105–4116.

44. Kim, S.J., Bang, I.C., Buongiorno, J., and Hu, L.W., "Study of pool boiling and critical heat flux enhancement in nanofluids." *Bulletin of the Polish Academy of Sciences* 55 no. 2, 2007: 211–216.

45. Kim, H., Kim, J., and Kim, M.H., "Effect of nanoparticles on CHF enhancement in pool boiling of nano-fluids." *International Journal of Heat and Mass Transfer* 49, 2006: 5070–5074.

46. Kim, H.D., Kim, J., and Kim, M.H., "Experimental studies on CHF characteristics of nano-fluids at pool boiling." *International Journal of Multiphase Flow* 333, 2007: 691–706.

47. Kim, H.D., and Kim, M.H., "Critical heat flux behavior in pool boiling of Water-TiO$_2$ nano-fluids." *Applied Physics Letters* 91, 2007: 014104.

48. Milanova, D., and Kumar, R., "Role of ions in pool boiling heat transfer of pure and silica nanofluids." *Applied Physics Letters* 87, 2005: 233107.

49. Xue, H.S., Fan, J.R., Hong, R.H., and Hu, Y.C., "Characteristic boiling curve of carbon nanotube nanofluid as determined by the transient calorimeter technique." *Applied Physics Letters* 90, 2007: 184107.

50. Park, K.J., Jung, D., and Shim, S.E., "Nucleate boiling heat transfer in aqueous solutions with carbon nanotubes up to critical heat fluxes." *International Journal of Multiphase Flow* 35, 2009: 525–532.

51. Bang, I.C., Chang, S.H., and Baek, W.P., "Direct observation of a liquid film under a vapor environment in a pool boiling using a nanofluid." *Applied Physics Letters* 86, 2005: 134107.

52. Liu, Z.H., Yang, X.F., and Guo, G.L., "Effect of nanoparticles in nanofluids on thermal performance in a miniature thermosyphon." *Journal of Applied Physics* 102, 2007: 013526.

53. Ma, H.B., Choi, S.U.S., Tirumala, M. et al., "An experimental investigation of heat transport capability in a nanofluid oscillating heat pipe." *Journal of Heat Transfer* 128, 2006: 1213–1216.

54. Liu, Z.H., Xiong, J.G., and Bao, R., "Boiling heat transfer characteristics of nanofluids in a flat heat pipe evaporator with micro-grooved heating surface." *International Journal of Multiphase Flow* 33, 2007: 1284–1295.

55. Liu, Z.H., and Qiu, Y.H., "Boiling heat transfer characteristics of nanofluids jet impingement on a plate surface." *Heat and Mass Transfer* 43, 2007: 699–706.

56. Lee, J., and Mudawar, I., "Assessment of the effectiveness of nanofluids for single-phase and two-phase heat transfer in micro-channels." *International Journal of Heat and Mass Transfer* 50, 2007: 452–463.

57. Park, Y., Sommers, A., Liu, L. et al., "Nanoparticles to enhance evaporative heat transfer." *The 22nd International Congress of Refrigeration*, August 21–26, 2007, Beijing, China, CD-Room, Paper number: ICR07-B1-309.

58. Henderson, K., Park, Y.-G., Liu, L., and Jacobi, A.M., "Flow boiling heat transfer of R-134a-based nanofluids in a horizontal tube." *International Journal of Heat and Mass Transfer* 53, 2010: 944–951.

59. Kim, S.J., McKrell, T., Buongiorno, J., and Hu, L.W., "Experimental study of flow critical heat flux in alumina-water, zinc-oxide-water, and diamond-water nanofluids." *Journal of Heat Transfer* 134, 2009: 043204.

60. Kim, T.I., Chang, W.J., and Chang, S.H., "Flow boiling CHF enhancement using Al$_2$O$_3$ nanofluid and an Al$_2$O$_3$ nanoparticle deposited tube." *International Journal of Heat and Mass Transfer* 54, 2011: 2021–2025.

61. Vafaei, S., and Wen, D., "Critical heat flux (CHF) of subcooled flow boiling of alumina nanofluids in a horizontal microchannel." *Journal of Heat Transfer* 132, 2010: 102404.

62. Vafaei, S., and Wen, D., "Flow boiling heat transfer of alumina nanofluids in single microchannels and the roles of nanoparticles." *Journal of Nanoparticle Research* 13, 2011: 1063–1073.

63. Ahn, H.S., and Kim, M.H., "The effect of micro/nanoscale structures on CHF enhancement." *Nuclear Engineering Technology* 43, 2011: 205–216.

64. Ahn, H.S., Kang, S.H., and Kim, M.H., "Visualized effect of alumina nanoparticles surface deposition on water flow boiling heat transfer." *Experimental Thermal and Fluid Science* 37, 2012: 154–163.

65. Lee, S.W., Park, S.D., Kang, S. et al., "Critical heat flux enhancement in flow boiling of Al_2O_3 and SiC nanofluids under low pressure and low flow conditions." *Nuclear Engineering Technology* 44, 2012: 429–436.

66. Wu, J.M., and Zhao, L., "A review of nanofluid heat transfer and critical heat flux enhancement—Research gap to engineering application." *Progress in Nuclear Energy* 66, 2013: 13–24.

67. Ahn, H.S., Kim, H., Jo, H., Kang, S., Chang, W., and Kim, M.H., "Experimental study of critical heat flux enhancement during forced convective flow boiling of nanofluid on a short heated surface." *International Journal of Multiphase Flow* 36, 2010: 375–384.

68. Ahn, H.S., Kang, S., Jo, H., Kim, H., and Kim, M.H., "Visualization study of the effects of nanoparticles surface deposition on convective flow boiling CHF from a short heated wall." *International Journal of Multiphase Flow* 37, 2011: 215–228.

69. Lee, T., Lee, J.H., and Jeong, Y.H., "Flow boiling critical heat flux characteristics of magnetic nanofluid at atmospheric pressure and low mass flux conditions." *International Journal of Heat and Mass Transfer* 56, 2013: 101–106.

70. Dewitt, G., Makrell, T., Buongiorno, J., Hu, L.W., and Park, R.J., "Experimental study of critical heat flux with alumina-water nanofluids in downward-facing channels for in-vessel retention applications." *Nuclear Engineering Technology* 45, 2013: 335–346.

71. Lee, S.W., Kim, K.M., and Bang, I.C., "Study on flow boiling critical heat flux enhancement of graphene oxide/water nanofluid." *International Journal of Heat and Mass Transfer* 65, 2013: 348–356.

72. Wojtan, L., Revellin, R., and Thome, J.R., "Investigation of critical heat flux in single, uniformly heated microchannels." *Experimental Thermal and Fluid Science* 30, 2006: 765–774.

73. Cheng, L., "Fundamental issues of critical heat flux phenomena during flow boiling in microscale-channels and nucleate pool boiling in confined spaces." *Heat Transfer Engineering* 34, 2013: 1011–1043.

74. Cheng, L., and. Mewes, D., "Review of two-phase flow and flow boiling of mixtures in small and mini channels." *International Journal of Multiphase Flow* 32, 2006: 183–207.

75. Cheng, L., Mewes, D., and Luke, A., "Boiling phenomena with surfactants and polymeric additives: A state-of-the-art review." *International Journal of Heat and Mass Transfer* 50, 2007: 2744–2771.

76. Cooper, M.G., "Saturation nucleate pool boiling—A simple correlation." *International Chemical Engineering Symposium Series* 86, 1984: 785–792.

77. Forster, H.K., and Zuber, N., "Dynamics of vapor bubbles and boiling heat transfer." *AIChE Journal* 1, 1955: 531–535.

78. Stephan, K., and Abdelsalam, M., "Heat transfer correlation for natural convection boiling." *International Journal of Heat and Mass Transfer* 23, 1980: 73–87.
79. Lienhard, J.H., and Dhir, V.K., "Peak pool boiling heat-flux measurements on finite horizontal flat plates." *Journal of Heat Transfer* 95, 1973: 477–482.
80. Kattan, N., Thome, J.R., and Favrat, D., "Flow boiling in horizontal tubes: Part 3. Heat transfer model based on flow pattern." *Journal of Heat Transfer* 120, 1998: 156–165.

Heat Pipes and Thermosyphons Operated with Nanofluids

Matthias H. Buschmann

CONTENTS

14.1 INTRODUCTION

Nanofluids rouse the hope that, if employed in heat pipes and thermosyphons, they could significantly improve the thermal performance of these devices. Two recent overviews (Liu and Li, 2012; Buschmann, 2013) compiling numerous related experimental and modeling investigations support this hope. However, due to the fact that thermosyphons and heat pipes utilize phase change heat transfer, the relevant physical mechanisms might be completely different than in standard heat exchangers operated with nanofluids. The general aim of replacing classical working fluids by nanofluids is that the transferred amount of heat under otherwise unchanged conditions can be increased. An aim more specific for thermosyphons and heat pipes is to lower the thermal resistance of the device.

14.2 WORKING PRINCIPLES OF THERMOSYPHONS AND HEAT PIPES

Thermosyphons and heat pipes are widely used as cooling devices where space is limited or high heat rates have to be led away. Because their design is properly described in several textbooks (i.e., Faghri, 1995; Reay and Kew, 2006), here only some main features are briefly discussed. Figure 14.1 shows the principal sketches of a cylindrical closed two-phase thermosyphon and a cylindrical heat pipe. The majority of experimental and modeling work connected with nanofluids follows these designs. A third type—oscillating heat pipe—is described in the work of Buschmann (2013).

Thermosyphons rely on gravity and can only be operated in an upright or moderately inclined position. The working fluid located in the lower part of the device is heated until nucleate boiling generates sufficient vapor to rise and reach the condenser region. Here it is cooled below the dew point. A condensate film develops, which runs down the inner wall of the thermosyphon and rejoins with the working fluid.

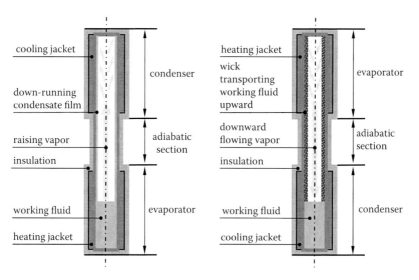

FIGURE 14.1 **(See color insert.)** Working principles of cylindrical closed two-phase thermosyphon (left) and cylindrical heat pipe (right).

Operation of heat pipes does not depend on orientation in space. Backflow of condensate is mostly organized via capillary force of a wick covering the inner wall of the device. Alternatives utilize centripetal, magnetic, or osmotic effects. The right plot of Figure 14.1 illustrates a heat pipe that is heated on its upper end. Working fluid is evaporated here and transported back to the lower part of the device where it condensates. Wicks can be manifold. They are mainly made of wire meshes or sintered material covering the inner wall of the heat pipe or microgrooves carved into this wall.

Besides the thermophysical properties of the working fluid, thermal performance of thermosyphons and heat pipes depends on a number of parameters, among them the filling ratio, the inner pressure, and the inclination angle. The filling ratio of the thermosyphon is defined as the amount of the evaporator region that is filled with working fluid. Filling of heat pipes is mostly sufficient when the wick is saturated.

14.3 NANOFLUIDS AS WORKING FLUIDS OF THERMOSYPHONS AND HEAT PIPES

We start with two simple questions: Why should nanofluids replace classical working fluids? and What are our expectations? To answer these questions, the thermal resistance R of the thermosyphon and heat pipe, which is defined as the ratio of the temperature difference between evaporator and condenser, and the transported heat, is discussed.

Figure 14.2 shows lumped models of the partial thermal resistances of a generic thermosyphon and heat pipe. Each partial thermal resistance is related to a distinct physical mechanism relevant for heat transfer. For the thermosyphon and the heat pipe, respectively, 13 and 15 serial and parallel partial thermal resistance components are identified. Not all of them are affected by the working fluid. This is especially true for the thermal resistances between the envelope and the heat source at evaporator R_{eo}, and between the envelope and the heat sink at condenser R_{co}, but also for the axial and radial thermal resistances, R_{pa} and R_{pr}, of the envelope itself. Only a few of the partial thermal resistances are affected definitively by a change of the working fluid. However, for the majority of the partial thermal resistances, the question of whether they are affected by nanoparticles or not is open and must be investigated experimentally.

Among the thermal resistances that are definitively affected by nanofluids is the inner heat transfer coefficient of the evaporator R_{ei}. The reason

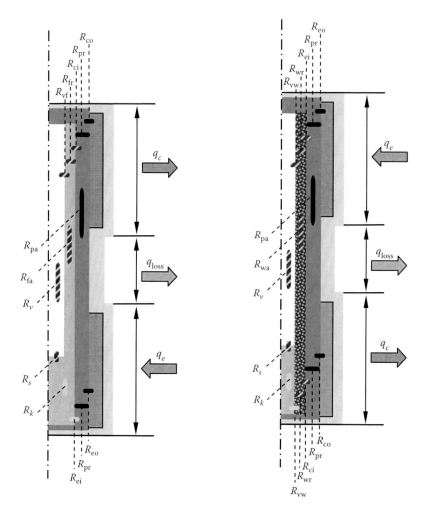

FIGURE 14.2 **(See color insert.)** Lumped model of thermal resistance of thermosyphon (left) and heat pipe (right). All partial thermal resistances R_i that are not affected by the working fluid are marked black, R_i that are possibly affected are marked yellow-black brindled, and R_i that are highly likely affected are marked yellow. The colors of design elements are as in Figure 14.1.

is mainly that nanoparticles interact with the evaporator wall. Either they form porous layers, as often reported by Grab et al. (2014), or they interact with vapor bubbles forming at the wall. In the first case, the structure of the evaporator surface, especially roughness, would be changed. In both cases, it is obvious that nanofluids cannot be described with an effective medium theory, but rather with a three-phase flow (solid: nanoparticles; liquid: base fluid; gaseous: bubbles of base fluid vapor) undergoing complex phase change and interaction processes. Additionally, specific thermophysical properties of a nanofluid following from chemical stabilizers (surfactants, dispersant agents, etc.) affect the local heat transfer at the evaporator inner wall. The same is true for heat conduction within the working fluid, which affects R_k. In heat pipes, R_k is only relevant if more working fluid is employed than needed for wick saturation.

A further phenomenon known from closed two-phase thermosyphons, the so-called geyser effect, related with superheating of the working liquid and unstable pressure oscillations (Tian et al., 2013), might be affected by nanoparticles directly. One may hypothesize that nanoparticles added to the working fluid could act as boiling stones and suppress geyser effects.

For the majority of the partial thermal resistances, it is open as to if they are indeed affected by nanoparticles dispersed in the working fluid. We start with the thermal resistance of the phase interface between the liquid working fluid and the vapor, R_s. The rate at which heat is transferred across the liquid/vapor interface depends on several thermophysical properties of both phases (Asselman and Green, 1973). Among them are vapor pressure and latent heat of vaporization enthalpy of the nanofluid. Unfortunately, these properties are rarely investigated and the few results are contradictory (Buschmann, 2013). Therefore, R_s must still be characterized as likely affected by nanoparticles suspended in the working liquid.

As already stated, nanofluids have to be understood as multiphase substances. Besides base fluid and solid nanoparticles, chemical stabilizers, if employed, have to be considered as an additional component. This is of special importance when it comes to evaporation. The question is whether all three components—base fluid, solid nanoparticles, and stabilizers—are part of the vapor. Base fluid will always be the main component. Whether stabilizers are vaporized depends on their specific thermophysical properties and has to be proven for each nanofluid separately. Due to the comparably low temperatures, nanoparticles cannot be evaporated. Therefore, it is rather a question of whether nanoparticles are transported by vapor.

The size of the smallest nanoparticles employed in thermosyphons so far is 4–5 nm (Huminic and Huminic, 2011) and in oscillating heat pipes, 2–3 nm (Cheng et al., 2010). By contrast, molecules of usual base fluids have sizes of several picometers (i.e., water about 0.2 nm). It seems to be rather implausible that these tiny molecules are able to rip out nanoparticles from the free surface of the working fluid and transport them over larger distances. Therefore, the thermal resistances of vapor R_v (both devices), the thermal resistance of the interface between the vapor and the condensate film R_{vf}, the thermal resistances in radial and axial directions of the condensate film R_{fr} and R_{fa}, respectively, and the thermal resistance between the condensate and the inner condenser wall R_{ci} in thermosyphons are rather not affected by nanoparticles. However, they might be influenced by evaporated stabilizers. If nanoparticles are found on the inner condenser wall, these depositions might not be caused by vapor transport but rather by violent boiling related to geyser effects (Buschmann and Franzke, 2014).

In heat pipes, further wick-related partial thermal resistances have to be considered. Nanoparticles might also be transported together with base fluid and stabilizer within the wick. In case the wick dries out, the latter two components would coat the wick surface and changes its capillarity (Do et al., 2010; Putra et al., 2012). The changing of capillarity affects especially radial and axial thermal resistances of wick, R_{wr} and R_{wa}, respectively, of which the latter is more important. Such a transport and deposition of nanoparticles in the wick would also affect thermal resistances between the pipe and the wick, R_{ei} and R_{ci}, and between the wick and the vapor phase, R_{vw}.

To summarize, the thermal resistances of the outer construction material of the devices are not affected by nanofluids. On the contrary, it is safe to say that for thermosyphons the thermal resistance between the evaporator wall and the working fluid and that of the working fluid itself are definitively influenced. All other partial thermal resistances are only affected by nanofluids if nanoparticles are transported either by vapor or by capillary forces inside the wick.

Based on the above analysis, the effects of nanofluids on the evaporator wall of thermosyphons will be discussed first. The thermal resistance of the working fluid is closely related to changed thermophysical properties. In Section 14.5, the wick of heat pipes is taken into focus. Finally, experimental and modeling results obtained by a variety of research teams are discussed.

14.4 EVAPORATOR SURFACE OF THERMOSYPHONS

At the inner evaporator surface of thermosyphons, nucleate boiling takes place. The phase of the working fluid changes from a liquid to a vapor state. Nanoparticles remain solid. Therefore, it is plausible that the nanoparticles stay behind on the evaporator surface and form a debris or porous layer. Boiling experiments carried out by Diao et al. (2013) on a microgrooved copper surface employing R141b-based Al_2O_3 nanofluid clearly revealed the effects of such deposits (Figure 14.3). Depending on operation pressure and

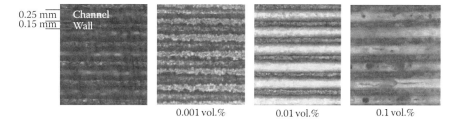

0.25 mm   Channel
0.15 mm   Wall

0.001 vol.%   0.01 vol.%   0.1 vol.%

FIGURE 14.3 **(See color insert.)** Evaporation experiments carried out on micro-grooved surfaces. Left photo shows fresh copper surface with microchannels. The other three images show the surface after boiling with R141b-based Al_2O_3 nanofluids. (Photos from Diao, Y.H. et al., *Int. J. Heat Mass Transf.*, 67, 183–193, 2013.)

superheating, an augmentation of the heat transfer coefficient was observed for the nanorefrigerant compared to the pure refrigerant. A maximum of about 100% enhancement was obtained for a concentration of 0.01 vol.%. As Figure 14.3 illustrates, the surface covering increased significantly with increasing nanoparticle concentration. Wen et al. (2011), boiling Al_2O_3–nanofluid (0.001 vol.%) on a smooth surface, showed also nearly a doubling of heat transfer rate compared to pure water. The same experiments carried out on a rough surface indicated no significant increase of heat transfer.

Nanoparticle coating of the lathed evaporator surface of a thermosyphon was reported by Buschmann and Franzke (2014) when employing water-based TiO_2 nanofluids (Figure 14.4). Surprisingly, the debris was limited to the valleys of lathing grooves. These authors found a minimal thermal resistance of their device for concentrations ranging between 0.2 and 0.3 vol.%.

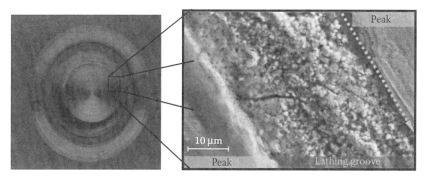

FIGURE 14.4 **(See color insert.)** Top view of evaporator surface of a thermosyphon. Left photo shows fresh copper surfaces with lathing grooves. Right photo shows scanning electron microscope image after experiment with 0.1 vol.% TiO_2 nanofluid, 20,000-fold enlarged. Orange arrows indicate the borders of a single lathing groove nearly completely filled with nanoparticles. (Data from Buschmann and Franzke 2014.)

Obviously, a complex interplay between original surface structure or roughness and nanoparticles leads to a new surface quality when nanofluids are employed as working fluids. This is in agreement with Harish et al. (2011), who argued that the surface interaction parameter—the ratio of the length scale of primary surface roughness to the nanoparticle size—is the relevant parameter, which indicates whether heat transfer is increased or not. Note that the *nanoparticle size* here denotes the actual size of the agglomerates, which primary nanoparticles form under the phase change processes.

Diao et al. (2013) repeated their experiments with pure R141b after coating the surfaces by boiling the nanofluids. The aim of these experiments was to prove whether the enhancement actually had to be attributed to the nanofluids or rather to the nanoparticle covered surface. Indeed, heat transfer coefficients were nearly the same for the uncoated surface with nanofluid R141b/Al_2O_3 and for Al_2O_3–nanoparticle-coated surface with pure base fluid R141b.

The experiments quoted above indicate that the number of nucleation sites is increased by nanoparticles settling down on the evaporator surfaces. This in turn augments boiling. If the concentration is too high, coating may become so thick that its thermal resistance suppresses heat transfer. In any case when nanoparticles are taken out of the working fluid and deposited at the evaporator surface enduringly, the concentration of the nanofluid is lowered. Therefore, modeling based on effective thermophysical properties of unspoiled suspension seems to be improper. Formation of nanoparticle deposits has to be seen in general as a time-dependent process. White (2010) experimentally investigated nanofluid boiling performance. The initially found increase of up to 25% had been significantly lowered after the layer of nanoparticle had been formed completely.

14.5 WICK OF HEAT PIPES

Besides others, the *wicking limit*—the ability of the wick to transport the liquid working fluid—is one of the limiting features of heat pipes. In general, capillary pumping pressure has to exceed the hydrostatic pressure in the liquid and the hydrodynamic pressure drops of vapor flow and liquid flow in the wick. A sufficient approximation of the capillary pressure Δp_c (VDI-Wärmeatlas, 2006) is

$$\Delta p_c = \frac{2\sigma\cos\theta}{R_{\text{effmin}}} \tag{14.1}$$

where:

σ is the surface tension of the liquid

θ is the contact angle between liquid and capillary

R_{effmin} denotes the minimal effective radius of the capillary

Tanvir and Qiao (2012) investigated aluminum, alumina, boron, and multiwalled carbon nanotubes (MCNTs) dispersed in deionized water and ethanol. In most cases, surface tension had increased significantly with increasing concentration and size of nanoparticles. This effect has to be attributed to increased van der Waals forces between nanoparticles at the nanofluid–gas interface.

The contact angle of nanofluids follows from a complex interplay of nanoparticles and surface. Kim et al. (2006) measured the contact angles of Al_2O_3–nanofluid droplets on differently treated surfaces and found increases as well as decreases depending on the surface quality. The remaining parameter to be discussed is R_{effmin}, which strongly depends on the geometry of the wick. If nanoparticles are deposited enduringly on the wick surface, the minimal effective radius is lowered and the capillary pressure increases. However, hydrodynamic loss of liquid flow increases with a decreasingly free cross section of the pores or microgrooves so that gains of capillary pressure might be compensated.

To summarize, increase of surface tension and lowering of minimal effective radius would increase capillary pressure accordingly. Therefore, employing nanofluids in wicked heat pipes seems to be favorable, but still has to be proven for any combination of wick and nanofluid experimentally.

Successful results of enhanced thermal performance of wicked heat pipes operated with different water and ethylene glycol-based nanofluids were presented by Putra et al. (2012). The employed wick was a single plaited stainless steel mesh with a wire diameter of 56.5 μm. Scanning electron microscopy (SEM) image and energy-dispersive X-ray spectroscopy analysis (Figure 14.4) indicated that after the experiments the wick was coated with nanoparticles that improved capillary structure. The best results were obtained with a water-based Al_2O_3 nanofluid with 5 vol.%. Operating the heat pipe with this nanofluid increased the evaporator heat transfer coefficient by a factor of about 4 ($q_e \approx 20\ kW/m^2$) compared with pure water.

Solomon et al. (2012) coated a screen mesh wick (wire diameter about 90 μm) by employing a simple immersing technology. After the sixth coating a pore size of about 4 μm had been created by deposited nanoparticles. According to Equation 14.1, the contact angle should not have changed because the materials of the screen mesh and the nanoparticles were both copper. Operating the heat pipe with the coated wick and pure water showed a significant reduction of the thermal resistance of the evaporator. Using the same heat pipe and operating it with a copper/water nanofluid (Shukla et al., 2012) indicated comparable reductions of the total thermal resistance. Similar to the results of Diao et al. (2013) for the heat transfer augmentation of microgrooved surfaces, enhancement of screen mesh wicked heat pipes seems rather due to the nanoparticles deposited on the surface than due to the nanofluid itself. Again, two things are clearly demonstrated: (1) a complex interplay between nanoparticles and surface is relevant for heat transfer deterioration or augmentation and (2) an effective medium theory describes these phenomena only in a very limited way.

14.6 EXPERIMENTAL RESULTS

The majority of the investigations related to nanoparticle-enforced working fluids employed in thermosyphons and heat pipes carried out worldwide are experimental. The overview by Buschmann (2013) counts 38 experimental studies addressing mostly characteristic operation parameters such as the amount of heat transferred, thermal resistance, and temperature distribution alongside the devices. Influences of the base fluid; particle size, shape, material, and suspension stabilization; nanoparticle concentration; and thermophysical properties relevant for phase change heat transfer were investigated. Although the inner diameter and inner height of thermosyphons and heat pipes ranged between 3 and 20 mm, the overall length of the devices was between 120 and 2000 mm. Oscillating heat pipes with inner capillary diameters as small as 0.76 mm and with up to 40 turns were investigated.

Investigations with respect to operation parameters, namely, filling ratio, inclination, and operation temperature, indicate that there are nearly no differences compared to the operation with the pure base fluid.

Virtually, all types of nanofluids have been tested in thermosyphons and heat pipes. The base fluid was mainly water, but also ethanol (C_2H_6O), acetone (C_3H_6O), the refrigerant R11 (trichlorofluoromethane, CCl_3F), and ethylene glycol mixtures were utilized. Nanoparticle material ranged from pure metals such as silver (Ag) and gold (Au) to oxides such as alumina (Al_2O_3), titania (TiO_2), and silica (SiO_2). Variations of carbon included diamond nanoparticles, CNTs (single- and multiwalled), and most recently graphene (Azizi et al., 2013). Actually, CNT and graphene are not particles that can be characterized as *nano*. CNTs have at least in one dimension, and graphene definitively in two dimensions, characteristic length scales that exceed the nanometer range significantly. The graphene pieces employed by Azizi et al. (2013) had a thickness of 4–20 nm and length scales of about 5–10 μm. Such outsized particles, but also large agglomerates of conventional nanoparticles, moving freely within the working fluid may provide an additional evaporation surface, a phenomenon probably not to be expected for well-stabilized nanofluids.

Only a few studies analyzed the influence of the base fluid on the thermal performance of gadgets. Putra et al. (2012) researched a screen mesh wicked heat pipe employing deionized water and ethylene glycol as base fluids. Thermal resistance of both fluids without nanoparticles indicated only small differences. When adding 5 vol.% alumina/titania nanoparticles, the decrease of the thermal resistance found for the base fluid deionized water was 78%/69%, but only 11%/2.4% for the base fluid ethylene glycol. The authors argued that these inequalities are due to the differently enhanced thermal conductivities of the working fluids. The increases amounted for the alumina/titania nanofluid to about 21.3%/14.0% for the base fluid deionized water and to about 13.2%/8.8% for the base fluid ethylene glycol. Analyzing just ethylene glycol experiments indicates that the reduction of the thermal resistance employing alumina nanoparticles was 4 times larger than for titania. On the contrary, the thermal conductivity increase of the alumina nanofluid was only about 50% higher than that of the titania nanofluid. Therefore, one may hypothesize that, besides increased thermal conductivity, other effects are responsible for the differently improved heat transfers.

Nanofluids are often chemically stabilized to keep their higher thermal conductivity under nonisothermal conditions stable for long periods of time. Some experiments analyzed the influence of surfactants on the thermal performance of thermosyphons (Paramatthanuwat et al., 2011) and screen mesh wicked heat pipes (Hajian et al., 2012). In the first case, experimenters changed the oleic acid surfactant concentration of a water-based silver nanofluids from 0.5 to 1.5 vol.%, keeping the nanoparticle concentration with 0.5 vol.% constant. The highest effectiveness of the device was found at a surfactant concentration of 1 vol.%. Hajian et al. (2012) employed silver nanoparticles suspended in water and a stainless steel screen wick. The contact angle between water and steel is about 50° and between water and silver about 90°. In cases where the nanofluid was not sufficiently stabilized and too much silver nanoparticles were deposited on the wick, the capillary pressure (Equation 14.1) went down due to the increased contact angle. In general, these experiments indicated that not only the amount of stabilization but also the nanoparticle material matter.

Kang et al. (2006) investigated the influence of the size of silver nanoparticles on the thermal performance of a heat pipe with microgrooves. One of the central outcomes of this study was that with nanoparticles with diameters of 35 nm a maximal reduction of the thermal resistance of 80% could be achieved. Nanoparticles of 10 nm diameter reduced thermal resistance by only 50%. The numerical simulations of the thermal performance of a flat micro heat pipe with rectangular grooved wick carried out by Do and Jang (2010) confirmed these findings. Based on their model, these authors explained the effect by a smaller increase of the hydraulic pressure drop of the porous layer when larger nanoparticles are employed. This increase was much less than the increase of the capillary pressure achieved. As a result, a net gain of force to transport vapor was obtained with larger nanoparticles. This in turn led to an enhancement of the evaporation rate.

Investigating large thermosyphons, Grab et al. (2014) found a reduction of thermal resistance for low heat flow rates provided at the evaporator up to 20% when gold nanofluids (5.2×10^{-4} vol.%) were used as working fluids. However, the reduction was 5%–10% less for gold nanoparticles having a mean size of 16 nm than for 66 nm gold nanoparticles. SEM images showed that the different sizes of the nanoparticles had changed the roughness structure of the evaporator surface differently.

Several experimental investigations have addressed the influence of nanoparticle concentration. The reason is simply that concentration is a key parameter to changing thermophysical properties of nanofluids. Grooved (Wei et al., 2005; Kang et al., 2006) and wicked heat pipes (Hajian et al., 2012) operated with silver nanofluids showed that thermal resistance can be lowered by up to 30% at an optimal concentration between 50 and 100 ppm. A screen mesh wicked heat pipe was investigated by Putra et al. (2012) employing several water and ethylene glycol oxide nanofluids. Increasing the concentration of Al_2O_3/H_2O nanofluid from 1 to 5 vol.% caused a lowering of the temperature in the entire heat pipe, except at the uppermost point of the condenser. Additionally, an enhancement of the evaporator heat transfer coefficient was observed when concentration was increased. This enhancement became the stronger, the higher the heat

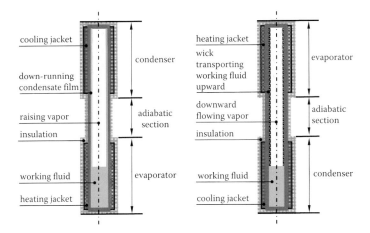

FIGURE 14.1   Working principles of cylindrical closed two-phase thermosyphon (left) and cylindrical heat pipe (right).

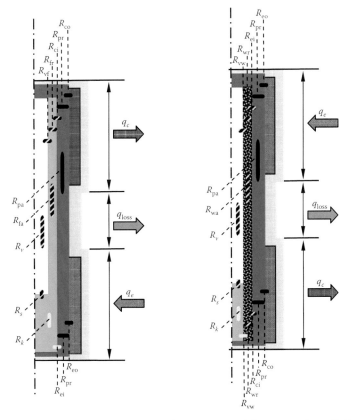

FIGURE 14.2   Lumped model of thermal resistance of thermosyphon (left) and heat pipe (right). All partial thermal resistances R_i that are not affected by the working fluid are marked black, R_i that are possibly affected are marked yellow-black brindled, and R_i that are highly likely affected are marked yellow. The colors of design elements are as in Figure 14.1.

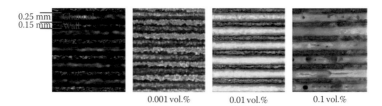

0.25 mm
0.15 mm

0.001 vol.% 0.01 vol.% 0.1 vol.%

FIGURE 14.3 Evaporation experiments carried out on microgrooved surfaces. Left photo shows fresh copper surface with microchannels. The other three images show the surface after boiling with R141b-based Al_2O_3 nanofluids. (Photos from Diao, Y.H. et al., *Int. J. Heat Mass Transf.*, 67, 183–193, 2013.)

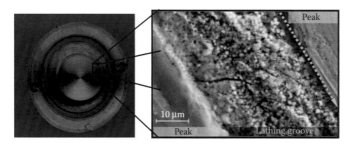

FIGURE 14.4 Top view of evaporator surface of a thermosyphon. Left photo shows fresh copper surfaces with lathing grooves. Right photo shows scanning electron microscope image after experiment with 0.1 vol.% TiO_2 nanofluid, 20,000-fold enlarged. Orange arrows indicate the borders of a single lathing groove nearly completely filled with nanoparticles. (Data from Buschmann and Franzke 2014.)

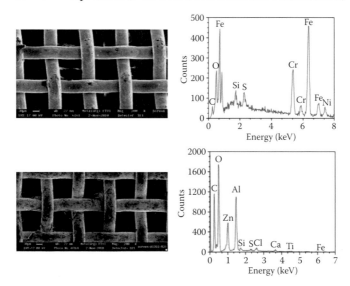

FIGURE 14.5 Scanning electron microscope image (left) and energy dispersive X-ray spectroscopy analysis (right) of the screen mesh wick of a heat pipe operated with water (above) and Al_2O_3 water nanofluid (below). (Data from Putra, N. et al., *Exp. Therm. Fluid Sci.*, 40, 2012, 10–17.)

flow rates provided at the evaporator. However, no optimum where effects reversed was found.

Thermosyphons with different nanoparticle concentrations of the working fluid have been investigated, for example, by Lu et al. (2011) and Buschmann and Franzke (2014). In the first case, an open thermosyphon operated with CuO/H_2O nanofluid of 1.2 wt.% concentration showed the highest evaporator heat transfer coefficient. Concentrations of 0.8, 1, and 1.5 wt.% and pure water showed less increases. Buschmann and Franzke (2014) measured a simple thermosyphon employing TiO_2/H_2O nanofluids with concentrations between 0.1 and 0.4 vol.%. A maximal decrease of the thermal performance was found at concentrations between 0.2 and 0.3 vol.%. However, long-term runs indicated that these improvements degenerated over time.

Looking for optima means to understand contradicting physical effects. The concentration-dependent minimum of the thermal resistance found for thermosyphons might be caused by a growing number of nucleation sites on the evaporator surface, which is compensated or even overcompensated for by an increasing thermal resistance of the porous layer itself. Both parameters depend on the concentration of nanoparticles in the working fluid. Accordingly, increased capillary pressure of screen mesh wicked heat pipes is compensated for by concentration-dependent hydrodynamic pressure caused by increasing debris on the wick. Currently, the acting and interacting of nanoparticles inside thermosyphons and heat pipes are not sufficiently understood to explain all these effects properly or to design tailor-made nanofluids for optimized heat transfer devices.

14.7 MATHEMATICAL MODELING AND NUMERICAL SIMULATION

Mathematical models of thermosyphons and heat pipes have to be split into three groups:

1. Semiempirical models based on similarity analysis

   (e.g., Nusselt number correlations)

2. Numerical simulations based on the effective medium theory

3. Numerical models based on the two-phase character of nanofluids

Models of type 1 are more or less only representations of experimental findings. Models of types 2 and 3, which are more complex, allow sensitivity analyses by varying, for example, approaches of thermophysical properties. Flexibility, physical correctness, and information gain increase from model 1 to model 3. Unfortunately, this has to be earned by exponentially increasing modeling and numerical efforts. Although the first outlay is measured in model complexity, numerical effort is characterized by increasing the CPU time.

Among others, Paramatthanuwat et al. (2010) and Buschmann and Franzke (2014) provided semiempirical approaches to describe closed two-phase thermosyphons. The first model is basically a correlation for the ratio of heat flux to critical heat flux (Kutateladze number) depending on 10 similarity numbers restricted to silver nanofluids with a concentration of 0.5 wt.%. Buschmann and Franzke (2014) provided a Nusselt number correlation for titania nanofluids with concentrations up to 0.2 vol.%. The Nusselt number that represents in this correlation the overall nondimensionalized amount of heat transferred depends on nondimensional temperature difference between the evaporator and the condenser and on the concentration of nanoparticles. Of course, both models represent experimental data from which they were derived properly. However, keeping in mind the complexity of nanofluids and the design manifold of thermosyphons and heat pipes, they lack generality. Despite this shortage, both models provide first insights into the relevant physics of thermosyphons operated with nanofluids. Both models see a strong influence of the thermophysical properties of the working fluid indicated by a dependency on the Prandtl number. Both models consider the latent heat of vaporization enthalpy of evaporation of the working fluid by employing the Jacob number or a comparable similarity number. Additionally, the level of operation temperature is considered.

Models of type 2 can be split into analytical and numerical approaches. Analytical models such as the Shafahi–Bianco–Vafai–Manca model (Shafahi et al., 2010a, 2010b) for flat-shaped heat pipes are based on a number of model equations describing different physical facts such as temperature distribution at the evaporator wall and pressure distributions of vapor and liquid. Additionally, thermophysical properties of the working fluid are modeled. A uniqueness of this special model is that it addresses the wick character as an independent parameter.

Numerical modeling is either based on commercially available computer fluid dynamics (CFD) codes or employs in-house codes. Examples employing standard CFD codes, namely, ANSYS-FLUENT and ANSYS CFX-12.0, are Gavtash et al. (2012) for heat pipes and Huminic and Huminic (2013) for thermosyphons. In both cases, effective medium correlations of thermophysical properties, in particular thermal conductivity and viscosity, are implemented in the CFD codes. Despite the fact that such approaches cannot represent phase separation as they obviously occur at the evaporator surface (Figures 14.3 and 14.4) and within the wick (Figure 14.5), the results especially obtained by Huminic and Huminic (2013) are in reasonable agreement with experiments. Numerical calculations indicate in both models a lowering of evaporator temperature. However, although Gavtash et al. (2012) see a monotonic decrease, the results of Huminic and Huminic (2013) rather show a falling and, above a certain concentration, again a rising evaporator temperature. Both models predict a decrease of thermal resistance with increasing nanoparticle concentration, an effect often shown also experimentally.

Do and Jang (2010) provided two complex models to study the effects connected with the formation of porous layers in a flat micro heat pipe

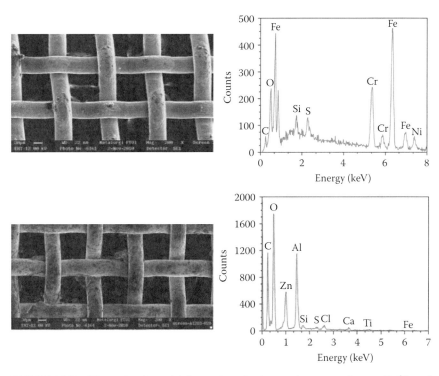

FIGURE 14.5 **(See color insert.)** Scanning electron microscope image (left) and energy dispersive X-ray spectroscopy analysis (right) of the screen mesh wick of a heat pipe operated with water (above) and Al_2O_3 water nanofluid (below). (Data from Putra, N. et al., *Exp. Therm. Fluid Sci.*, 40, 2012, 10–17.)

with rectangular grooved wick. The first model considered thermophysical properties of nanofluids and the second model a thin porous coating as major reasons for increasing thermal performance. Model 1 indicated that the higher thermal conductivity of the nanofluid increased evaporation heat transfer in the liquid film. Model 2 showed that top and side walls of the fins were covered with nanoparticles originally suspended in the nano-fluid, which is in agreement with the boiling experiments of Diao et al. (2013). Depending on nanoparticle concentration in the working fluid, this debris increased the heat transfer area but also the pressure drop of the wick. These two effects counteract each other. Hence, the optimum of low-ering thermal resistance depending on the concentration as found by many experimenters is confirmed by the model of Do and Jang (2010).

The model by Do and Jang (2010) is already close to the need of type 3 models. Such approaches would describe the interaction of nanopar-ticles with evaporator wall and/or wick strictly based on a two-phase approach for nanofluids. Future models should consider an additional transport equation for the nanoparticle volume fraction as provided, for example, by Avramenko et al. (2012) considering thermophoresis and Brownian motion explicitly. However, to the best knowledge of the author, such models have not yet been employed to model thermosyphons and heat pipes operated with nanofluids.

14.8 APPLICATION

The advantage of thermosyphons and heat pipes is that they are closed and the contact of the working fluid with the environment is avoided. Exceptions are filling, undesirable leaks, and removal of nanofluid after use. However, during operation no additional safety measures or changed maintenance procedures are needed to protect the environment from uncontrolled transported nanoparticles. Hence, thermosyphons and heat pipes operated with nanofluids appear comparatively safe and easy to handle.

Applied research with respect to thermosyphons and heat pipes oper-ated with nanofluids are reported, for example, for the following:

- Thermoelectric cooling of electronic equipment (Putra et al., 2011)

- Preheating of air employing a thermosyphon heat exchanger (Leong et al., 2012)

- Open thermosyphon for high-temperature evacuated tubular solar collectors (Lu et al., 2011)

These examples indicate cooling, heat recovery, and renewable energies as main application fields. However, currently several physical phenomena related to the interaction between nanoparticles and walls/wicks are still not sufficiently understood to expect large-scale industrial applications soon. Besides solving the problem of long-term stability of nanofluids, further research has to answer these questions.

REFERENCES

Asselman, G.A.A., Green, D.B., Heat pipes. *Phillips Technical Review* 33, 1973, 104–113.

Avramenko, A., Blinov, D.G., Shevchuk, I.V., Kuznetsov, A.V., Symmetry analysis and self-similar forms of fluid flow and heat-mass transfer in turbulent boundary layer flow of a nanofluid. *Physics of Fluids* 24, 2012, 092003. http://dx.doi.org/10.1063/1.4753945.

Azizi, M., Hosseini, M., Zafarnak, S., Shanbedi, M., Amiri, A., Experimental analysis of thermal performance in a two-phase closed thermosyphon using graphene/water nanofluid. *Industrial Engineering and Chemical Research* 52, 2013, 10015–10021. doi:dx.doi.org/10.1021/ie401543n.

Buschmann, M.H., Nanofluids in thermosyphons and heat pipes: Overview of recent experiments and modelling approaches. *International Journal of Thermal Sciences* 72, 2013, 1–17. http://dx.doi.org/10.1016/j.ijthermalsci.2013.04.024.

Buschmann, M.H., Franzke, U., Improvement of thermosyphon performance by employing nanofluid. *International Journal of Refrigeration* 40, 2014, 1–13. http://dx.doi.org/10.1016/j.ijrefrig.2013.11.022.

Cheng, P., Thompson, S., Boswell, J., Ma, H.B., An investigation of flat-plate oscillating heat pipes. *Journal of Electronic Packaging* 132, 2010, 041009. doi:10.1115/1.4002726.

Diao, Y.H., Liu, Y., Wang, R., Zhao, Y.H., Guo, L., Tang, X., Effects of nanofluids and nanocoatings on the thermal performance of an evaporator with rectangular microchannels. *International Journal of Heat and Mass Transfer* 67, 2013, 183–193. http://dx.doi.org/10.1016/j.ijheatmasstransfer.2013.07.089.

Do, K.H., Ha, H.J., Jang, S.P., Thermal resistance of screen mesh wick heat pipes using the water-based Al_2O_3 nanofluids. *International Journal of Heat and Mass Transfer* 53, 2010, 5888–5894. doi:10.1016/j.ijheatmasstransfer.2010.07.050.

Do, K.H., Jang, S.P., Effect of nanofluids on the thermal performance of a flat micro heat pipe with a rectangular grooved wick. *International Journal of Heat and Mass Transfer* 53, 2010, 2183–2192. doi:10.1016/j.ijheatmasstransfer.2010.07.050.

Faghri, A., *Heat Pipe Science and Technology, Mechanical Engineering*, Taylor & Francis Group, New York, 1995.

Gavtash, B., Hussain, K., Layeghi, M., Lafmejani, S.S., Numerical simulation of the effects of nanofluid on a heat pipe thermal performance. *World Academy of Science, Engineering and Technology* 68, 2012, 549–555.

Grab, T., Gross, U., Franzke, U., Buschmann, M.H., Operation performance of thermosyphons operated with titania and gold nanofluids. *International Journal of Thermal Sciences* 86, 2014, 352–364.

Hajian, R., Layeghi, M., Sani, K.A., Experimental study of nanofluid effects on the thermal performance with response time of heat pipe. *Energy Conversion Management* 5, 2012, 63–68. doi:10.1016/j.enconman.2011.11.010.

Harish, G., Emlin, V., Sajith, V., Effect of surface particle interactions during pool boiling of nanofluids. *International Journal of Thermal Sciences* 50, 2011, 2318–2327. doi:10.1016/j.ijthermalsci.-2011.06.019.

Huminic, G., Huminic, A., Heat transfer characteristics of a two-phase closed thermosyphons using nanofluids. *Experimental Thermal and Fluid Science* 35, 2011, 550–557. doi:10.1016/j.expthermflusci.2010.12.009.

Huminic, G., Huminic, A., Numerical study on heat transfer characteristics of thermosyphon heat pipes using nanofluids. *Energy Conversion and Management* 76, 2013, 393–399. http://dx.doi.org/10.1016/j.enconman.2013.07.026.

Kang, S.W., Wei, W.C., Tsai, S.H., Yang, S.Y., Experimental investigation of silver nano-fluid on heat pipe thermal performance. *Applied Thermal Engineering* 26, 2006, 2377–2382. doi:10.1016/j.applthermaleng.2006.02.020.

Kim, S.J., Bang, I.C., Buongiorno, J., Hu L.W., Effects of nanoparticle deposition on surface wettability influencing boiling heat transfer in nanofluids. *Applied Physics Letters* 89, 2006, 153107. doi:10.1063/1.2360892.

Leong, K.Y., Saidur, R., Mahlia, T.M.I., Yau, Y.H., Performance investigation of nanofluids as working fluid in a thermosyphon air preheater. *International Communications Heat and Mass Transfer* 39, 2012, 523–529. doi:10.1016/j. icheatmasstransfer.2012.01.014.

Liu, Z.H., Li, Y.Y., A new frontier of nanofluid research—Application of nanofluids in heat pipes. *International Journal of Heat Mass Transfer* 55, 2012, 6786–6797. http://dx.doi.org/10.1016/-j.ijheatmasstransfer.2012.06.086.

Lu, L., Liu, Z.H., Xiao, H.S., Thermal performance of an open thermosyphon using nanofluids for high-temperature evacuated tubular solar collectors. Part 1: Indoor experiment. *Solar Energy* 85, 2011, 379–387. doi:10.1016/j. solener.2010.11.008.

Paramatthanuwat, T., Rittidech, S., Pattiya, A., A correlation to predict heat-transfer rates of a two-phase closed thermosyphon (TPCT) using silver nanofluid at normal operating conditions. *International Journal of Heat Mass Transfer* 53, 2010, 4960–4965. doi:10.1016/j.ijheatmasstransfer.2010.05.046.

Paramatthanuwat, T., Rittidech, S., Pattiya, A., Ding, Y., Witharana, S., Application of silver nanofluid containing oleic acid surfactant in a thermosyphon economizer. *Nanoscale Research Letters* 6, 2011, 315. doi:10.1186/1556-276X-6-315.

Putra, N., Ferdiansyah, Y., Iskandar, N., Application of nanofluids to a heat pipe liquid-block and the thermoelectric cooling of electronic equipment. *Experimental Thermal and Fluid Science* 35, 2011, 1274–1281. doi:10.1016/j. expthermflusci.2011.04.015.

Putra, N., Septiadi, W.N., Rahman, H., Irwansyah, R., Thermal performance of screen mesh wick heat pipes with nanofluids. *Experimental Thermal and Fluid Science* 40, 2012, 10–17. doi:10.1016/j.expthermflusci.2012.01.007.

Reay, D.A., Kew, P.A., *Heat Pipes*, 5th ed., Elsevier, Oxford, 2006.

Shafahi, M., Bianco, V., Vafai, K., Manca, O., An investigation of the thermal performance of cylindrical heat pipes using nanofluids. *International Journal of Heat Mass Transfer* 53, 2010a, 376–383. doi:10.1016/j.ijheatmasstransfer.2009.09.019.

Shafahi, M., Bianco, V., Vafai, K., Manca, O., Thermal performance of flat-shaped heat pipes using nanofluids. *International Journal of Heat Mass Transfer* 53, 2010b, 1438–1445. doi:10.1016/j.ijheatmasstransfer.2009.12.007.

Shukla, K.N., Solomon, A.B., Pillai, B.C., Singh, B.J.R., Kumar, S.S., Thermal performance of heat pipe with suspended nanoparticles. *Heat Mass Transfer* 48, 2012, 1913–1920. doi:10.1007/s00231-012-1028-4.

Solomon, B., Ramachandran, K., Pillai, B.C., Thermal performance of a heat pipe with nanoparticles coated wick. *Applied Thermal Engineering* 36, 2012, 106–112. doi:10.1016/j.applthermaleng.2011.12.004.

Tanvir, S., Qiao, L., Surface tension of nanofluid-type fuels containing suspended nanomaterials. *Nanoscale Research Letters* 7, 2012, 226. doi:10.1186/1556-276X-7-226.

Tian, F.Z., Xin, G.M., Wang, X.Y., Cheng, L., An investigation of the unstable oscillation phenomenon of two-phase closed thermosyphon. *Advanced Materials Research* 668, 2013, 608–611. doi:10.4028/www.scientijic.net/AlvfR. 668. 608.

Verein Deutscher Ingenieure, *VDI-Wärmeatlas: Berechnungsblätter für den Wärmeübergang*, Springer-Vieweg, Berlin, Germany, 2006.

Wei, W.C., Tsai, S.H., Yang, S.Y., Kang, S.W., Effect of nanofluid concentration on heat pipe thermal performance. *IASME Transaction* 2, 2005, 1432–1439.

Wen, D., Corr, M., Hua, X., Lin, G., Boiling heat transfer of nanofluids: The effect of heating surface modification. *International Journal of Thermal Sciences* 50, 2011, 480–485. doi:1 0.1016/j.ijthermalsd.2010.10.017.

White, S.B., Enhancement of boiling surfaces using nanofluid particle deposition, PhD Thesis, The University of Michigan, Ann Arbor, MI, 2010.

CHAPTER 15

Entropy Generation Minimization in Nanofluid Flow

Omid Mahian, Clement Kleinstreuer, Ali Kianifar, Ahmet Z. Sahin, Giulio Lorenzini, and Somchai Wongwises

CONTENTS

15.1 INTRODUCTION

Entropy generation due to irreversibilities results in the destruction of exergy. In other words, entropy generation in a thermal system causes loss of useful work. Therefore, it is essential to carry out entropy generation analysis for investigating the location and sources of irreversibilities that are responsible for exergy destruction. The irreversibilities in a thermal system are mainly due to heat transfer and fluid friction. Minimizing the loss of exergy and improving the performance of thermal systems are possible through entropy generation minimization (EGM) procedures. Techniques used for decreasing the entropy generation due to heat transfer may often cause increases in the entropy generation due to the fluid friction. Therefore, an optimum set of operating parameters may be found to minimize overall entropy generation. As the size of system components such as computer chips is reduced, local temperatures may significantly rise because of considerable local heat generation. Thus, efficient cooling techniques are needed to prevent overheating. The thermal conductivities of ordinary fluids are rather low, and therefore the cooling capacity of such fluids is limited. In order to enhance heat transfer through nanofluids, that is, dilute suspensions of nanoparticles in liquids, various kinds of nanoparticles, some with high thermal conductivities, have been employed. Because of the elevated thermal conductivities and heat transfer performance characteristics, nanofluids have become a very attractive choice for practical applications in thermal engineering systems. Thus, nanofluids are found to be suitable not only for micro applications but also for large size thermal systems such as nuclear reactors.

Extensive studies can be found in the literature on the development of nanofluids and improvements of their thermofluid characteristics both theoretically and experimentally. On the contrary, most of the studies on nanofluids have been conducted from the first law of thermodynamics perspective. The number of investigations on the treatment of nanofluids is less from the second law of thermodynamics (entropy generation analysis) point of view. Nanoparticles available in nanofluids enhance the heat transfer characteristics of the fluid, but at the same time the viscous dissipation due to the increase of the viscosity may be affected considerably. Therefore, entropy generation in nanofluid flows due to both heat transfer and viscous dissipation should be considered to optimize a thermal system's operating conditions. In addition, complications such as agglomeration and settlement of the nanoparticles introduce new challenges for the thermal and entropy generation analyses.

In this chapter, first the general formulations related to entropy generation are presented. Next, some results of the works conducted on nanofluids are outlined to illustrate the current status of the literature on the subject. The studies are classified based on the system's geometry. Finally, a conclusion is given and some suggestions for future works are proposed.

15.2 ENTROPY GENERATION FORMULATIONS

Before presenting the works conducted on the entropy generation of nanofluid flows in different geometries and regimes, it is useful to provide some background information on the methods used to calculate the entropy generation. To determine the entropy generation, one should know the fluid flow and heat transfer characteristics of the thermal system. Pumping power and friction factor are used to show the fluid flow characteristics in a system, whereas the heat transfer characteristics are represented by the heat transfer coefficient and Nusselt number. Thus, these parameters should be calculated first, considering two different cases. In the first case, in lieu of solving the momentum and energy equations to obtain the flow and heat transfer fields, available correlations in the literature are used to calculate the friction factor and Nusselt number. This approach does not deal with the local distribution of entropy generation in the system. For example, Bejan [1] derived a relation to obtain the entropy generation in a duct of diameter D, the mass flow rate $\dot{m}$, and the density ρ as follows:

$$\dot{S}'_{gen} = \frac{q''^2 \pi D^2}{kT^2 \text{Nu}\left[(\text{Re})_D, \text{Pr}\right]} + \frac{32\dot{m}^3}{\pi^2 \rho^2 T_{ave}} \frac{f\left[(\text{Re})_D\right]}{D^5}$$

$$= (\dot{S}'_{gen})_{\text{heat transfer}} + (\dot{S}'_{gen})_{\text{fluid friction}}$$

(15.1)

where:
$\dot{S}'_{gen}$ is the entropy generation rate per unit length
k is the thermal conductivity of the fluid

The Nusselt number is a function of the Reynolds and Prandtl numbers, and the friction factor is a function of the Reynolds number. The duct is under a constant wall heat flux of q''. It should be noted that in the above relation, f is the Fanning friction factor. If one wants to use the Darcy friction factor, the factor 32 has to be replaced by 8. In Equation 15.1, the term containing the Nusselt number represents the heat transfer contribution to entropy generation, whereas the term with the friction factor implies the contribution of fluid friction to entropy generation. As mentioned in the first approach, the Nusselt number and friction factor are obtained using available correlations.

In the second case, first the momentum and energy equations have to be solved. Then, using the flow and temperature fields, the distribution of entropy can be obtained at any point. The total entropy generation can be estimated simply by integrating the local entropy generation over the volume.

The local entropy generation in different coordinate systems is given by the following relations [1]:

Cartesian coordinates (x, y, z)

$$\dot{S}_{gen}''' = \frac{k}{T^2}\left[\left(\frac{\partial T}{\partial x}\right)^2 + \left(\frac{\partial T}{\partial y}\right)^2 + \left(\frac{\partial T}{\partial z}\right)^2\right]$$
$$+ \frac{\mu}{T}\left\{ \begin{array}{l} 2\left[\left(\frac{\partial v_x}{\partial x}\right)^2 + \left(\frac{\partial v_y}{\partial y}\right)^2 + \left(\frac{\partial v_z}{\partial z}\right)^2\right] \\ + \left(\frac{\partial v_x}{\partial y} + \frac{\partial v_y}{\partial x}\right)^2 + \left(\frac{\partial v_x}{\partial z} + \frac{\partial v_z}{\partial x}\right)^2 + \left(\frac{\partial v_x}{\partial z} + \frac{\partial v_z}{\partial y}\right)^2 \end{array} \right\} \quad (15.2)$$

Cylindrical coordinates (r, θ, z):

$$\dot{S}_{gen}''' = \frac{k}{T^2}\left[\left(\frac{\partial T}{\partial r}\right)^2 + \left(\frac{1}{r}\frac{\partial T}{\partial \theta}\right)^2 + \left(\frac{\partial T}{\partial z}\right)^2\right]$$
$$+ \frac{\mu}{T}\left\{ \begin{array}{l} 2\left[\left(\frac{\partial v_r}{\partial r}\right)^2 + \frac{1}{r^2}\left(\frac{\partial v_\theta}{\partial \theta} + v_r\right)^2 + \left(\frac{\partial v_z}{\partial z}\right)^2\right] + \left(\frac{\partial v_\theta}{\partial z} + \frac{1}{r}\frac{\partial v_z}{\partial \theta}\right)^2 \\ + \left(\frac{\partial v_z}{\partial r} + \frac{\partial v_r}{\partial z}\right)^2 + \left[\frac{1}{r}\frac{\partial v_r}{\partial \theta} + r\frac{\partial}{\partial r}\left(\frac{v_\theta}{r}\right)\right]^2 \end{array} \right\} \quad (15.3)$$

Spherical coordinates (r, θ, φ):

$$\dot{S}'''_{gen} = \frac{k}{T^2}\left[\left(\frac{\partial T}{\partial r}\right)^2 + \left(\frac{1}{r}\frac{\partial T}{\partial \theta}\right)^2 + \left(\frac{1}{r\sin\theta}\frac{\partial T}{\partial \varphi}\right)^2\right]$$

$$+ \frac{\mu}{T}\left\{\begin{array}{l}2\left[\left(\frac{\partial v_r}{\partial r}\right)^2 + \left(\frac{1}{r}\frac{\partial v_\theta}{\partial \theta} + \frac{v_r}{r}\right)^2 + \left(\frac{1}{r\sin\theta}\frac{\partial v_\varphi}{\partial \varphi} + \frac{v_r}{r} + \frac{v_\theta\cot\theta}{r}\right)^2\right] \\[2ex] + \left[r\frac{\partial}{\partial r}\left(\frac{v_\theta}{r}\right) + \frac{1}{r}\frac{\partial v_r}{\partial \theta}\right]^2 + \left[\frac{1}{r\sin\theta}\frac{\partial v_r}{\partial \varphi} + r\frac{\partial}{\partial r}\left(\frac{v_\varphi}{r}\right)\right]^2 \\[2ex] + \left[\frac{\sin\theta}{r}\frac{\partial}{\partial \theta}\left(\frac{v_\varphi}{\sin\theta}\right) + \frac{1}{r\sin\theta}\frac{\partial v_\varphi}{\partial \varphi}\right]^2\end{array}\right\} \quad (15.4)$$

where $\dot{S}'''_{gen}$ is the entropy generated per unit of volume (W/m³K). The above equations show the role of temperature and velocity gradients in entropy generation. For problems dealing with nanofluid flow, special models describing a nanofluid's thermal conductivity, heat capacity, viscosity, and density should be employed. Such models and their applications are discussed by Kleinstreuer and Xu in Chapter 8.

15.3 ENTROPY GENERATION IN NANOFLUID FLOW: A LITERATURE REVIEW

In this section, a literature review is presented, where the papers are grouped based on the geometry of the problem. It should be noted that Mahian et al. [2] had published a review paper on entropy generation in nanofluid flow; thus, the focus here is on new contributions.

15.3.1 Entropy Generation in Microchannels

Channels based on the magnitude of hydraulic diameter can be classified into three groups: conventional channels, minichannels, and microchannels. In this section, first the entropy generation of nanofluids in microchannels is reviewed. One of the earliest papers on this topic was that of Singh et al. [3]. They performed an analytical analysis of entropy generation due to Al_2O_3–water nanofluid flow for three different tube sizes, including a microchannel ($D_h = 0.1$ mm), a minichannel ($D_h = 1$ mm), and a conventional channel ($D_h = 10$ mm), where the hydraulic diameter is $D_h = 4A/P$. They considered both laminar and turbulent flow regimes and predicted the entropy generation rate by using an order-of-magnitude approach. In their work, Equation 15.1 was used to obtain the entropy generation in conjunction with the following well-known relations to estimate the Nusselt number and friction factor, assuming laminar fully developed flow:

$$Nu = \frac{48}{11} \qquad (15.5a)$$

$$f = \frac{64}{Re} \qquad (15.5b)$$

For turbulent flow, they calculated the Nusselt number and friction factor from the Dittus–Boelter and Blasius equations, respectively:

$$Nu = 0.023 \, Re^{0.8} \, Pr^{0.4} \qquad (15.6a)$$

$$f = 0.316 \, Re^{-1/4} \qquad (15.6b)$$

Singh et al. [3] solved the problem by using an order-of-magnitude analysis for the three tubes and both flow regimes. To study the effects of using both theoretical and experimental relations to estimate the viscosity and thermal conductivity, they introduced two models. The results showed that for laminar and turbulent flows the use of Al_2O_3–water nanofluid in both the microchannels and conventional channels increases the entropy generation ratio. They also concluded that the two models used to calculate the thermophysical properties may give opposite predictions. They obtained a general relation for an optimal diameter for which entropy generation is minimized.

Li and Kleinstreuer [4] investigated numerically entropy generation of CuO–water nanofluid flow with volume fractions less than 4% in trapezoidal microchannels, assuming steady laminar developing flow. The longest trapezoidal side was on the top and adiabatic, whereas the other sides were subject to a constant wall heat flux. The authors concluded that adding nanoparticles to the base fluid reduces entropy generation, although there is an optimum volume fraction for which entropy generation is minimized. Tabrizi and Seyf [5] investigated numerically entropy generation in Al_2O_3 nanofluid flow in a tangential micro heat sink (TMHS). The schematic of the problem is shown in Figure 15.1. The cold nanofluid enters

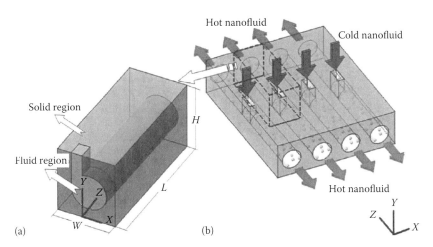

FIGURE 15.1 Schematic of the TMHS assembly and coordinate system considered by Tabrizi and Seyf (a) general view of the microchannels, (b) a close view of a channel. (Reprinted from *Int. J. Heat Mass Transf.*, 55, Tabrizi, S.A. and Seyf, H.R., Analysis of entropy generation and convective heat transfer of Al_2O_3 nanofluid flow in a tangential micro heat sink, 4366–4375, Copyright 2012, with permission from Elsevier.)

the TMHS and after cooling the system exits it. In addition to particle size, the effects of Reynolds number and volume fraction on the entropy generation rate were examined. They found that the contribution of viscous effects to entropy generation is negligible compared to the heat transfer contribution. Thus, an overall change in total entropy generation was greatly determined by thermal effects. It was found that with a decrease in nanoparticle size from 47 to 29 nm, the entropy generation decreased, whereas with an increase in the Reynolds number the entropy generation decreased. The reasons are that decreasing the particle size and increasing the Reynolds number lead to higher Nusselt numbers within the TMHS, and hence the entropy generation decreases (see Equation 15.1). They also concluded that with an increase in the volume fraction of nanoparticles, the entropy generation decreases. Mah et al. [6] studied analytically the effects of viscous dissipation on entropy generation induced by Al_2O_3–water nanofluids in a circular microchannel. They concluded that when the viscous effects are considered in the entropy generation analysis, the entropy generation increases with an increase in the volume fraction of nanoparticles.

Sohel et al. [7] considered entropy generation in nanofluid flow in mini- and microchannels. They used Cu/water, Cu/ethylene glycol (EG), Al_2O_3/water, and Al_2O_3/EG combinations as nanofluids, where three

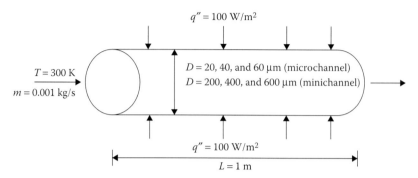

FIGURE 15.2 Schematic diagram of a circular microchannel and minichannel considered by Sohel et al. (Reprinted from *Int. Commun. Heat Mass*, 46, Sohel, M.R. et al., Analysis of entropy generation using nanofluid flow through the circular microchannel and minichannel heat sink, 85–91, Copyright 2013, with permission from Elsevier.)

different sizes were considered for both mini- and microchannels. The schematic of the problem is shown in Figure 15.2. They found that with an increase in the size of mini- and microchannels the entropy generation increases considerably. The results also showed that water-based nanofluids give lower values of entropy generation than EG-based nanofluids because of the higher thermal conductivity of water; also, Cu nanoparticles are preferable compared to Al_2O_3 nanoparticles. In both types of channels, the entropy decreased by increasing the nanofluid concentration.

Ting et al. [8] studied the entropy generation due to alumina/water nanofluid flow in a circular microchannel for low-Peclet number flow regimes by considering the effect of streamwise conduction. Based on this, they considered two models. In model 1, streamwise conduction was not considered, whereas in model 2, the effect of streamwise conduction was taken into account. Their findings indicated that at a Peclet number Pe = 1 and volume fractions of 10%, the total entropy generation estimated by model 2 was about 43% higher than that of model 1. With a decrease in the volume fraction, the effect of streamwise conduction decreased. They considered the problem for the Peclet numbers between 1 and 50, and found an optimal Peclet number in which the total entropy was minimized.

Mohammadian et al. [9] studied the entropy generation due to laminar flow in a counterflow microchannel heat exchanger where Al_2O_3/water nanofluids with concentrations between 1% and 4% were used to cool down the hot water flowing in the heat exchanger. Besides the volume concentration, the effects of three different particle sizes (29, 38.4, and 47 nm) and the Reynolds number (between 10 and 480) on the flow and entropy generation were considered. They found that with an increase in particle size, the entropy generation due to fluid friction decreases, although the heat transfer contribution increases. By contrast, increases in the volume fraction resulted in an inverse trend. It was concluded that to reduce the entropy generation, nanofluids of high volume fraction and small nanoparticle size are needed.

15.3.2 Entropy Generation in Cavities

Extensive studies on natural and mixed convection in cavities of different geometries have been performed. Such problems are important in the design of nuclear reactors, solar collectors, storage tanks, electronic cooling, and so on. In the following, studies conducted on entropy generation for nanofluid flow in cavities are reviewed. Shahi et al. [10] examined numerically the entropy generation due to Cu–water nanofluids in a square cavity with a heat source mounted inside it, considering four different designs. The distance of the heat source from the side walls was changed as well. The dimensionless distances were 0.2, 0.4, 0.6, and 0.8, where the distance 0.2 was near the left wall. The authors concluded that the use of the Case 4 configuration, where the heat source was mounted at the dimensionless distance of 0.2, led to maximum entropy generation. Thus, they suggested installing the heat source on the bottom wall to achieve the maximum heat transfer rate and the minimum entropy generation rate.

Cho et al. [11] studied the entropy generation due to natural convection heat transfer of three different nanofluids, including Cu/water, Al_2O_3/water, and TiO_2/water in a wavy cavity, where the schematic view is shown in Figure 15.3. The results indicated that independent of the value of the Rayleigh number, the total entropy generation reduces with an increase in nanofluid concentration. The authors concluded that Cu/water nanofluids provide the highest heat transfer performance and a minimum in total entropy generation. It was also mentioned that through an expedient design of the cavity, the total entropy generation can be minimized.

Mahmoudi et al. [12] presented a numerical solution to elucidate the influence of magnetohydrodynamic (MHD) flow on the entropy generation in a trapezoidal cavity filled with Cu/water nanofluid. The cavity was heated at the bottom. The results showed that adding nanoparticles decreased the entropy generation, whereas increasing the magnetic field intensity led generally to an increase in the total entropy generation.

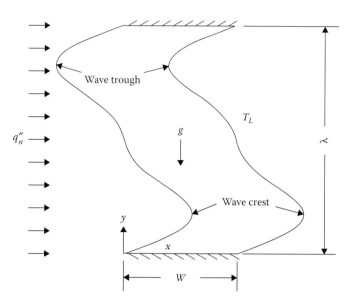

FIGURE 15.3 Schematic illustration of wavy wall enclosure considered by Cho et al. (Reprinted from *Int. J. Heat Mass Transf.*, 61, Cho, C.C. et al., Natural convection heat transfer and entropy generation in wavy-wall enclosure containing water-based nanofluid, 749–758, Copyright 2013, with permission from Elsevier.)

The results were presented for value ranges of $10^4 \leq Ra \leq 10^7$ and $0 \leq Ha \leq 100$ as well as volume fractions less than 5%, considering different positions of the heat source at the bottom wall. Sheikhzadeh and Nikfar [13] investigated numerically the aspect ratio (AR) effects of a centered adiabatic rectangular obstacle on natural convection and entropy generation in a differentially heated enclosure filled with either water or nanofluid (Cu–water). The governing equations were solved numerically by a finite-volume method using the SIMPLER algorithm. The study was done for Rayleigh numbers between 10^3 and 10^6 and with ARs of 0.33, 0.5, 1, 2, and 3. It was found that using the nanofluid led to an increase in flow strength, average Nusselt number, and entropy generation, whereas the Bejan number decreased, especially at high Rayleigh numbers. In low Rayleigh numbers, the entropy generation was very low. It was observed that the viscous entropy generation was larger than the thermal entropy generation. The maximum entropy generation occurred at AR = 0.33 and 3, and the minimum entropy generation occurred at AR = 1 and 0.5. It was observed that the effect of the AR on the Nusselt number, the entropy generation, and the Bejan number depends on the Rayleigh number. Kashani et al. [14] solved the entropy generation problem in a two-dimensional wavy cavity filled with Cu/water nanofluid near the point when the density of water has a maximum, using a pressure-based finite-volume method. The effects of nonlinear temperature dependence of water density and its extreme near 4°C on the entropy generation are investigated through the comparison between the results obtained from Boussinesq and non-Boussinesq approximations. Also the influence of adding nanoparticles to pure fluid, different volume fractions of nanoparticle, surface waviness, AR, and various wavy patterns on the entropy generation was studied. The main results can be expressed as follows:

- The role of adding nanoparticles in reducing the entropy generation is more pronounced at higher Ra numbers when enforcing the Boussinesq approximation.

- The results indicate that the average Nu number and entropy generation decrease with an increase of the volume fraction in all cases.

- The effect of density inversion is to reduce the total heat transfer rate and entropy generation.

- Increasing the Ra number results in an increase in average Nu number and entropy generation.

Parvin and Chamkha [15] investigated the entropy generation in an odd-shaped cavity using Cu/water nanofluids for the Rayleigh numbers between 10^3 and 10^6. The schematic of the problem is shown in Figure 15.4. They reported that using nanofluids increases the entropy generation and Bejan number. For example, at Ra = 10^4 the total entropy generation increased about 10%, when the nanofluid concentration changed from 0% to 5%. Their results showed that to minimize the total entropy generation the Rayleigh number should be 10^5. Mehrez et al. [16] investigated the entropy generation due to mixed convection of various nanoparticles (Cu, Al_2O_3, CuO, and TiO_2) suspended in water inside an open cavity (see Figure 15.5). The results were presented for different Reynolds numbers ($100 \leq Re \leq 500$), Richardson numbers ($0.05 \leq Ri \leq 1$), and volume concentrations less than 10%. It was shown that the entropy generation increases with an increase in Reynolds and Richardson numbers and nanofluid concentration. They found that CuO/water nanofluid gives the highest heat transfer

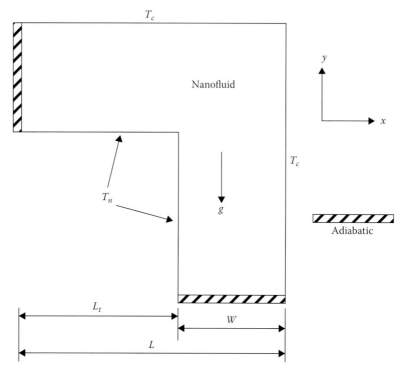

FIGURE 15.4 Schematic of odd-shaped cavity considered by Parvin and Chamkha. (Reprinted from *Int. Commun. Heat Mass Transf.*, 54, Parvin, S. and Chamkha, A.J., An analysis on free convection flow, heat transfer and entropy generation in an odd-shaped cavity filled with nanofluid, 8–17, Copyright 2014, with permission from Elsevier.)

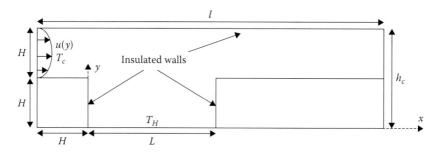

FIGURE 15.5 Schematic of the problem considered by Mehrez et al. (Reprinted from *Comput. Fluids*, 88, Mehrez, Z. et al., Heat transfer and entropy generation analysis of nanofluids flow in an open cavity, 363–373, Copyright 2013, with permission from Elsevier.)

rate and entropy generation, whereas TiO_2/water nanofluid produces minimum heat transfer and entropy. It was concluded that the density and thermal conductivity of nanoparticles played an important role in this problem solution.

Mahmoudi and Hooman [17] studied the effect of localized heat sources on the entropy generation owing to mixed convection flow in a vented square cavity. Laminar steady forced convection flow of copper–water nanofluid through the cavity has been affected by density variations as a result of heat input from a wall-mounted heat source. To investigate the effect of the heat source location, three different placement configurations of the heat source were considered. To generalize the results, three different nonuniform heat flux conditions were also examined. The entropy generation rate was analyzed for Richardson numbers $0 \leq Ri \leq 10$ and for solid volume fractions within $0 \leq \varphi \leq 0.05$. With either uniform or nonuniform wall heating, the entropy generation rate was found to be minimal when the heat source and cavity exit were located at the same wall. The maximum heat transfer rate, however, corresponded to the case when the main flow was parallel to the heated wall.

15.3.3 Entropy Generation in Different Ducts

Bianco et al. [18] studied the entropy generation for turbulent Al_2O_3–water nanofluid flow in a conduit with square cross section, subject to a constant wall heat flux. They applied Equation 15.1 in a somewhat different form; however, they used the following relations to calculate the Nusselt number and the friction factor for turbulent flow:

$$Nu = 0.021 Re^{0.8} Pr^{0.5} \tag{15.7}$$

$$f = 0.184 Re^{-0.2} \tag{15.8}$$

They obtained an optimal Reynolds number for which the entropy generation is minimized. They found that with an increase in the volume fraction, the Reynolds number decreases for a minimum entropy generation to occur.

Moghaddami et al. [19] used Equation 15.1 to estimate the entropy generation for Al_2O_3–water and Al_2O_3–EG flows in a circular tube, subject to a constant wall heat flux, considering both laminar and turbulent flows. They found that adding nanoparticles decreases the entropy generation for any Reynolds number in the laminar flow, whereas in the turbulent flow there is an optimum Reynolds number for which entropy generation is minimized. To analyze the impact of changes in key parameters, the increase in the entropy generation number ($N_{S,a}$) is defined as follows:

$$N_{S,a} = \frac{\dot{S}'_{\text{gen,nf}}}{\dot{S}'_{\text{gen},f}}$$ (15.9)

where $\dot{S}'_{\text{gen,nf}}$ and $\dot{S}'_{\text{gen},f}$ are the entropy generation rates for the nanofluid and base fluid, respectively. The authors found that for laminar flow of Al_2O_3–water nanofluids, the entropy generation for any volume fraction ($0 \leq \varphi \leq 5\%$) is minimized at Re = 853. They also perceived that the entropy generation decreases with an increase in the volume fraction. For turbulent flow of Al_2O_3–water nanofluids, particle loading is helpful when the Reynolds number is less than 40,000, where the optimal point is Re = 13,500. For Al_2O_3–EG nanofluids, it was found that adding nanoparticles is only advantageous at the Reynolds numbers less than 11. They concluded that adding nanoparticles is useful when the fluid friction contribution to entropy generation is adequately less than the contribution of heat transfer to entropy generation. Moghaddami et al. [20] investigated the entropy generation for tubular flow under constant heat flux using Al_2O_3–water nanofluids for both laminar and turbulent flow regimes. The difference between their work in Ref. [19] and that in Ref. [20] is the method to obtain entropy generation results. In the work of Moghaddami et al. [20], the velocity and temperature fields were obtained, and then the entropy generation distribution was computed locally. However, in the work of Moghaddami et al. [19], Equation 15.1 was used. Moghaddami et al. [20] concluded that with an increase in the Reynolds number and volume fraction of nanoparticles the contribution of heat transfer to entropy generation decreases, whereas the viscous dissipation contribution increases. The reason is that at higher Reynolds numbers and volume fractions, nanofluids are very effective in distributing the heat within the whole fluid and hence decreasing the local temperature gradients. As a result, the entropy generation induced by heat transfer decreases. On the contrary, viscous dissipation becomes more significant at higher Reynolds numbers and volume fractions because of the increase in velocity gradients as well as generated fluctuations and disturbances due to the addition of the nanoparticles. Consequently, the fluid friction contribution to entropy generation increases. They also found that using nanofluids is advantageous for laminar flow, whereas for turbulent flow there is an optimal Reynolds number for which the entropy generation can be minimized.

The optimal Reynolds number value decreases with an increase in the volume fraction of nanoparticles. Leong et al. [21] examined the entropy generation due to flow of titanium dioxide (TiO_2) and alumina (Al_2O_3) suspended in water in a circular tube where the wall of the tube was of constant temperature. They considered both laminar and turbulent flows. They found that the use of TiO_2 was more beneficial than Al_2O_3 nanoparticles. They also showed that the entropy generation is reduced in both laminar and turbulent flow regimes by adding nanoparticles. Whereas it was observed that for the constant wall heat flux condition (see Refs. [19,20]), adding nanoparticles in the turbulent flow does not always reduce the entropy generation. Karami et al. [22] investigated numerically the effects of Al_2O_3–water nanofluids in a circular tube under constant heat flux on the entropy generation, pumping power, and wall temperature. They found that nanoparticle loadings decrease the generated entropy and the temperature of the wall, whereas they increase the required pumping power. In light of these findings, they concluded that using nanofluids at high concentrations is not beneficial.

Falahat and Vosough [23] computed the entropy generation in a coiled tube, assuming a constant wall heat flux, for both laminar and turbulent flows of Al_2O_3–water nanofluids. They used Equation 15.1 and suitable correlations for the Nu number and friction factor to obtain entropy generation values.

Their results showed that by adding 1% volume fraction of nanoparticles to the base fluid, entropy generation decreased about 3% in the laminar flow. For the turbulent flow, they concluded that adding nanoparticles at the Reynolds numbers greater than 140,000 would be disadvantageous, whereas for the Reynolds numbers lower than 40,000 adding nanoparticles reduces the entropy generation. Also, they obtained an optimal Reynolds number (~41,500) for which the entropy generation was minimized.

Ahadi and Abbassi [24] considered the entropy generation for fully developed laminar forced convection of Al_2O_3–water nanofluid through helical coiled tubes. The nanofluid was assumed to be a single-phase homogeneous fluid. The entropy generation rates due to heat transfer and friction loss and the total entropy generation rate were evaluated, and the effects of various coil and flow parameters on the entropy generation rates were analyzed. The results showed that increasing the Reynolds number of the base fluid, nanoparticle concentration and curvature ratio caused thermal entropy generation to decrease and frictional entropy generation to increase. Furthermore, the variation of the dimensionless pitch had a minor effect on the total entropy generation distribution of nanofluids, whereas the effect of nanoparticle concentration on the total entropy generation rate was significant.

Abdin et al. [25] investigated the flow of three different nanofluids including CuO/water, Al$_2$O$_3$/water, and ZnO/water in a helical tube. The entropy generation through the tube was obtained for volume fractions between 1% and 4%, and the volumetric flow rates between 3–6 L/min. They found that particle loading leads to a reduction in the entropy generation. In addition, they showed that using CuO/water produced maximum heat transfer enhancement and minimum entropy generation rate. The authors used Equation 15.1 to obtain the entropy generation through the helical tube.

Bianco et al. [26] investigated the entropy generation due to flow of Al$_2$O$_3$/water nanofluid (volume concentrations up to 6%) in a circular tube with a constant wall heat flux. Different criteria were applied to describe the flow intensity, including the inlet velocity, the Reynolds number, and the mass flow rate. To solve the problem, they used Equation 15.1. They found that considering the constant inlet velocity condition, adding nanoparticles may decrease the entropy generation only for very low solid volume fractions, that is, 0.5%, but using higher concentrations led to an increase in the entropy generation. For the case of constant Reynolds number, an optimal concentration may be found in which the entropy generation is minimized. Finally, for the case of constant mass flow rate, the entropy generation increased with particle loading. In this case, to keep the constant mass flow rate, it was necessary to reduce the velocity. By decreasing the velocity the thermal conductance decreases; hence, the contribution of heat transfer to entropy generation increases, so that the total entropy generation increases. An interesting note in their results is that, for volume fraction of 4%, the entropy generation increases with an increase in particle size, whereas the Reynolds number is less than 40×10^3, but for higher Reynolds number, that is, 80×10^3, the entropy decreases with an increase in particle size. Figure 15.6 shows this scenario.

Sarkar et al. [27] studied the entropy generation due to natural convection heat transfer of Al$_2$O$_3$/water nanofluids between two infinite horizontal parallel plates under a magnetic field. The governing equations were simplified so that an analytical solution was obtained for the problem. Two different thermal boundary conditions were assumed: (1) Both plates are insulated and (2) the temperature of the bottom plate is zero, whereas the top plate has a higher temperature. The results are presented for volume fractions up to 4% with particle sizes between 10 and 50 nm, where $0 \leq Ha \leq 50$ and $10^3 \leq Gr \leq 10^5$. The findings show that with an increase in the particle

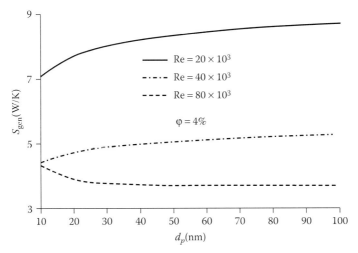

FIGURE 15.6 Impact of particles' dimension on the total entropy generation for various Re and for $\varphi = 4\%$. (Reprinted from *Energ. Convers. Manage.*, 77, Bianco, V. et al., Entropy generation analysis of turbulent convection flow of Al_2O_3-water nanofluid in a circular tube subjected to constant wall heat flux, 306–314, Copyright 2014, with permission from Elsevier.)

size the total entropy generation increases. The increase was attributed to the decreasing Brownian motion of particles, and hence the decrease of the effective thermal conductivity of nanofluids. A lower thermal conductivity leads to a high temperature gradient, and consequently higher entropy generation. They also showed that there is a Hartmann number in which the Bejan number is minimized. Finally, it was concluded that with an increase in the Grashof number the entropy generation increases (especially when Gr increases from 10^4 to 10^5), whereas the entropy production decreases with increasing volume fraction of nanoparticles.

Mahian et al. [28] investigated the entropy generation in mixed convection flow inside a vertical annulus using TiO_2/water nanofluids with volume concentrations less than 2% where a magnetic field was applied to the annulus. They simplified the momentum and energy equations to obtain an analytical solution for the problem. To calculate the thermal conductivity and viscosity, they used two sets of models including correlations based on the experimental data and the well-known theoretical models. The results showed that at a Hartmann number of 5, the theoretical model predicts the entropy generation 4% higher than experimental models. In general, it was concluded that adding the nanoparticles reduces the entropy generation, whereas boosting the magnetic field will increase the entropy generation.

Matin et al. [29] studied the impact of physical parameters on the entropy generation rate for a nanofluid flow inside an MHD channel, taking into account the thermal radiation effects. The entropy was minimized in terms of nanoparticle volume fraction, radiation parameter, as well as Hartmann and Peclet numbers. The results imply that there are conditions in which entropy generation can be minimized.

15.3.4 Entropy Generation between Two Rotating Cylinders

In this section, contributions related to the entropy generation between two rotating cylinders are reviewed. This type of flow is important in electrical motors, turbojet turbines, rotating heat pipes, swirl nozzles, rotating disks, and standard commercial rheometers. For example, Mahian et al. [30] investigated the entropy generation in nanofluid flow between two horizontally rotating, isoflux cylinders. They considered two nanofluids, that is, Al_2O_3–EG and TiO_2–water, in their analytical analysis. They solved the problem analytically by neglecting the radial velocity in the momentum equation, whereas for the energy equation they considered viscous dissipation but disregarded convection. It was found that the entropy generation can be reduced with an increase in the nanofluid concentration. Specifically, the heat transfer contribution to entropy generation was dominant in the annulus compared to the fluid friction contribution. They also concluded that TiO_2–water nanofluid is more suitable than Al_2O_3–EG nanofluid as the working fluid at low Brinkman numbers. Mahian et al. [31] studied the effects of uncertainties in thermophysical models for Al_2O_3–EG nanofluids ($0 \leq \varphi \leq 6\%$) when predicting the entropy generation between two rotating cylinders. The inner cylinder was subjected to a constant wall heat flux, whereas the outer cylinder was at a constant temperature. They compared the amounts of entropy generation, using six different models for thermophysical properties. They found that there is a critical radius ratio for the annulus in which the entropy generation based on all models decreases with an increase of the volume fraction of nanoparticles. This work shows the importance of using suitable thermophysical models. Later on, Mahian et al. [32] considered the effects of MHD flow on the entropy generation between two cylinders using TiO_2–water nanofluids. In this study, besides the heat transfer and fluid friction effects, the contribution of the magnetic field was considered in the entropy generation equation. They suggested using nanofluids between two cylinders in the presence of MHD flow only for low values of the Brinkman number.

15.3.5 Entropy Generation in Heat Exchangers

Leong et al. [33] investigated the performance of three types of shell-and-tube heat exchangers, including helical baffles of 25°, 50°, and segmental, using Cu–water nanofluids with volume fractions below 2%. Apparently, nanofluids at this low concentration have no considerable effect on the total entropy generation. It was also found that, in the presence of nanofluids, the entropy generation for the helical heat exchanger of 50° was the lowest.

Elias et al. [34] investigated the effects of various shapes of γ-AlOOH nanoparticles, including cylindrical, bricks, blades, and platelets as well as different baffle angles (20°, 30°, 40°, and 50°) on the performance and entropy generation of a shell-and-tube heat exchanger. As shown in Figure 15.7, to minimize the entropy generation it can be suggested to use the nanoparticles with platelet shape where the baffle angle and volume fraction are the

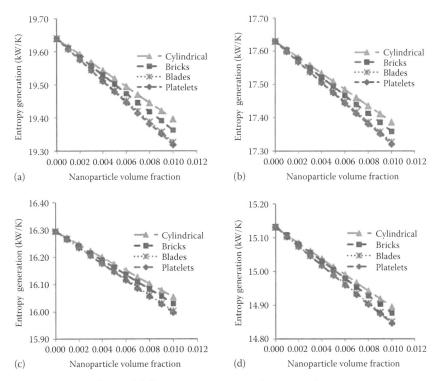

FIGURE 15.7 Effects of different nanoparticle shapes on the entropy generation of nanofluid at baffle angles of 20° (a), 30° (b), 40° (c), and 50° (d). (Reprinted from *Int. J. Heat Mass Transf.*, 70, Elias, M.M. et al., Effect of different nanoparticle shapes on shell and tube heat exchanger using different baffle angles and operated with nanofluid, 289–297, Copyright 2014, with permission from Elsevier).

highest. As the authors showed that the platelet nanoparticles produce the lowest heat transfer rate, it seems that in future works one can obtain an optimal shape considering both first and second law analyses.

15.3.6 Entropy Generation of Flow over Different Types of Plates

Matin et al. [35] presented a numerical study of MHD mixed convection flow of SiO_2–water nanofluids (with volume fractions up to 30%) over a nonlinear stretching sheet. They concluded that the entropy generation decreases with an increase of the volume fraction of nanoparticles. The Brinkman and Maxwell models were selected to calculate the thermal nanofluid properties, although at nanoparticle concentrations as high as 30% non-Newtonian fluid flow effects should be considered.

Noghrehabadi et al. [36] analyzed the boundary layer heat transfer and entropy generation of a nanofluid over an isothermal linear stretching sheet with heat generation/absorption. In the nanofluid model, the development of nanoparticles' concentration gradient due to slip mechanisms, the effects of Brownian motion and thermophoresis, is taken into account. The dependency of the local Nusselt number and the entropy generation number of the nondimensional parameters is numerically investigated. The results show that the increase of heat generation, Brownian motion, or thermophoresis parameters decreases the entropy generation number in the vicinity of the sheet.

Makinde et al. [37] considered the problem of entropy generation and inherent irreversibility in the steady boundary layer shear flow of nanofluids over a moving flat plate. The governing partial differential equations are transformed into ordinary differential equations using a similarity transformation and then solved numerically by a Runge–Kutta–Fehlberg method with the shooting technique. Two types of nanofluids, namely, Cu–water and TiO_2–water, were used. The effects of nanoparticle volume fraction, the type of nanoparticles, the group parameter, the local Reynolds number on the entropy generation rate, the irreversibility ratio, and the Bejan number are discussed. It is found that the entropy generation rate at the plate surface decreases with increasing nanoparticle volume fraction and the group parameter. Moreover, the heat transfer irreversibility at the plate surface with TiO_2–water nanofluid is slightly higher than that at the plate surface with Cu–water nanofluid.

Sheikholeslami et al. [38] investigated the entropy generation due to nanofluid flow over a permeable stretching sheet in a porous medium. Various water-based nanofluids containing copper, silver, alumina, and titanium oxide nanoparticles are considered in this analysis. They found that with an increase in the volume fraction of nanoparticles the entropy generation decreases; this decrease has a direct relation with the thermal conductivity of particles so that Ag nanoparticles with the highest thermal conductivity give the minimum entropy generation.

15.3.7 Entropy Generation in Solar Collectors

Alim et al. [39] studied the entropy generation and pressure drop due to nanofluid flow in a flat plate solar collector using different nanofluids. They investigated the performance for four different nanoparticles, including alumina, copper oxide, titanium oxide, and silicon dioxide with volume fractions between 1% and 4%, and volumetric flow rates of 1–4 L/min. They estimated analytically the entropy generation in the collector and found that CuO nanoparticles, having the highest density, provide the lowest entropy generation, decreasing the entropy generation by 4.34%.

Parvin et al. [40] studied the entropy generation in a direct absorption solar collector filled with Cu/water nanofluid. Figure 15.8a shows the variations of average entropy generation with volume fraction at Re = 600. Clearly, the entropy generation due to nanofluid flow is higher than for pure water. It is also seen that the entropy generation increases with an increase in the nanofluid concentration by volume fraction of 3%, but

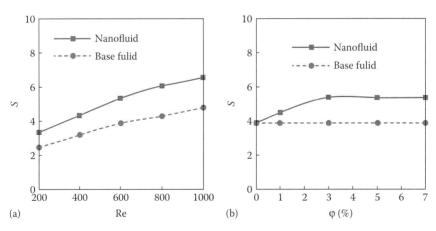

FIGURE 15.8   Variations of average entropy generation with volume fraction for Re = 600 (a) and with Reynolds number for φ = 3% (b). (Reprinted from *Int. J. Heat Mass Transf.*, 71, Parvin, S. et al., Heat transfer and entropy generation through nanofluid filled direct absorption solar collector, 386–395, Copyright 2014, with permission from Elsevier.)

it does not change for volume fractions higher than 3%. This may happen because for higher volume fractions the positive effects of thermal conductivity enhancement compensate the negative effects of viscosity enhancement; therefore, the entropy generation does not grow anymore. Figure 15.8b displays the variations of average entropy generation with the Reynolds number for volume fraction of 3%. It was observed that the entropy generation increases with increasing Reynolds number.

15.3.8 Entropy Generation in Nanofluid Flow Past a Barrier

Boghrati et al. [41] examined numerically the entropy generation due to the flow of Al_2O_3 and carbon nanotubes suspended in water through two horizontal parallel plates. A rectangular barrier was mounted between the two plates so that the nanofluid can flow up and down the barrier. They found that the entropy generation increases by adding nanoparticles. They also observed that the entropy generation due to carbon nanotubes is more than 5 times when using Al_2O_3 nanoparticles. Sarkar et al. [42] investigated the entropy generation due to mixed water-based Al_2O_3 and Cu nanofluids that were flowing past an inserted square barrier in the middle of two vertical parallel plates. They employed the finite-element method for the transient analysis of the problem, considering the highest value of volume fraction of nanoparticles to be equal to 20%. Their results show that the total entropy generation decreased by about 25% when the volume fraction increased from 0% to 20%. They also found that the entropy generation due to fluid friction is negligible compared to the irreversibilities due to heat transfer. There is no considerable difference between the total entropy generation for both nanoparticles (Al_2O_3 and CuO) when the volume fraction increases from 0% to 20%, except for the negative values of the buoyancy parameter Gr/Re^2. It should be noted that Gr/Re^2 is negative when the inlet temperature of nanofluid to the channel is higher than the temperature of the channel walls. The Brinkman and Maxwell models were used to determine the viscosity and thermal conductivity of the nanofluids.

15.3.9 Entropy Generation in Other Systems

Feng and Kleinstreuer [43] studied numerically the entropy generation in Al_2O_3–water nanofluid flow between parallel disks. The schematic of the problem is shown in Figure 15.9. They assumed steady, laminar, axisymmetric, developing flow of a dilute nanoparticle suspension. The authors [43] concluded that the total entropy generation rate in the system reduces by adding nanoparticles. Specifically, whereas adding nanoparticles

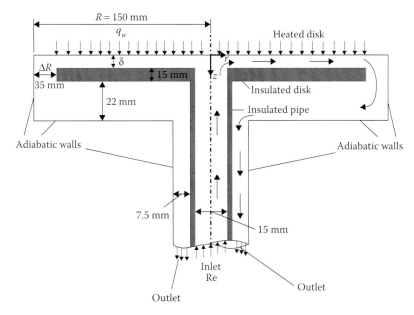

FIGURE 15.9 Schematic of the parallel disks considered by Feng and Kleinstreuer. (Reprinted from *Int. J. Heat Mass Transf.*, 53, Feng, Y. and Kleinstreuer, C., Nanofluid convective heat transfer in a paralleldisk system, 4619–4628, Copyright 2010, with permission from Elsevier.)

increases the entropy generation due to fluid friction, because of the elevated mixture viscosity, particle loading reduces the entropy generation induced by heat transfer because of smaller temperature gradients.

Rashidi et al. [44] examined the second law of thermodynamics for the flow of three different water-based nanofluids containing Cu, CuO, and Al_2O_3 nanoparticles over a horizontal porous disk. A similarity solution is presented where a magnetic field is applied to the disk perpendicularly. The results showed that the entropy generation decreases with increasing nanofluid concentration. The maximum and minimum values of the entropy generation were reported for Cu and Al_2O_3 nanoparticles, respectively. It was also found that an increase in the magnetic field intensity and the suction flow parameter leads to an increase in the entropy generation.

15.4 CONCLUSIONS

EGM through nanofluid flow is the subject of this chapter. First, the two general methods for formulating the problems are presented. In one approach to estimate the entropy generation, the available correlations in the literature were outlined to calculate the Nusselt number and friction factor. In another approach, the momentum and energy equations have to be solved. Then, the gradients of velocity and temperature are utilized to obtain the entropy generation. A literature review of the studies on nanofluids was conducted. In general, the conclusions as well as suggestions for future work are as follows:

- Adding nanoparticles is advantageous when the heat transfer contribution to the entropy generation is adequately higher than any fluid friction contribution.

- Selecting suitable thermophysical models is very important in order to accurately predict trends and total values of entropy generation.

- In natural convection problems, the entropy generation in nanofluid flow increases with an increase in the buoyancy parameters, as indicated by the Rayleigh number and Grashof number.

- The Bejan number increases with an increase in nanofluid concentration.

- Water-based nanofluids show a lower entropy generation compared to EG-based nanofluids with similar nanoparticles because of lower viscosity and higher thermal conductivity of water.

- In general, using nanoparticles with a smaller size can reduce the entropy generation due to thermal conductivity enhancements; but, in some situations, it was reported that at high Reynolds numbers increasing the nanoparticle size leads to a reduction in the entropy generation (see Bianco et al. [26]).

- For future works, the entropy generation due to nanofluid flow should consider the effects of radiation and porous media.

- In addition, the EGM technique should be expanded to classic problems where the working fluid is a nanofluid.

NOMENCLATURE

Variables

A	surface area (m^2)
D	inner diameter (m)
f	friction factor
k	thermal conductivity (W/m K)

$\dot{m}$ mass flow rate (kg/s)

Nu   Nusselt number

Pr   Prandtl number

q'' heat flux (W/m²)

Re   Reynolds number

$\dot{S}'_{gen}$ entropy generation rate in length unit (W/m K)

$\dot{S}'''_{gen}$ entropy generation rate in volume unit (W/m³ K)

T temperature (K)

v velocity component

Greek symbols

μ   dynamic viscosity (kg m/s)

ρ   density (kg/m³)

υ   kinematics viscosity (m²/s)

Subscripts

ave   average

REFERENCES

1. Bejan, A. *Entropy Generation through Heat and Fluid Flow*, Wiley, New York, 1982.
2. Mahian, O., Kianifar, A., Kleinstreuer, C., Al-Nimr, M.A., Pop, I., Sahin, A.Z., and Wongwises, S. "A review of entropy generation in nanofluid flow." *International Journal of Heat and Mass Transfer* 65, 2013: 514–532.
3. Singh, P.K., Anoop, K.B., Sundararajan, T., and Das, S.K. "Entropy generation due to flow and heat transfer in nanofluids." *International Journal of Heat and Mass Transfer* 53, 2010: 4757–4767.
4. Li, J., and Kleinstreuer, C. "Entropy generation analysis for nanofluid flow in microchannels." *Journal of Heat Transfer* 132, 2010: 122401.1–122401.8.
5. Tabrizi, S.A., and Seyf, H.R. "Analysis of entropy generation and convective heat transfer of Al_2O_3 nanofluid flow in a tangential micro heat sink." *International Journal of Heat and Mass Transfer* 55, 2012: 4366–4375.
6. Mah, W.H., Hung, Y.M., and Guo, N. "Entropy generation of viscous dissipative nanofluid flow in microchannels." *International Journal of Heat and Mass Transfer* 55, 2012: 4169–4182.
7. Sohel, M.R., Saidur, R., Hassan, N.H., Elias, M.M., Khaleduzzaman, S.S., and Mahbubul, I.M. "Analysis of entropy generation using nanofluid flow through the circular microchannel and minichannel heat sink." *International Communications in Heat and Mass Transfer* 46, 2013: 85–91.
8. Ting, T.W., Hung, Y.M., and Guo, N. "Entropy generation of nanofluid flow with streamwise conduction in microchannels." *Energy* 64, 2014: 979–990.

9. Mohammadian, S.K., Seyf, H.R., and Zhang, Y. "Performance augmentation and optimization of aluminum oxide-water nanofluid flow in a two-fluid microchannel heat exchanger." *Journal of Heat Transfer* 136, 2013: 021701.

10. Shahi, M., Mahmoudi, A.H., and Raouf, A.H. "Entropy generation due to natural convection cooling of a nanofluid." *International Communications in Heat Mass Transfer* 38, 2011: 972–983.

11. Cho, C.C., Chen, C.L., and Chen, C.K. "Natural convection heat transfer and entropy generation in wavy-wall enclosure containing water-based nanofluid." *International Journal of Heat and Mass Transfer* 61, 2013: 749–758.

12. Mahmoudi, A.H., Pop, I., Shahi, M., and Talebi, F. "MHD natural convection and entropy generation in a trapezoidal enclosure using Cu-water nanofluid." *Computers & Fluids* 72, 2013: 46–62.

13. Sheikhzadeh, G.A., and Nikfar, M. "Aspect ratio effects of an adiabatic rectangular obstacle on natural convection and entropy generation of a nanofluid in an enclosure." *Journal of Mechanical Science and Technology* 27, 2013: 3495–3504.

14. Kashani, S., Ranjbar, A.A., Mastiani, M., and Mirzaei, H. "Entropy generation and natural convection of nanoparticle-water mixture (nanofluid) near water density inversion in an enclosure with various patterns of vertical wavy walls." *Applied Mathematics and Computation* 226, 2014: 180–193.

15. Parvin, S., and Chamkha, A.J. "An analysis on free convection flow, heat transfer and entropy generation in an odd-shaped cavity filled with nanofluid." *International Communications in Heat and Mass Transfer* 54, 2014: 8–17.

16. Mehrez, Z., Bouterra, M., El Cafsi, A., and Belghith, A. "Heat transfer and entropy generation analysis of nanofluids flow in an open cavity." *Computers & Fluids* 88, 2013: 363–373.

17. Mahmoudi, A.H., and Hooman, K. "Effect of a discrete heat source location on entropy generation in mixed convective cooling of a nanofluid inside the ventilated cavity." *International Journal of Exergy* 13, 2013: 299–319.

18. Bianco, V., Nardini, S., and Manca, O. "Enhancement of heat transfer and entropy generation analysis of nanofluids turbulent convection flow in square section tubes." *Nanoscale Research Letters* 6, 2011: 252.

19. Moghaddami, M., Mohammadzade, A., and Varzane Esfehani, S.A. "Second law analysis of nanofluid flow." *Energy Conversion and Management* 52, 2011: 1397–1405.

20. Moghaddami, M., Shahidi, S.E., and Siavashi, M. "Entropy generation analysis of nanofluid flow in turbulent and laminar regimes." *Journal of Computational and Theoretical Nanoscience* 9, 2012: 1–10.

21. Leong, K.Y., Saidur, R., Mahlia, T.M.I., and Yau, Y.H. "Entropy generation analysis of nanofluid flow in a circular tube subjected to constant wall temperature." *International Communications in Heat and Mass Transfer* 39, 2012: 1169–1175.

22. Karami, M., Shirani, E., and Avara, A. "Analysis of entropy generation, pumping power, and tube wall temperature in aqueous suspensions of alumina particles." *Heat Transfer Research* 43, 4, 2012: 327–342.

23. Falahat, A., and Vosough, A. "Effect of nanofluid on entropy generation and pumping power in coiled tube." *Journal of Thermophysics and Heat Transfer* 26, 1, 2012: 141–146.
24. Ahadi, M., and Abbassi, A. "Exergy analysis of laminar forced convection of nanofluids through a helical coiled tube with uniform wall heat flux." *International Journal of Exergy* 13, 2013: 21–35.
25. Khairul, M.A., Saidur, R., Rahman, M.M., Alim, M.A., Hossain, A., and Abdin, Z. "Heat transfer and thermodynamic analyses of a helically coiled heat exchanger using different types of nanofluids." *International Journal of Heat and Mass Transfer* 67, 2013: 398–403.
26. Bianco, V., Manca, O., and Nardini, S. "Entropy generation analysis of turbulent convection flow of Al_2O_3-water nanofluid in a circular tube subjected to constant wall heat flux." *Energy Conversion and Management* 77, 2014: 306–314.
27. Sarkar, S., Ganguly, S., and Biswas, G. "Buoyancy driven convection of nanofluids in an infinitely long channel under the effect of a magnetic field." *International Journal of Heat and Mass Transfer* 71, 2014: 328–340.
28. Mahian, O., Pop, I., Sahin, A.Z., Oztop, H.F., and Wongwises, S. "Irreversibility analysis of a vertical annulus using TiO_2/water nanofluid with MHD flow effects." *International Journal of Heat and Mass Transfer* 64, 2013, 671–679.
29. Matin, M.H., Hosseini, R., Simiari, M., and Jahangiri, P. "Entropy generation minimization of nanofluid flow in a MHD channel considering thermal radiation effect." *Mechanika* 19, 4, 2013: 445–450.
30. Mahian, O., Mahmud, S., and Heris, S.Z. "Analysis of entropy generation between co-rotating cylinders using nanofluids." *Energy* 44, 2012: 438–446.
31. Mahian, O., Mahmud, S., and Heris, S.Z. "Effect of uncertainties in physical properties on entropy generation between two rotating cylinders with nanofluids." *Journal of Heat Transfer* 134, 2012: 101704.
32. Mahian, O., Mahmud, S., and Wongwises, S. "Entropy generation between two rotating cylinders in the presence of magnetohydrodynamic flow using nanofluids." *Journal of Thermophysics and Heat Transfer* 27, 2013: 161–169.
33. Leong, K.Y., Saidur, R., Khairulmaini, M., Michael, Z., and Kamyar, A. "Heat transfer and entropy analysis of three different types of heat exchangers operated with nanofluids." *International Communications in Heat and Mass Transfer* 39, 2012: 838–843.
34. Elias, M.M., Shahrul, I.M., Mahbubul, I.M., Saidur, R., and Rahim, N.A. "Effect of different nanoparticle shapes on shell and tube heat exchanger using different baffle angles and operated with nanofluid." *International Journal of Heat and Mass Transfer* 70, 2014: 289–297.
35. Matin, M.H., Nobari, M.R.H., and Jahangiri, P. "Entropy analysis in mixed convection MHD flow of nanofluid over a non-linear stretching sheet." *Journal of Thermal Science and Technology* 7, 1, 2012: 104–119.
36. Noghrehabadi, A., Saffarian, M.R., Pourrajab, R., and Ghalambaz, M. "Entropy analysis for nanofluid flow over a stretching sheet in the presence of heat generation/absorption and partial slip." *Journal of Mechanical Science and Technology* 27, 2013: 927–937.

37. Makinde, O.D., Khan, W.A., and Aziz, A. "On inherent irreversibility in Sakiadis flow of nanofluids." *International Journal of Exergy* 13, 2013: 159–174.

38. Sheikholeslami, M., Ellahi, R., Ashorynejad, H.R., Domairry, G., and Hayat, T. "Effects of heat transfer in flow of nanofluids over a permeable stretching wall in a porous medium." *Journal of Computational and Theoretical Nanoscience* 11, 2014: 486–496.

39. Alim, M.A., Abdin, Z., Saidur, R., Hepbaslid, A., Khairul, M.A., and Rahim, N.A. "Analyses of entropy generation and pressure drop for a conventional flat plate solar collector using different types of metal oxide nanofluids." *Energy and Buildings* 66, 2013: 289–296.

40. Parvin, S., Nasrin, R., and Alim, M.A. "Heat transfer and entropy generation through nanofluid filled direct absorption solar collector." *International Journal of Heat and Mass Transfer* 71, 2014: 386–395.

41. Boghrati, M., Ebrahim nia Bajestan, E., and Etminan, V. "Entropy generation minimization of confined nanofluids laminar flow around a block." *Proceedings of the 10th Biennial Conference on Engineering Systems and Analysis*, ESDA 2010, July 12–14, 2010, Istanbul, Turkey.

42. Sarkar, S., Ganguly, S., and Dalal, A. "Analysis of entropy generation during mixed convection heat transfer of nanofluids past a square cylinder in vertically upward flow." *Journal of Heat Transfer* 134, 2012: 122501.

43. Feng, Y., and Kleinstreuer, C. "Nanofluid convective heat transfer in a parallel-disk system." *International Journal of Heat and Mass Transfer* 53, 2010: 4619–4628.

44. Rashidi, M.M., Abelman, S., and Freidooni mehr, N. "Entropy generation in steady MHD flow due to a rotating porous disk in a nanofluid." *International Journal of Heat and Mass Transfer* 62, 2013: 515–525.

CHAPTER 16

Gas-Based Nanofluids (Nanoaerosols)

Wesley C. Williams

CONTENTS

16.1 INTRODUCTION AND OVERVIEW

16.1.1 Introduction to Gas-Based Nanofluids (Nanoaerosols)

Gas-based nanofluids or nanoaerosols are defined as any aerosol created with nanometer-scale particulates, typically 1–100 nm. Similar to their liquid counterparts, gas-based nanofluids have the potential to increase the thermal conductivity as well as the heat capacity of the base gas and therefore enhance the heat transfer performance of the gas. Initial investigation of gas-based nanofluids came from their potential application in gas-cooled nuclear power plants, particularly for advanced gas-reactor cooling such as the gas-cooled fast reactor or the high-temperature gas-cooled reactor. Gas nanofluids are also of great interest in the fusion reactor diverter cooling system. Improved heat transfer performance of gas coolants would have a significant impact on the performance and safety of gas-cooled nuclear power reactors, as well as, many other technologies.

As part of a National Laboratory Directed Research and Development project, the Idaho National Lab (INL) took the lead in the design and construction of a gas–nanoparticle suspension convective cooling experiment as shown in Figure 16.1. This chapter will cover some initial theoretical background work on gas–particulate flows, which were performed to guide the experimental and future theoretical investigations of gas nanofluids.

16.1.2 Overview of This Chapter

This chapter theoretically investigates the potential for gas-based nanofluids to be utilized as a cooling medium in nuclear reactors. This investigation comes from the development of an experimental apparatus for performing gas-based nanofluid convective heat transfer performance. The analysis discusses the validity of the continuum assumption for gas-based nanofluids, transient temperature response of the particles inside of the gas, local thermal equilibrium between the particles and the gas, and potential heat transfer enhancement and turbulence strengthening due to particle dispersion. It also discusses the potential for thermal conductivity increase due to nanoparticle

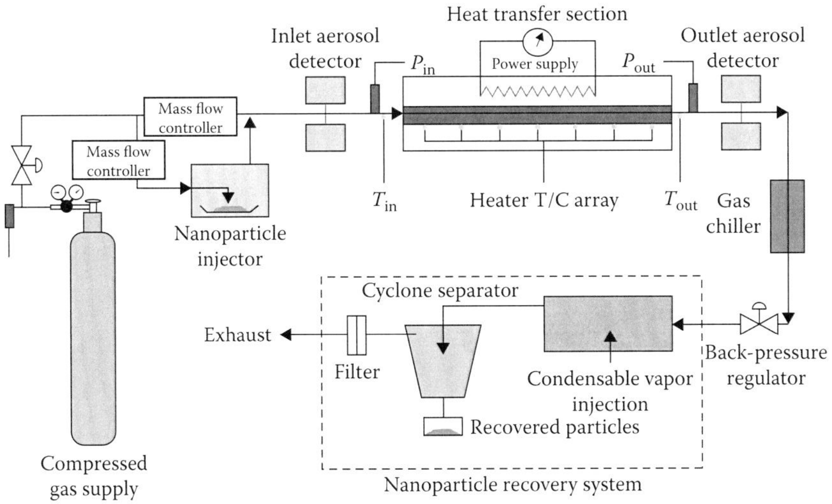

FIGURE 16.1   Schematic of the INL gas loop experiment. P_{in}, inlet pressure; T_{in}, inlet temperature sensors; P_{out}, outlet pressure; T_{out}, outlet temperature sensors; *T/C*, thermocouple array.

dispersion, the radiant heat transfer between the particles and the wall, and the development of thermal conductivity and viscosity gradients in the boundary layer due to particle migration. This discussion hopes to provide a theoretical basis for further experimental and theoretical studies of gas-based nanofluids.

The author spent a three-week period at the INL and contributed this work to the gas nanofluid project under way there. A determination of the key transport phenomena in the INL gas nanofluid experiment is made through a semiquantitative approach. The current specifications of the experiment (i.e., temperature, pressure, flow, ID, etc.) are used in the analysis. It is important to note that, due to the sensitivity of gas properties to temperature and pressure, gas nanofluid systems can be regarded as continuous, free molecular, or transitional in behavior; thus, the results of this analysis are applicable to the described experimental situation only. The insights found lead to the recommendation of certain focal points for the experimental and future theoretical work with gas nanofluids. This chapter is divided into three sections, which are described subsequently.

16.2 ASSUMPTIONS

The main assumption is that the flow situation considered is for the INL gas nanofluid experiment. The experiment is a once-through heated pipe convection heat transfer system. The system has the means to inject, measure, and remove nanometer-sized particles at various concentrations. The purpose of the experiment is to determine the potential for using gas nanoparticle dispersions (nanofluids) to enhance convective heat transfer. The experiment uses helium gas as the base fluid and, for the sake of analysis, carbon-based nanoparticles. For the analysis, the following parameters are set:

Helium gas state and properties

- Pressure $(P) = 0.2$ MPa
- Temperature $(T) = 873$ K
- Viscosity $(\mu) \cong 4 \times 10^{-5}$ Pa·s
- Thermal conductivity $(k) \cong 0.3$ W/m·K
- Density $(\rho) \cong 0.11$ kg/m^3
- Heat capacity $(c) \cong 5193$ J/kg·K
- Ideal gas constant $(R) = 8.31$ J/mol·K
- Avogadro's number $(N_{av}) = 6.022 \times 10^{23}$ atoms/mol
- Effective diameter of single molecule $(d_{He}) \cong 0.6 \times 10^{-10}$ m

Nanoparticle properties

- Material—Pyrocarbon spheres
- Particle diameter $(d_p) = 1{-}100$ nm
- Thermal conductivity $(k_p) \cong 4$ W/m·K or greater
- Density $(\rho_p) \cong 1900$ kg/m^3
- Heat capacity $(c_p) \cong 709$ J/kg·K

Operating conditions

- Reynolds number (Re) $= 4000$
- Pipe diameter $(D) = 17.3$ mm

Furthermore, in order to have an obtainable starting point to the study, the nanoparticles are assumed to be ideally dispersed and perfectly spherical.

This, however, is likely a weak assumption, which will be further discussed at the end of Section 16.3.

16.3 ANALYSIS

The analysis is composed of five parts as follows: (1) assessment of the continuum assumption, (2) evaluation of the internal particle temperature response, (3) evaluation of the particle/fluid heat transfer response, (4) evaluation of nanoparticle dispersion and turbulence intensification as heat transfer enhancement mechanisms, and (5) discussion of other possible heat transfer enhancement mechanisms.

16.3.1 Continuum Assumption

The common way of determining whether the fluid–particle interaction can be treated as a continuum phenomenon or not is to calculate the Knudsen number (Kn). The Knudsen number is the ratio of the mean free path (λ) of the fluid particles to the characteristic length of the system with which they interact. Here, it is chosen as the mean free path of the gas atoms divided by the diameter of the nanoparticles as shown in Equation 16.1, where the mean free path for helium is calculated as shown in Equation 16.2:

$$\text{Kn} = \frac{\lambda}{d_p} \tag{16.1}$$

$$\lambda = \frac{RT}{\sqrt{2}\pi d_{\text{He}} P N_{\text{av}}} \tag{16.2}$$

Knudsen values less than 0.01 are considered to be in the continuum; values from 0.01 to 1.0 can be considered near-continuum; values from 1.0 to 10 can be considered transitional; and values above 10 are free molecular. Navier–Stokes equations are valid for the continuum regime below 0.01 and can be extended, through the use of slip conditions, up to 0.1. Figure 16.2 shows the dependence of the Knudsen number on the diameter of the nanoparticle. From the figure, it is determined that the nanoparticle will see the surrounding gas as a free molecular interaction and therefore not a continuum. Further analysis determined the Knudsen number to be very weakly dependent on the base gas composition. However, the analysis of another work (Willeke 1976) demonstrates the pressure and temperature dependence of the gas mean free path,

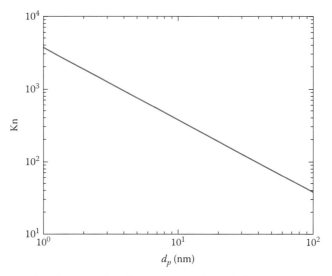

FIGURE 16.2 Knudsen number for nanoparticles in helium gas.

which can be calculated proportional to the initial mean free path values with subscript of 0:

$$\lambda = \lambda_0 \left(\frac{T}{T_0} \right) \left(\frac{P_0}{p} \right) \left[\frac{1+(S/T_0)}{1+(S/T)} \right] \tag{16.3}$$

The Sutherland constant (S), which accounts for the attractive potentials between molecules, in the above equation can be assumed to have a typical value of 113 K, as is the case for air (Chapman and Cowling 1958). Therefore, it is determined that at lower temperatures and higher pressures, the latter being typical of gas-cooled reactor applications, the interaction can be moved to the transitional regime.

16.3.2 Determination of Internal Particle Temperature Response

A comparison of the timescale (τ_p) of energy redistribution within the nanoparticle and the timescales of the energy transfer from the particle to the surrounding gas determines whether the temperature distribution inside the nanoparticle can be regarded as uniform. Energy is transferred to and from the particle by conduction and radiation, their timescales being indicated here as τ_{con} and τ_{rad}, respectively.

16.3.2.1 Energy Redistribution within Particle

The timescale of the energy transfer can be calculated using Equation 16.4 utilizing the thermal diffusivity as calculated using Equation 16.5:

$$\tau_p \cong \frac{d_p^2}{0.12\alpha_p} \tag{16.4}$$

$$\alpha_p = \frac{k_p}{\rho_p c_p} \tag{16.5}$$

This analysis follows from the transient conduction solution for spherical systems with low Biot number (Incropera and DeWitt 1990).

16.3.2.2 Heat Conduction from Particle to Gas

An energy balance for the particle is developed as shown in Equation 16.6. The equation utilizes the helium atom concentration in atoms/m³ defined by Equation 16.7, the average helium speed in m/s in Equation 16.8, and the mass of the helium atom in kg in Equation 16.9:

$$\rho_p C_p \frac{\pi}{6} d_p^3 \frac{dT_p}{dt} = -\frac{\pi}{2} d_p^2 n_g \bar{c} k_b (T_p - T) \tag{16.6}$$

$$n_g = \frac{PN_{av}}{RT} \tag{16.7}$$

$$\bar{c} = \sqrt{\frac{8k_b T}{\pi n_g}} \tag{16.8}$$

$$n_g = \frac{0.004}{N_{av}} \tag{16.9}$$

The right-hand term in Equation 16.6, the particle surface heat flux, is drawn from Filippov and Rosner's paper (Filippov and Rosner 2000). If one assumes an accommodation coefficient equal to 1, $\gamma = 5/3$, and T as the bulk gas temperature from the energy balance equation, the timescale for heat conduction to the gas is derived as shown in the following equation:

$$\tau_{con} \cong \frac{\rho_p c_p d_p}{3n_g \bar{c} k_b} \tag{16.10}$$

16.3.2.3 Heat Radiation from Particle to Surroundings

An energy balance for the particle with radiation heat transfer is developed and shown in the following equation:

$$\rho_p c_p \frac{\pi}{6} d_p^3 \frac{dT_p}{dt} = -\pi d_p^2 \sigma \left(T_p^4 - T^4 \right) \tag{16.11}$$

Assuming the black-body behavior, the Stefan–Boltzmann constant (σ) is selected to be 5.67×10^{-8} W/m^{2}K^4. From the energy balance equation, the timescale for radiative heat transfer is derived as

$$\tau_{rad} \cong \frac{\rho_p c_p d_p}{6\sigma T^3} \tag{16.12}$$

Comparison of these timescales defined in Equations 16.4, 16.10, and 16.12 demonstrates the relative importance of each mechanism in the overall heat transfer process. Figure 16.3 shows a plot of the timescales of the above three derived mechanisms and how these vary with the prescribed nanoparticle diameter.

Two conclusions can be drawn from Figure 16.3. First, the energy redistribution within the particle is much faster than the energy transfer

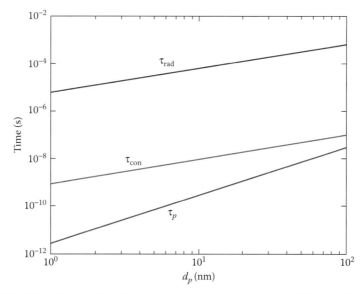

FIGURE 16.3 Timescales for heat transfer mechanisms within and around nanoparticles in helium gas.

between particle and gas due to the large particle conductivity. Thus, the temperature distribution within the particle is uniform. Second, the dominant heat transfer mechanism from the particle to the gas is conduction. However, due to the long-range nature of radiation, particle-to-particle and particle-to-wall radiation can still play a significant role in the overall heat transfer enhancement in the channel.

16.3.3 Determination of Local Thermal Equilibrium between the Particles and the Gas

A comparison of the timescale of the heat transfer from the particle to the gas with the timescale of the particle slip motion in the gas can be used to determine the thermal equilibrium characteristics of the mixture. The heat transfer timescale has been determined in Section 16.3.3. In turbulent flow, the dominant slip mechanism is determined from assuming an inertial flight of the nanoparticle following an abrupt stop of the eddy carrying the particle. The timescale for this process is the so-called relaxation time of the particle and is quantified by Equation 16.13 utilizing Equation 16.14, the Cunningham corrective factor, which accounts for the noncontinuum nature of the particle–gas interaction:

$$\tau_{\mathrm{rel}} = \frac{\rho_p d_p^2}{18\mu} C_c \qquad (16.13)$$

$$C_c = 1 + \mathrm{Kn}\left(\alpha + \beta e^{-\gamma/\mathrm{Kn}}\right) \qquad (16.14)$$

The coefficients inside the Cunningham corrective factor α, β, and γ are determined experimentally for exact fluid/particle combinations (which is not done here). For the calculation, the values are assumed as $\alpha = 2.34$, $\beta = 1.05$, and $\gamma = -0.39$. It was later found from a recent experiment of a similar system using polystyrene latex particles from 20 to 270 nm that $\alpha = 1.165$, $\beta = 0.483$, and $\gamma = 0.997$ (Kim et al. 2005). The Cunningham factors were recalculated with these new values and found to be on the same order as the original calculation; therefore, the original values were kept for this analysis. The timescales τ_{rel} and τ_{con} are shown in Figure 16.4. Because $\tau_{\mathrm{con}} \ll \tau_{\mathrm{rel}}$, it is concluded that particles can exchange energy very effectively as they fly within the gas; therefore, we can assume that there is local particle/gas thermal equilibrium.

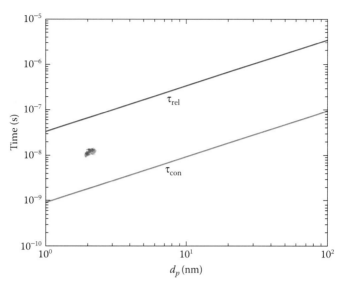

FIGURE 16.4 Comparison of timescales for energy transfer and nanoparticle slip motion in helium gas.

16.3.4 Heat Transfer Enhancement Due to Particle Dispersion, Turbulence Strengthening, and Turbulent Deposition

Due to their small size, nanoparticles can be entrained in both large and small turbulent eddies. This is shown by the size (ℓ), velocity (V), and timescales (τ) of the eddies as follows:

Scales for large eddies

- $\ell_o \cong 0.07D \cong 1.2\ \mathrm{mm}$

- $V_o \cong V_{\mathrm{shear}} \cong 6\ \mathrm{m/s}$

- $\tau_o \cong \ell_o/V_o \cong 2 \times 10^{-4}\ \mathrm{s}$

Scales for small eddies (using Kolomogorov's scaling laws)

- $\ell_s/\ell_o \cong \mathrm{Re}^{-3/4} \rightarrow \ell_s \cong 2.4\ \mu\mathrm{m}$

- $\tau_s/\tau_o \cong \mathrm{Re}^{-1/2} \rightarrow \tau_s \cong 3 \times 10^{-6}\ \mathrm{s}$

- $V_s \cong \ell_s/\tau_s \cong 0.7\ \mathrm{m/s}$

It is thus evident that the particle sizes (1–100 nm) are much smaller than the eddy sizes (1.2 and 2.4 μm).

The next step is to determine the stopping distance (S) for the particles entrained in the small and large eddies in order to see how far they can be thrown by the eddy velocities. The following equations can be used to determine the stopping distances for particles entrained in large and small eddies, respectively.

$$S_o = \frac{\rho_p d_p^2}{18\mu} C_c V_o \tag{16.15}$$

$$S_s = \frac{\rho_p d_p^2}{18\mu} C_c V_s \tag{16.16}$$

The calculated stopping distances for each eddy scale are shown in Figure 16.5. The results in Figure 16.5 suggest that nanoparticles do not significantly project out of the eddies with the possible exception of large nanoparticles, that is, $d_p > 100$ nm, entrained by small eddies. Therefore, one can conclude that for this case the nanoparticles move with the turbulent eddies. Thus, the contribution of inertial slip to nanoparticle dispersion is probably negligible. Regarding the question of turbulence intensification, the presence of the nanoparticles will likely increase the viscosity, which will delay the onset of turbulence for a given mean velocity.

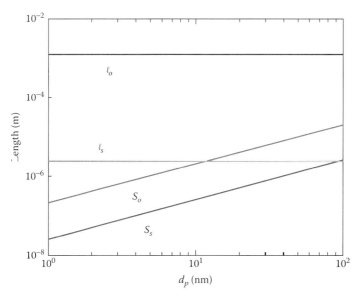

FIGURE 16.5 Comparison of eddy sizes and particle stopping distances.

Moreover, Figure 16.5 suggests that particles entrained by large eddies may interact with (i.e., breakup) small eddies, because their length scales are comparable. Based on the order-of-magnitude estimates of the kinetic energy carried by each eddy and using Kolomogorov's scaling laws, one finds that the ratio of the number of small eddies generated per unit time and volume to the number of large eddies generated per unit time and volume is proportional to $Re^{11/4}$, which in our case is about 1010.

Therefore, small eddies are much more numerous than large eddies. This would mean that nanoparticles being projected out of large eddies will affect only a very limited number of small eddies. In sum, any effect on the turbulence due to nanoparticle dispersion, beyond the obvious effect via the change in viscosity, seems unlikely.

Turbulent fluctuations can also impart enough inertia for the nanoparticles to potentially penetrate the boundary layer. This is known as turbulent deposition of particles. The intensification of the particle velocity follows parallel to the intensification of the turbulence. Per the work of Friedlander (2000), no turbulent deposition of particles occurs until local Reynolds numbers reach 10^5. The assumed Reynolds number for this study is 4000. Therefore, it is determined that there will be no turbulent deposition of the nanoparticles.

16.3.5 Discussion of Other Possible Heat Transfer Enhancement Mechanisms

16.3.5.1 Thermophoresis

The thermophoretic effect can be visualized as an imbalance in the forces of the gas molecules colliding on the opposite sides of the nanoparticle. Per kinetic gas theory, gas molecules on the hotter side of the nanoparticle would collide with a higher velocity than the colder side molecules. This force imbalance tends to drive the nanoparticle from hotter regions toward colder regions of the gas, an effect known as thermophoresis. Thermophoresis is used to create "dust-free" zones or layers near heated surfaces. The system being analyzed here is a heated tube surface being cooled by the nanoaerosol. Obviously, thermophoresis would drive the particles out and away from the heated surface and potentially out of the boundary layer. Due to the molecular nature of thermophoresis, its effects are highly dependent on the size of the particle. Therefore, selection of the correct model for thermophoresis depends on the analysis of the continuum assumption from Section 16.3.1. The work of Mädler and Friedlander (2007) thoroughly covers the theory of thermophoresis in nanoparticle aerosols. The preceding work states that the thermophoretic velocity in the free molecular regime, where the particle size is much smaller than the mean free path of the gas, does not depend on the particle size, except for potentially particles smaller than 5 nm. The analysis of Section A showed that the system under investigation is in the free molecular regime or potentially in the transitional regime at low temperatures and high pressures. The dimensionless thermophoretic velocity should be approaching a value of 0.55, as derived from the kinetic gas theory.

Thermophoresis is driven by the temperature gradient in the gas. The system in this study is a low-pressure turbulent gas flow. It can be assumed that the largest temperature gradients are in the laminar boundary layer near the wall. This would create an area of high thermophoretic velocities. It can be assumed that thermophoretic effects would be far overshadowed by turbulent mixing in the bulk flow of gas, and therefore is inconsequential in the bulk flow. For this experimental system, it is determined that the particles would be pushed out of the boundary layer by the creation of a dust-free zone by the large temperature gradient. For this reason, we can assume that particles likely only reside in the bulk flow due to thermophoresis and that it would likely not contribute to the heat transfer in the boundary layer. However, further analysis is required to determine the effects on heat transfer.

16.3.5.2 Brownian Diffusion

Brownian motion of the gas in the nanoaerosol would have the tendency to evenly distribute the nanoparticles inside the fluid. The movement of the nanoparticles can be viewed as a diffusion process that depends on the kinetic nature of the gas and the coefficient of friction of the particles. The general theory that covers this behavior is the Stokes–Einstein expression for the coefficient of diffusion as seen in Equation 16.17 using the corrected Stokes friction factor defined by Equation 16.18, where μ is the gas viscosity:

$$D = \frac{kT}{f} \tag{16.17}$$

$$f = \frac{3\pi\mu d_p}{C_c} \tag{16.18}$$

A table of typical values for the Cunningham correction factor can be found on page 34 of the book of Friedlander (2000). This diffusion would effectively push particles into the areas of lower concentration, that is, the boundary layer. This diffusion force would contradict the thermophoretic force. In turbulent flows similar to the system under investigation, it is found that Brownian diffusion and turbulent diffusion are the two strongest actors in particle migration in the flow.

16.3.5.3 Agglomeration

As the particles move inside the gas, they collide and/or come into close proximity. At these close ranges, the particles could coagulate or coalesce. Obviously, Brownian and turbulent diffusion and shear can create enough force to push particles into each other. Once the particles come into close range, attractive van der Waals forces can pull the particles temporarily or permanently together. These forces can cause the particles to form larger agglomerations. If the particles chemically react with one another, then coalescence can occur. It is assumed that the nanoparticles in this system will be nonreactive; therefore, coalescence is not a possibility. However, agglomeration will occur. Agglomeration sizes are determined by the equilibrium between agglomerate formation due to the particle collisions and van der Waals forces that bring them together, and the shear and Brownian forces that can break the agglomerates apart. In the work of Friedlander (2000) on page 209, there is a comparative table of coagulation mechanisms as a function of particle size. It can be seen for particle diameters at 100 nm (and presumably below) that Brownian motion is the primary mechanism for coagulating particles. For this reason, further studies should look at the size distribution of the particle in the flow. How the particles are distributed in the gas flow will greatly impact all of the theoretical analysis of the experimental results.

16.3.5.4 Thermal Conductivity Increase

It is assumed that the thermal conductivity of the gas will be greatly increased through the addition of solid particles. It is obvious that addition of a solid with an order of magnitude higher conductivity would create and increase in the gas's mixture conductivity. An estimation of the conductivity increase can be made using existing models for liquid particle conductivity similar to those of Hamilton–Crosser (Choi and Eastman 1995), or other methods that involve modifications for agglomeration shapes and size distributions. However, similar to nanofluids, the existing models may not completely predict the mixture thermal conductivity. The only way to definitively know the thermal conductivity would be through experimental measurement. The full measurement of thermophysical properties of the nanoaerosol would allow for a solid theoretical formulation and analysis of the nanoaerosol heat transfer performance. To the author's knowledge, no such experimental measurements of nano-aerosol thermal conductivities exist in the literature.

16.3.5.5 Radiative Heat Transfer from Particle to Particle and from Particle to Wall

Radiative heat transfer is possible from the heated wall directly to the nanoparticles or between the particles themselves. Direct thermal radiation heat transfer could occur between the wall and the particles, thus enhancing the amount of heat that is transferred by convection alone. Quantification of this heat transfer could be pursued using the vast amount of theoretical works on the subject similar to that of Chandrasekhar (1960). For this study, a full analysis was not pursued. The current system will be operated at lower wall temperature and then measured at higher wall temperatures in order to experimentally verify if radiative heat transfer is occurring. However, it is nearly impossible to decouple the potential heat transfer enhancements due to all the different potential mechanisms. For this reason, further evaluation of the radiative heat transfer in nano-aerosol systems would be a lucrative area of investigation.

16.4 RECOMMENDATIONS AND FURTHER STUDY

16.4.1 Recommendations

In view of the above considerations, the following recommendations were made for the INL experimental system:

1. Measure thermal conductivity and viscosity in static gas–particle systems.

2. Run experiments at temperatures for which the effect of radiative heat transfer is expected to be small and then compare with higher temperatures where the radiative heat transfer would be greater.

3. Attempt to measure nanoparticle concentration and particle size distribution inside and outside the test section in stagnant systems.

16.4.2 Further Study

One can gain great insight into the thermal behavior of nanofluids through a study of the sister field of colloidal and surface science. Similarly, nano-aerosols can gain insight into the thermal behavior through further study of the field of aerosols. Historically, the heat transfer behavior of aerosols was of little importance. More recently with the advent of potential applications such as those found in nuclear reactors or computational cooling, the heat transfer behavior of nanoaerosols is becoming more interesting as a research topic. One such computational work exploring the theoretical application of nanoaerosols to electronic cooling is that of Hudson (2013). This work tries to formulate a computational model that encompasses many of the heat transfer mechanisms described in this chapter. The results show the importance of thermophoresis and diffusion of the particles inside a natural convection laminar flow system. Further works along these lines will help to refine the needs of experimental data in the area of nanoaerosols.

REFERENCES

Chandrasekhar, S. 1960. *Radiative Transfer*. New York: Dover.
Chapman, S. and T.G. Cowling. 1958. *The Mathematical Theory of Non-Uniform Gases*. London: Cambridge University Press.
Choi, S.U.S. and J.A. Eastman. 1995. Enhancing thermal conductivity of fluids with nanoparticles. *ASME International Mechanical Engineering Congress & Exposition*, ASME, San Francisco, CA, November 12–17.
Filippov, A.V. and D.E. Rosner. 2000. Energy transfer between an aerosol particle and gas at high temperature ratios in the Knudsen transition regime. *International Journal of Heat and Mass Transfer*, 43(1):127–138.
Friedlander, S.K. 2000. *Smoke, Dust, and Haze: Fundamentals of Aerosol Dynamics*. New York: Oxford University Press.
Hudson, A. 2013. *Computational Analysis to Enhance Laminar Flow Convective Heat Transfer Rate in an Enclosure Using Aerosol Nanofluids*. Electronic Theses & Dissertations. Paper 38. Georgia Southern University, Statesboro, GA.
Incropera, F.P. and D.P. DeWitt. 1990. *Fundamentals of Heat and Mass Transfer*. New York: John Wiley & Sons.
Kim, J.H., G.W. Mulholland, S.R. Kukuck, and D.Y.H. Pui. 2005. Slip correction measurements of certified PSL nanoparticles using a nanometer differential mobility analyzer (Nano-DMA) for Knudsen number from 0.5 to 83. *Journal of Research of the National Institute of Standards and Technology* 110(1):31–54.
Mädler, L. and S.K. Friedlander. 2007. Transport of nanoparticles in gases: Overview and recent advances. *Aerosol and Air Quality Research* 7(3):304–342.
Willeke, K. 1976. Temperature dependence of particle slip in a gaseous medium. *Journal of Aerosol Science* 7:381–387.

Index

Printed and bound by CPI Group (UK) Ltd, Croydon, CR0 4YY

22/10/2024

01777647-0005